The Official Guide

Robot Sumo

The Official Guide

Robot Sumo

Pete Miles

Library Media Center
Kittatinny Regional High School
77 Halsey Rd.
Newton, New Jersey 07860

McGraw-Hill Osborne

New York   Chicago   San Francisco   Lisbon
London   Madrid   Mexico City   Milan   New Delhi
San Juan   Seoul   Singapore   Sydney   Toronto

McGraw-Hill/Osborne
2600 Tenth Street
Berkeley, California 94710
U.S.A

To arrange bulk purchase discounts for sales promotions, premiums, or fund-raisers, please contact **McGraw-Hill**/Osborne at the above address. For information on translations or book distributors outside the U.S.A., please see the International Contact Information page immediately following the index of this book.

Robot Sumo: The Official Guide

Copyright © 2002 by The McGraw-Hill Companies. All rights reserved. Printed in the United States of America. Except as permitted under the Copyright Act of 1976, no part of this publication may be reproduced or distributed in any form or by any means, or stored in a database or retrieval system, without the prior written permission of publisher, with the exception that the program listings may be entered, stored, and executed in a computer system, but they may not be reproduced for publication.

1234567890 CUS CUS 0198765432

ISBN 0-07-222617-X

Publisher	Brandon A. Nordin
Vice President &	
Associate Publisher	Scott Rogers
Acquisitions Editor	Marjorie McAneny
Project Editor	Lisa Wolters-Broder
Acquisitions Coordinator	Tana Allen
Technical Editors	Bill Harrison, Jim Frye
Copy Editor	Marilyn Smith
Proofreader	Robin Small
Indexer	James Minkin
Computer Designer	Lucie Ericksen
Illustrators	Michael Mueller, Lyssa Wald
Series Design	Lucie Ericksen, Mickey Galicia
Cover Design	Theresa Hines

This book was composed with Corel VENTURA™ Publisher.

Information has been obtained by **McGraw-Hill**/Osborne from sources believed to be reliable. However, because of the possibility of human or mechanical error by our sources, **McGraw-Hill**/Osborne, or others, **McGraw-Hill**/Osborne does not guarantee the accuracy, adequacy, or completeness of any information and is not responsible for any errors or omissions or the results obtained from the use of such information.

About the Author

Pete Miles is a senior research engineer who develops advanced abrasive waterjet machining technologies and automated CNC machine tools for Ormond LLC, in Kent, Washington. After serving in the U.S. Marine Corps as a tank killer using guided missiles, he went to college and obtained both bachelor's and master's degrees in mechanical engineering. He then went on to developing new advanced machining technologies and capabilities that have become enabling technologies to several different companies for manufacturing difficult-to-machine structures.

Pete is an avid competitor in robot sumo, and prefers to build unusual styles of sumo robots to spark the interest of others. He is the first person who built and competed a bi-ped robot in robot sumo. His sumo robots have traveled to many robotics events in the U.S. and in Japan. He has built mini sumo robots that weigh as little as 6 ounces and very large sumo robots that have weighed as much as 198 pounds for robot combat events such as *Robotica*. He is also the co-author of *Build Your Own Combat Robot* (McGraw-Hill/Osborne, 2001).

Pete is an active member of the Seattle Robotics Society, one of the largest amateur robotics societies in the world. He's also the organizer of the Robothon, one of the largest amateur robotics events held in North America. In his free time, he goes to various community activities to promote amateur robotics, and to show that it is one of the best ways for people to learn that math, science, and engineering can be fun, exciting, and easy to do.

About the Technical Editors

Bill Harrison, President of Sine Robotics, has a degree in mechanical engineering and is the president of a small research and development company that specializes in miniature robotics. He is a veteran robot sumo designer, builder, and competitor, and has competed in—and won—many sumo contests throughout the U.S., Canada, and Japan. He has devoted much of his time to promoting and teaching robot sumo and has assisted many clubs throughout North America in setting up robot sumo events. He is the inventor of the popular Mini Sumo class of robot sumo, and is the founder and organizer of the Northwest Robot Sumo Tournament, the oldest robot sumo event in the U.S.

Jim Frye, President of Lynxmotion, has a degree in electronics engineering and also has a passion for making cool robots. His company is one of the largest manufacturers of hobby robot kits in the world. He has designed and built more than 50 different robots that walk, crawl, roll, push, or just move things around. He also founded the Central Illinois Robotics Club, and in his spare time, is heavily involved in the activities and competitions hosted by the club.

Dedication

I would like to dedicate this book to the most wonderful woman in the world, my wife Kristina, for all of her support and help with writing this book and her willingness in helping me promote amateur robotics to the general public. Without all of her support, I wouldn't be writing this book.

Contents At A Glance

1	Introducing Robot Sumo	1
2	The Rules of the Game	11
3	Building the Sumo Ring	27
4	Robot Materials and Construction Techniques	47
5	Making Your Robot Move	61
6	Power Transmission Fundamentals	89
7	Batteries: The Heart of a Robot	119
8	It's All About Power: Selecting the Right Motor	137
9	Motor Control Fundamentals	165
10	Remotely Controlling Your Robot	187
11	Introduction to Microcontrollers	211
12	Sensors: Letting Your Robot See the World	233
13	Building the Robot	271
14	Offensive and Defensive Strategies	297
15	Improving Your Robot's Pushing Capability	309
16	Real-Life Sumo Robots: Lessons from Robot Builders	325
A	Robot Resources and References	339
B	Robot Power Supplies	353
	Index	357

Contents

Acknowledgments xxi
Introduction xxiii

CHAPTER 1 **Introducing Robot Sumo** 1
    Why Participate in Robot Sumo? 1
    What Is Robot Sumo? 4
        Sumo Robot Sizes 4
        Sumo Robot Design 5
    A Short History of Robot Sumo 7
    What You Need to Know and What You'll Learn 8

CHAPTER 2 **The Rules of the Game** 11
    The Official Japanese Rules 11
        Section 1: Definition of a Match 12
        Section 2: Dohyo Specifications 12
        Section 3: Robot Specifications 13
        Section 4: Game Principles 14
        Section 5: Game Procedure 15
        Section 6: Yuko (Effective) Points 16
        Section 7: Violations and Penalties 17
        Section 8: Injury and Accidents 18
        Section 9: Objections 18
        Section 10: Specifications of Robot Markings 18
        Section 11: Others 19

	Interpretation of the Rules	19
	Game Points	19
	The Weigh-in and Sizing Check	19
	The Beginning of the Game	20
	Robot Specifications	21
	Robot Restrictions	22
	Restarting a Match	23
	Modification of the Rules	23
	Robot Sumo Weight Classes	24
	Tournament Styles	25
	A Simple Game	26
CHAPTER 3	**Building the Sumo Ring**	**27**
	Specifications for the Sumo Ring	28
	Color and Size Specifications	28
	The Surface of the Dohyo	29
	Building a Practice Sumo Ring	30
	Building a Full-Sized Wooden Sumo Ring	30
	Wooden Sumo Ring Design	30
	Wooden Sumo Ring Construction	31
	Building a Full-Sized Metal Dohyo	36
	Metal Sumo Ring Design	36
	Metal Sumo Ring Construction	38
	Building a Mini Sumo Dohyo	38
	Applying the Top Surface to the Dohyo	40
	Preparing a Wooden Surface for a Practice or Mini	
	Sumo Ring	40
	Adding a Metal Surface	40
	Adding a Vinyl Surface	41
	Painting Your Dohyo	42
	Painting the Wooden Surface	42
	Painting Vinyl and Steel Surfaces	46
	Final Comments on Building Sumo Rings	46
CHAPTER 4	**Robot Materials and Construction Techniques** ..	**47**
	Robot Weight	48
	Robot Materials	48
	Wood	48
	Plastic	49

Aluminum	50
Brass	51
Material Comparison	51
Fasteners for Robot Assembly	53
Adhesives	53
Screws and Bolts	53
Tools for Robot Construction	55
Hand Tools	55
Power Tools	56
Electronic Tools	56
Final Comments on Robot Materials and Construction Techniques	59

CHAPTER 5 Making Your Robot Move — 61

Choosing a Locomotion Style for Your Robot	61
The Wheeled Robot	62
The Tank-Treaded Robot	64
The Walking Robot	67
Selecting a Steering Method	68
Ackerman Steering	69
Differential Steering	70
The Synchro Drive System	71
Holonomic Motion Control	71
Selecting the Right Wheel for Your Robot	73
Wheel Sources	73
Custom Wheels	77
Mini Sumo Robot Wheels	79
Mounting Wheels to Drive Shafts	83
Set Screw Mounting	84
The Pinned Shaft Method	85
Hex Nut-Style Mounting	86
Wheel Mounting Alignment	87
Quick-Disconnect Coupling	87
Closing Comments on Locomotion	88

CHAPTER 6 Power Transmission Fundamentals — 89

Power Transmission Choices	90
Kinematic Fundamentals	91
Differential Steering	91
Pushing Force	93

Speed Reduction	95
Increasing the Wheel Torque	98
Power Transmission System Components	99
Chain-Drive Systems	100
Belt-Drive Systems	105
Gears	107
Bearings	109
Gearboxes	112
Closing Comments on Power Transmission	118

CHAPTER 7   Batteries: The Heart of a Robot   119

Battery Types	120
Alkaline Batteries	120
Nickel-Cadmium Batteries	121
Nickel-Metal-Hydride Batteries	121
Sealed-Lead-Acid Batteries	122
Lithium-Ion Batteries	122
Other Types of Batteries	123
Battery Basics	123
Battery Capacity	123
Internal Resistance	126
Battery Summary	128
Battery Packs	128
Battery Recharging	130
Battery and Wire Installation	132
Battery Placement	132
Electrical Wire Requirements	133
Terminal Blocks	134
Multiple Power Supplies	135
Closing Comments on Batteries	135

CHAPTER 8   It's All About Power: Selecting the Right Motor   137

Motor Types	138
Rare-Earth and Ferrite Magnet Motors	138
Brushless PMDC Motors	139
Electric Motor Basics	139
Motor Speed	140
Motor Torque	140

Stall Current	141
Power and Efficiency	142
Determining the Motor Constants	143
Power and Heat	146
Battery Effects	148
Pushing the Limits of Your Motors	149
Effects of Increasing Voltage	149
Cooling Motors	150
How to Select a Motor	151
Defining Your Robot's Performance Specifications	151
Finding the Motor	152
R/C Servos for Gearmotors	154
Using Servos as Sumo Robot Motors	155
R/C Control Signals	156
Modifying a Servo	158
Motor Sources	163
Final Comments on Motors	163

CHAPTER 9 Motor Control Fundamentals — 165

Motor-Direction Control	166
Switch Combinations for Motor Actions	166
Relay Control	167
Transistor Control	168
Solid-State H-Bridges	171
Integrated-Circuit-Based Solutions	173
Variable-Speed Control	175
Variable Resistors	175
Pulse Width Modulation	176
Commercial Electronic Speed Controllers	178
Commercial H-Bridges	179
Radio-Control Electronic Speed Controllers	179
Radio-Control-Compatible Electronic Speed Controllers	183
Advanced Motor Controllers	184
Final Comments on Motor Speed Controllers	185

CHAPTER 10 Remotely Controlling Your Robot — 187

Radio-Control System Basics	188
Radio-Control Channels	189
Radio-Control Frequencies	189

	Radio Transmission Methods	192
	R/C Control Signals	194
	Radio Interference	194
	Reducing Electrical Noise from Motors	194
	Reducing Interference from Other Radios	196
	R/C Antenna Setup	196
	R/C Configurations for Sumo Robots	198
	Mini Sumo Robot R/C Configuration	200
	Servo Mixing for Steering Controls	201
	Tether-Control Systems	206
	General Specifications for Tether-Controlled Robots	207
	Simple Tether-Control Circuits	207
	Failsafe Control	208
	Final Comments on Remote-Control Robots	209
CHAPTER 11	**Introduction to Microcontrollers**	**211**
	What Is a Microcontroller?	212
	Basic Microcontroller Features	213
	Special Microcontroller Features	214
	Programming Environments	215
	Types of Microcontrollers	216
	Parallax Basic Stamp Modules	217
	The Parallax Javelin Stamp Module	221
	Basic Micro Basic Atom Modules	223
	The Basic Micro MBasic Compiler	224
	The Acroname BrainStem	225
	The LEGO RCX Brick	226
	Handy Board	228
	The BotBoard Plus	228
	Single-Chip Solutions	229
	Closing Comments on Microcontrollers	230
CHAPTER 12	**Sensors: Letting Your Robot See the World**	**233**
	Edge-Detection Sensors	234
	Mechanical Edge-Detection Sensors	234
	Optical Edge-Detection Sensors	238
	Opponent-Detection Sensors	242
	Mechanical Opponent-Detection Sensors	242
	Infrared Reflective Sensors	243

Contents

Infrared Sensor Variations	248
Sharp Infrared Proximity Detectors	249
Sharp Infrared Range Sensors	253
Ultrasonic Range Sensors	258
Using Ultrasonic Sensors	259
Sensor Placement	261
Protecting Your RobotÆs Rear	262
Sensor Range Issues	262
Using Multiple Sensors	262
Discovering Your Opponent's Location	265
Sensor Accuracy Issues	266
Drawbacks to Reflective Sensors	266
It's a Noisy World Out There	267
Calibrating Sensors	268
Closing Remarks on Sensors	269

CHAPTER 13 Building the Robot ... 271

Full-Sized Sumo Robots	271
Viper Sumo Robot	272
Terminator Sumo Robot	284
Mini Sumo Robots	286
A Mini Sumo Robot Built from Scratch	286
Mini Sumo Kits	294
Closing Remarks on Building Robots	296

CHAPTER 14 Offensive and Defensive Strategies ... 297

General Competition Tips	297
Knowing Your Competition	298
Showing Confidence	298
The Luck Factor	299
Strategies for Remote-Control Robots	299
Programming Tips	299
Using Subroutines	300
Subsumptive Programming	300
Using Multiple Programs	301
Competition Strategies	302
Offensive Strategies	302
Defensive Strategies	306
Closing Remarks on Strategy	308

CHAPTER 15	**Improving Your Robot's Pushing Capability** **309**
	Improving Wheel Traction 310
	Coating the Wheels 310
	Casting Wheels 311
	Increasing the Apparent Weight of Your Robot 313
	Vacuum Systems 313
	Magnetic Systems 320
	Closing Remarks on Traction and Pushing Capabilities 323
CHAPTER 16	**Real-Life Sumo Robots: Lessons from Robot Builders** **325**
	Mini Sumo Robots 326
	LEGO Sumo Robots 331
	Middleweight Sumo Robots 331
	Full-Sized Sumo Robots 332
	Now Go Build a Robot! 338
APPENDIX A	**Robot Resources and References** **339**
	Sumo Robot Parts Suppliers 340
	Sumo Ring Construction Materials 340
	Wheels, Hubs, and Mounting Hardware 340
	Motor and Gearmotor Sources 341
	Surplus Motors and Gearmotors 342
	R/C Servo Motors 342
	Gears and Drive Components 342
	Batteries 343
	Motor Controllers 343
	Remote-Control Equipment 344
	Microcontrollers 344
	Robot Sensors 345
	Electronic Parts Suppliers 346
	Mechanical Parts Suppliers 346
	Vacuum Pump Suppliers 346
	High-Strength Magnets 347
	Silicon and Polyurethane High Traction Wheel
	Casting Compounds 347
	Printed Circuit Board Manufacturers 347
	Hobby Stores 348
	Tools 348

	References	348
	Robot Building	348
	Electronics	349
	Machine Design	349
	Programming	349
	Other Robot References and Resources	350
	Major Robot Sumo Tournaments	350
	Robotics Clubs	351
APPENDIX B	**Robot Power Supplies**	**353**
	Index	**357**

Acknowledgments

Nobody gets anywhere without help from many different people along the way. I am no exception to this. First off, I would like to thank my wife, Kristina Lobb Miles, for all of her hard work in preparing all of the graphics that went into this book. Kristina took many of the photographs in this book, and she edited all of the graphics so that the primary point of interest is well-defined and separated from the background. Without her help, I would not have been able to complete this book.

I would like to thank Bill Harrison for introducing robot sumo to me. He was able to use the right words that sparked the sumo flame inside me. He has been, and will always be, my mentor in robot sumo. Much of what I know about how to build sumo robots is due to his teachings, and now I am teaching others about robot sumo and promoting the robot sumo event. Without his tireless efforts in promoting robot sumo here in the United States and Canada, robot sumo would not be where it is today.

I would like to thank Mr. Hiroshi Nozawa, chairman of Fujisoft ABC, Inc., for creating robot sumo, and for all of his efforts in organizing and promoting the All Japan Robot Sumo Tournaments, and the International Robot Sumo Tournaments, along with his encouragement to see this book written. Without all of his efforts, robot sumo would not exist today. I would also like to thank Mr. Tetsuya Chiba for all of his help in obtaining and translating information between me and the robot sumo tournament committee in Tokyo. I would also like to thank Mr. Mato Hattori for introducing robot sumo to the United States by creating a video about robot sumo and distributing it to various robot clubs in the United States.

I would also like to thank Ms. Mary Ohno, Mato Hattori, and Tetsuya Chiba for all of their help in translating the current Japanese robot sumo rules into English so that they can be presented in this book. Mr. Mato Hattori's personal experience in competing in robot sumo has been very helpful in providing clearer interpretations of the rules.

I would like to thank Daryl Sandberg, Pete Burrows, Tom Dickens, Carlo Bertocchini, Jim Sincock, Steve Richards, Jeff Loitz, Peter Campbell, Dave Hylands, Steve Hassenplug, Lim Beng Soon, Dana and David Weesner, Steve Richards, and Jim Frye for providing photographs and descriptions of their sumo robots for this book.

I would like to thank Jim and Annette Frye of Lynxmotion, Ken Gracey and John Cunningham of Parallax, Inc., Robert Bonner and Craig Justice of Basic Micro, and Lon Glazner of Solutions Cubed for all of their test materials and technical support for this book.

I would like to thank the crew at McGraw-Hill/Osborne for all of their hard work in managing and preparing this book—Margie McAneny and Tana Allen for managing this project and keeping me on track; Lisa Wolters-Broder, Marilyn Smith, and Robin Small for all of their editing, proofreading, and shepherding; Lucie Ericksen for designing the page layout; and Theresa Hines for designing the jacket cover.

Finally, I would like to thank Bill Harrison and Jim Frye for all of their enthusiastic willingness to help me write this book by being the technical editors. Their insightful comments really helped form the content of this book, and their keen eyes were able to catch many subtle errors. Without their help, the quality of this book would not be where it is now.

Introduction

Robot sumo is the fastest growing robotics sport on this planet, a sport where two robots try to push each other out of a sumo ring. Robot sumo is not to be confused with combat robot events like *BattleBots, Robot Wars, BotBash,* or *Antweights,* to name a few. The goal of robot sumo is to push your opponent out of a sumo ring. Points are awarded based on whether or not your robot is successful in doing this, unlike combat robot events where you earn points by damaging your opponent. The main similarity between the two types of events is that you get to pit your creation up against another person's creation to see who has the better robot.

The goal of this book is to teach you all about robot sumo so that you can start building your own sumo robots and creating local sumo contests. What's so nice about sumo robots is that anyone can build them. They are small and lightweight, and you can go from contest to contest competing with the same robot. Contests don't need to be formal regional events; they can be local clubs competing against each other, local schools competing against each other, or simply neighbors competing against each other. In Japan, the country that invented robot sumo, these robots are part of the educational programs in many schools. The students learn about electronics, mechanics, and computer programming, and use sumo robots as the end goal to apply their knowledge, and then compete against each other, school versus school.

I hope that teachers and parents who read this book will see that building sumo robots is one of the best ways to learn that science, mathematics, and engineering can be fun and exciting, and that you can use your brains and creativity to compete in a sport. One of the unique things about robotics is that it is an equalizer. Men and women compete on equal terms. Athletes and handicapped people compete on equal grounds. Old versus young compete equally. Ingenuity, creativity, and luck rule the day. Robot sumo events in the schools are one of the few events where every student can compete equally.

This book is divided into a series of logical steps for building sumo robots. First, I introduce the event and its rules. Then I present a set of instructions for building an actual sumo ring. Next, a series of robot basics are presented: materials and construction steps, locomotion, wheels and gears, transmissions, batteries and motors, and how to control the motors and motion. This is followed by a series of chapters on how to remotely and autonomously control sumo robots, and how to give robots the ability to see and feel within their environment. Several real-world robot construction examples will be presented to tie all this information together. Once your robot is built, I'll give you a set of offensive and defensive strategies and a series of advanced techniques to improve the robot's competitive advantages. Finally, there is a gallery of many different robots to give you lots of ideas about what other people are doing in the field of robot sumo.

Most people build their robots from the ground up. Why? Because this is one of the most exciting parts of robot sumo. The rules are so simple; you are open to doing many different things. Because the robots are small, you can use common materials and household tools to build them. Usually the hardest part for people, when they first get started in building sumo robots, is finding parts. Throughout this book, there are links to many different off-the-shelf solutions and low-cost parts, so that the mystery around how to find parts will go away. The information presented in this book does not represent the whole subject. There are so many different ways to build a robot that I could write an entire book on each approach.

What *is* presented here are a bunch of working examples, and they can be combined together to build complete working robots. Or, you can use parts of this book along with your own ideas. A lot of the information that is in this book can be applied to any robot that you might want to build, so this book may become a valuable resource in your collection.

A famous line from the movie *Field of Dreams* is, "If you build it, they will come." This is very true about sumo robots. If you build a sumo ring and a sumo robot, someone *will* build a robot to compete against your robot. Take your robot to your local school, workplace, church, library, club, or community center, and show off how it works. When people see your excitement, and how much fun sumo robots are, they will soon be bitten by the sumo bug, and will build robots to compete, as well.

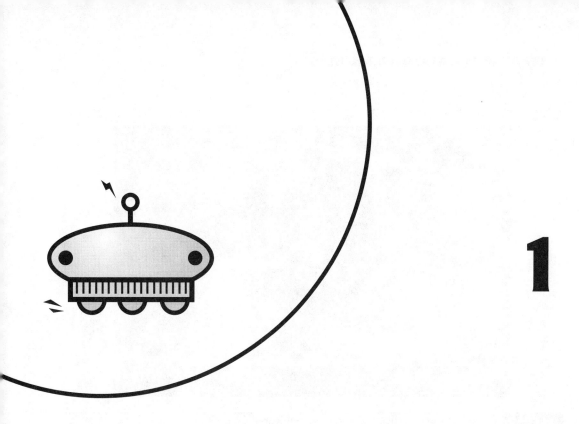

Introducing Robot Sumo

This book aims to provide all the information you need to become part of the thrilling world of robot sumo. It contains details for building sumo rings and competitive sumo robots, as well as planning game strategies. But before we get to sumo ring and robot assembly, let's take a broad view of robot sumo to see what it's all about.

Why Participate in Robot Sumo?

I quietly walk up to the sumo arena and set my robot down for its first contest. I look around and notice everyone in the crowd is quiet and watching everything I do. I begin to think that I should not be here. I was up until 4 A.M. trying to get my robot to work, so I got only a couple hours of sleep. The other guy's robot looks really cool. There is no way my robot can beat his. I am going to lose and be totally embarrassed.

FIGURE 1-1 Robot sumo at the 2002 International Robot Sumo Tournament

I set my robot down on the sumo ring and position it so that it will run straight at the other robot. My opponent picks up his robot and repositions it so that it will run right past my robot. I begin to wonder why he did this, since I thought the goal of robot sumo was to push the other robot out of the sumo ring.

The referee signals start, and my heartbeat rate increases as I press my robot's start button. I stand back to mentally count down the five seconds my robot must remain stationary until it can move. In my excitement, I mentally reach the five seconds before my machine starts to move. I begin to panic, thinking I must have forgotten something; I suddenly don't remember testing the five-second subroutine the night before. But then, both sumos start moving, and I exhale a sigh of relief. Both robots are moving forward, but I can't tell if my robot sees the other robot. Each robot is moving forward and looking for the other. My robot approaches its opponent, and I smile as the crowd starts to cheer, thinking, "I've got him now!"

But the robots just pass each other, as if the other one wasn't even there, and the crowd sighs and goes silent. My robot is now heading full steam ahead toward the edge of the ring, and I suddenly think, "What if the edge sensors aren't working?!" As soon as my machine gets to the white edge of the sumo ring, it stops, backs up a bit, and then spins around. I breathe another sigh of relief that all seems to be working correctly. This time, my robot is heading toward the side of the other guy's robot, and the moment of truth is about to occur. Will my robot be able to push the other robot out of the ring?

As my robot approaches its foe, it makes a couple of quick course corrections and is now zeroing in on the opposition. I'm thrilled to see my object-detection sensors are working. My robot is closing in on the adversary, and the crowd starts cheering again. Just as it is about to broadside its opponent, the other robot suddenly turns toward mine. The crowd gets louder, and the robots crash into each other. My breathing almost stops as both machines halt near the center of the ring. The wheels of both robots are spinning on the surface of the ring.

My robot slowly starts pushing its challenger steadily backwards. Just as I start thinking I'm going to win this bout, the other robot somehow gets better traction and begins pushing mine backwards. The crowd goes wild. I bite my lower lip. Then my robot's traction improves, and it begins pushing its foe backwards. But this soon turns into a back-and-forth dance for both robots. I begin to notice that our robots are starting to slip sideways with respect to each other, and as soon as the corners of the robots come apart, they shoot toward each other. But this time, their wheels get caught, and they start a do-se-do spinning dance. The crowd quiets down. This is turning out to be a much tougher battle than I had anticipated.

The referee stops the match to separate the robots. I take a couple deep breaths, and we restart. This time, when my robot goes to the sumo ring edge and backs up, it turns only 90 degrees. I think to myself that I'm glad my random turning method is working. Now my robot is approaching the rear of its rival. My robot is following its opponent and the crowd goes wild. I glance up at my opponent's face, and I can see in his eyes that he is concerned. As the other robot stops at the white edge of the sumo ring, my robot slams right into its back. The crowd screams as my robot launches its opponent out of the ring.

As the crowd is cheering on my robot, I hear the referee indicate the match is over and that I can pick up my robot. I spend a moment watching the referee and the other guy catch the robot that is running away on the floor. I can't stop smiling from ear to ear because my baby worked as planned and won its first bout. I look up at the large display game timer and notice only 30 seconds have passed. It seemed like hours. Now the butterflies are returning because this was only the first bout in a best of two out of three game. I think, "Can I do it again?"

The feelings I've just described are what everyone experiences when they compete in one of the fastest growing and most popular robot contests in the world: robot sumo. Since its creation, robot sumo has found its way into most robotic clubs, as well as many universities and schools throughout the world.

What Is Robot Sumo?

Robot sumo is a robotic version of one of Japan's most popular sports, *sumo*. Instead of two humans trying to push each other out of a sumo ring, two robots try to attempt the same feat.

In robot sumo, there are three bouts per match, and a robot must win two out of three bouts to win a match. All three bouts must be completed in a three-minute time frame. Robot sumo tournaments are conducted using either a single-elimination, double-elimination, or round-robin style competition. There are not that many restrictions on what the robots can and cannot do, so there is a lot of room for creativity in your robot design.

NOTE *The proper term is* sumo, *not sumo wrestling.* Sumo *is a Japanese word for wrestling. Saying "sumo wrestling" is like saying "wrestling wrestling."*

Sumo Robot Sizes

The size constraints for a sumo robot are that it must fit inside a 20 centimeter square (7.87 inch) box (no height limit) and not weigh more then 3 kilograms (6.6 pounds). The robots must push their opponents out of a 154-centimeter (60.6-inch) diameter ring within the three-minute time limit.

Along with "full-sized" robots, there are two other popular sumo weight classes: lightweight and mini sumo. The lightweight class has the same size specifications as described above, but the maximum weight is 1 kilogram (2.2 pounds). The popular mini sumo class robot must fit inside a 10 centimeter (3.94 inch) square box (again, there are no height restrictions) and weigh no more than 500 grams (1.1 pounds). The sumo ring for the mini sumo class is 77 centimeters (30.31 inches).

For the most part, sumo robot matches held throughout the world follow the basic Japanese set of rules, with only a few local modifications. The modifications are generally in regard to the weight and size limitations of the robots. However, most competitions are migrating to full compliance with the Japanese rules for 3-kilogram (6.6-pound) and mini sumo robots, since robot builders are now traveling from contest to contest. Some competitors even travel around the world to compete in different sumo events.

Sumo Robot Design

There are two different robot sumo classifications: autonomous and remote control. An *autonomous* sumo robot must run completely on its own, with no human control except for turning on the robot. The *remote-control* class usually uses standard radio-control equipment, or tethers, to remotely control the robot.

Since the rules of the event are not complicated (Chapter 2 provides the details on the rules of the game), robot builders are free to use any number of innovative designs to give their robots either a competitive edge or a high coefficient of coolness to make them crowd favorites. Some innovative mini sumo robot designs are shown here.

One of the factors that makes building sumo robots so attractive is the relative low cost. Many sumo robots are made from parts scavenged from toys, cordless power tools, and other types of electronic household gizmos. Due to their small size, robots can be carried around easily, and they do not require any significant repair costs after a contest. You can choose to spend a lot of money building your robot with top-of-the-line components, but it still might not get past the first round in a tournament.

When building a sumo robot (or any type of robot), the selection of all of the components that go into the robot has an effect on how the other components work. Many times, you will need to make compromises when selecting components. A well-thought-out robot that doesn't use any high-performance components can beat a poorly thought-out robot that has high-performance components. Good upfront planning will help you build a top-quality robot.

Because sumo robots are small, space is always a premium. When thinking about different parts to use, also consider how they will physically fit in the robot. Before purchasing parts, make many sketches of what the robot should look like and how the parts will fit inside the robot. It is best to design the robot at full scale using graph paper and a ruler. Sometimes, before cutting parts out of metal, making a wooden, plastic, cardboard, or paper model is very helpful.

Typically, a big problem most robot sumo builders face is determining where to get parts for their robots. One of the main goals of this book is to tell you where to find good-quality robot parts. Throughout the book, I will refer to sources (including websites) for the various parts being discussed. At the end of the book, you'll find a comprehensive list of robot parts suppliers, focusing on companies that sell parts that are ideal for sumo robots.

Robot Safety

While you're building your sumo robot, and after you've finished, keep safety in mind. Some of the components used to construct robots can be dangerous. Sumo robots can be very fast, and they can cause injury and property damage. Here are some guidelines for robot safety:

- The small parts used to build a robot could be a choking hazard. Keep them away from small children.
- When constructing the robot, wear safety glasses.
- When dealing with chemicals, use the appropriate gloves and respirators.
- Always keep your fingers away from moving parts.
- Only use power tools with correct supervision.
- Clamp your parts before cutting and drilling.
- Soldering irons are hot and should not be left unattended.
- Motors, transistors, and motor controllers also can get very hot, so don't touch them after use.
- Make sure that your wiring is correct before applying power to the circuit. A shorted nickel-cadmium (NiCd) battery pack can deliver a few hundred amps of current for a short time. This is enough to melt most wires and wire insulation and can be very dangerous.
- Never leave a powered robot unattended.
- Robots should have a manual power-kill switch to quickly shut off the robot.

With all of this said, building sumo robots is fun and safe, as long as you take proper safety precautions.

A Short History of Robot Sumo

Robotic sumo was originally invented in Japan in the late 1980s by Hiroshi Nozawa, Chair-man of Fuji Software Inc. The first exposition game was held in August 1989, with 33 robots. The first official tournament was held in 1990, with 147 robots. Since then, robot sumo in Japan has grown steadily. More than 4,000 robots competed in the four-month-long, country-wide tournament season of 2001.

> **NOTE** *Japan is divided into nine sumo-robot regions. Within each region, there is a high school division and an open division for everyone else to compete in. A series of tournaments narrows the total number of robots down to 128 robots (64 autonomous and 64 remote control) for the final championship, the All Japan Robot Sumo Tournament, which is held each year on December 23 at the Kokugikan Stadium in Tokyo. The Kokugikan Stadium is the famous stadium for human sumo in Japan.*

In the early 1990s, Mato Hattori (a robot sumo competitor from Japan) introduced robot sumo to the United States. He prepared a videotape showing highlights of the third robot sumo tournament season in Japan, with his BigTX sumo robot, shown here.

After receiving a copy of the videotape that Mato Hattori prepared, Bill Harrison took the robot sumo bull by the horns and became the number one supporter and organizer of robot sumo in the United States. Bill Harrison is head of the Northwest Robot Sumo Tournament, which is the oldest and longest running robot sumo tournament in the United States. He also helped create several other robot sumo tournaments in the United States and Canada.

Bill Harrison and Robert Jorgenson invented the mini sumo class, which has become so popular there are more than twice as many mini sumo robots than 3-kilogram sumo robots in North America. Because of the popularity of mini sumo robots, there are now four different companies in the United States selling mini sumo kits.

Since the early 1990s, robot sumo tournaments have been occurring around the world, and the number of sumo robots is growing at an exponential rate.

What You Need to Know and What You'll Learn

If you're reading this book, I assume that you are familiar with basic electronics and can assemble electronic circuits, that you have experience with standard shop tools (which are used to fabricate robot sumo parts), and that you are familiar with computers and can program microcontrollers.

If you are not familiar with basic electrical circuits and microcontrollers, I recommend that you purchase one of the starter kits from Parallax Inc. (www.parallaxinc.com), such as the BoeBot robotics kit. Parallax has done an excellent job of preparing a set of documents and tutorials to help people get up to speed with electronics and microcontroller programming.

This book is not a step-by-step, hold-your-hand type of how-to book. The goal of this book is to teach you what it takes to build a sumo robot. If you understand how all of the various components work and how they relate to each other, building a robot becomes easy.

If everyone just followed step-by-step instructions, every tournament would have identical robots that had the same capabilities. The only differences between them would be the names of the robots and their paint colors. This would be very boring. It is the innovative robot designs that make robot sumo so much fun. I encourage you to use your imagination and be creative. Throughout this book, you will find photos of many different types of robots, to give you ideas on the various approaches you can take to building a sumo robot. For example, here is a photo of a sumo robot with all of its control circuits external to the robot.

CHAPTER 1: Introducing Robot Sumo

By the time you have finished reading this book, you will have learned all this:

- How to build a sumo ring, both for full-sized robots and for mini sumo robots
- How to select different types of wheels to use in your robot, along with how to mount them to a motor/drive shaft and where to purchase these critical parts
- How to make a robot move using different locomotion methods
- How gears, belts, chains, and bearings are used to reduce motor speed so you can build your own gearbox, and where to purchase different types of gearboxes and gearmotors for your robot
- How different battery technologies will work in a robot and how to select the right battery for your robot
- How motors work, how to select the right type of motor for your robot, and how all of this information about transmission components is related
- How to control the direction and speed of motors and where to purchase low-cost motor-controller solutions
- How to make your robot into a remote-control robot, how the various radio-control components work, and what components work best in sumo robots
- How to use different microcontrollers in a sumo robot
- How to build and use a wide variety of sensors that are well suited for sumo robots
- How to use different programming strategies to gain both offensive and defensive skills in your robot
- How to use different tricks to give your robot that extra competitive edge

With all of this information, you will be able to build top-quality sumo robots, and this book will become a valued reference for all of your robot-creation projects.

When you're designing your robot, you will need to know if any restrictions apply for the matches you plan on participating in. You will need to keep up-to-date on the local rules for the events that you are considering entering. The next chapter discusses the rules of game.

The Rules of the Game

As with every game, robot sumo has a set of rules that needs to be followed. This chapter will present the rules of the game. Because rule changes occur from time to time, you should periodically check the website for the official Japanese rules at www.fsi.co.jp/sumo. Also note that the English version of the rules on their web site may not be the current translated rules.

This chapter begins with a presentation of the official Japanese rules, followed by interpretations of some of the less straightforward rules. Then it discusses some of the different weight classes. Finally, you'll learn about the various tournament styles.

The Official Japanese Rules

The following rules are an English translation of the official FSI All Japan Robot Sumo Tournament rules, as of July 4, 2001. They also include additional information about the new dohyo specifications and changes to the Japanese radio-control frequencies to legal United States radio frequencies.

The Japanese translation has 11 chapters and 21 articles to the rules. To avoid confusion between rule chapters and book chapters, I've called the rule chapters *sections* instead.

Section 1: Definition of a Match

Article 1–Definition

A match involves two contestants (one operator per robot can be registered, but a mechanic may assist the operator) who operate robots that they have made themselves (either autonomous or remote control) in the sumo ring (*dohyo*) according to the game rules (*rules*). The contest continues until a (*yuko*) point is scored by one of the contestants. The decision on when a point is scored will be made by the referee.

Section 2: Dohyo Specifications

Article 2–Definition of the Dohyo Interior

The dohyo interior is defined as the dohyo area surrounded by and including the border line.

Article 3–Dohyo

A dohyo is an aluminum cylinder with a height of 5 centimeters (cm) and a diameter of 154 cm (including the border line), consisting of aluminum base and black, steel sheet metal on top (see Figure 2-1).

- The top surface is a 154-cm diameter, cold-rolled steel sheet (ASTM A366) with a thickness of 3.2 millimeters (mm). The surface will be flat and smooth using a nonglossy black Melamine coating that was baked on at 130 degrees Celsius for 30 minutes.

- The starting lines (*shikiri-sen*) are indicated as two parallel brown (color ratio—blue : red : yellow = 4 : 4 : 2) lines with a width of 2 cm and a length of 20 cm. The outside edge of each line is 20 cm apart, and they are centered on the dohyo.

- The border line is indicated as a white circle ring with a width of 5 cm and an outside diameter of 154 cm. "On the border" is defined as being within the interior of the dohyo.

- During the games, it is up to the referee to decide whether the dohyo can continue to be used or whether it should be replaced.

Article 4–Dohyo Exterior

The exterior area of a dohyo extends at least 100 cm from the border line. The color of the exterior can be any color except white. There are no restrictions on the type of material that can be used or the shape of the exterior, as long as they do not violate the spirit of the rules.

CHAPTER 2: The Rules of the Game

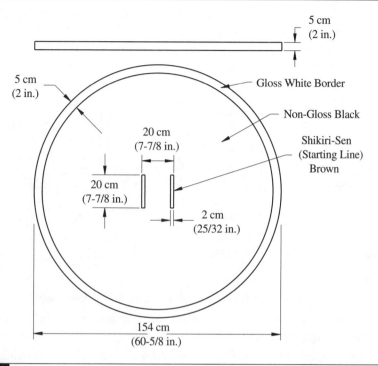

FIGURE 2-1 Dohyo specifications

Section 3: Robot Specifications

Article 5–Specifications

- The robot must be able to fit in a box with a width and depth of 20 cm. There are no restrictions on height (see Figure 2-2).
- Weight (including accessories) must not exceed 3 kilograms (kg). However, the wireless remote control unit of a remote-controlled robot is excluded.
- Usable frequencies for remote control robots are 27 megahertz (MHz) (wide bands 1–6), 50 MHz (bands 00–09), and 75 MHz (bands 11–60)
- Usable remote-control units are limited to the following brands: Futaba, Hitec, Airtronics (Sanwa), JR, and Kondo Kagaku (KO Propo). Multiple remote-control units cannot be used. When you use 27 MHz, use a remote-control unit that is compatible with the wide bands.
- A remote-controlled robot must use the frequency crystals prepared by the tournament officials for the receiver and transmitter. Consequently, the robot's receiver's radio crystals must be detachable.

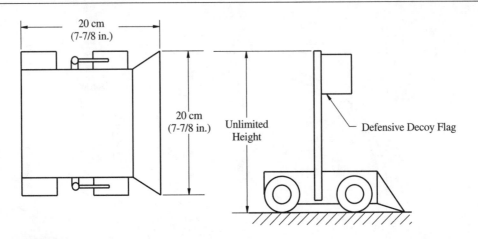

FIGURE 2-2 Sumo size specifications

- There are no restrictions on the type of control method used with autonomous robots.
- An autonomous robot should be designed to begin action no earlier than five seconds after the contestant presses the robot's start button.
- There are no restrictions on the type of microprocessor or the amount of memory used in the robot.

Article 6–Restrictions on Robot Design

- The robot will not include a device that obstructs the control of the opponent's operation, such as a jamming device or strobe light.
- The robot will not include any parts that might damage or deface the dohyo.
- The robot will not include a device that insufflates any liquid, powder, or gas.
- The robot will not include an inflaming device.
- The robot will not include a throwing device.
- The robot will not include any part that fixes the robot to the dohyo surface and prevents it from moving (such as suckers, glue, and so on).

Section 4: Game Principles

Article 7–Game Principles

- A game consists of three matches of three minutes each. The first contestant to win two yuko points is the winner of the game.

- The contestant who has the most yuko points at the end of the game will be judged as the winner.
- When neither contestant receives any yuko points, or both contestants have one yuko point, the winner will be decided by the judges. However, if no obvious superiority exists and a winner cannot be determined, an extra three-minute match can be played.

Section 5: Game Procedure

Article 8–Beginning of the Game

Before the match, the contestants bow to each other outside the dohyo following the chief referee's instructions, then enter the dohyo. After that, the contestants put their robots on or behind their starting lines. No part of the robot can be placed in front of the starting line before the match begins (see Figure 2-3).

- With a remote-controlled robot, the match begins at the referee's signal. Then the contestant can begin to operate the robot with a remote-control unit.
- With an autonomous robot, the contestant can press the start button on the robot at the referee's signal. The match begins five seconds after the referee's signal. The contestant exits the dohyo when the match begins.

Article 9–End of the Game

The match ends when the referee calls the winner. Both contestants bow after removing their robots.

Article 10–Game Cancellation and Rematches

A match will be stopped and a rematch will be started under the following conditions:

- The robots are locked together in such a way that no more action appears to be possible, or they rotate in circles several times.
- Both robots touch the exterior of the dohyo at the same time.
- Any other conditions under which the referee judges that no winner can be decided.
- In case of a rematch, maintenance of competing robots is prohibited until a yuko is observed, and the robots must be immediately put back to the location specified in Article 8.
- If neither of the competing robots win or lose after a rematch, the referee may reposition both robots to a specified location and restart. If even that does not yield a winner, the match may continue at any location decided by the referee, until the time limit is reached.

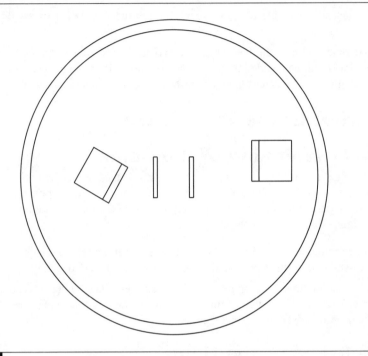

FIGURE 2-3 Start of a sumo match

Section 6: Yuko (Effective) Points

Article 11–Yuko

The following conditions are determined as yuko (effective) points:

- When a robot ejects its opponent from the dohyo with a fair action.
- When the opponent's robot steps out of dohyo on its own (for any reason).
- When the opponent's robot is disqualified or has had more than one violation or warning.
- When two yusei points are given.

Article 12–Yusei

The following condition is determined as yusei (advantage) points:

- When the opponent's robot gets stuck on the border line and cannot move off the border line on its own.

A yuko point will be given when two yusei points are given. A yuko point will also be given when a yusei point is given and the opponent has already been given a warning.

Section 7: Violations and Penalties

Article 13–Warnings

A contestant who takes any of the following actions will receive a warning:

- The operator or any part of the operator (remote control, for example) enters the dohyo before the referee's call ends the match.
- Preparation for the restart of a match takes more than 30 seconds.
- A remote-controlled robot begins action (physical expansion or moving) before the chief referee's start signal.
- An autonomous robot begins action (physical expansion or moving) within five seconds after the chief referee's start signal.
- Any other actions that may be deemed unfair occur.

When a contestant receives two warnings, the contestant's opponent will be awarded one yuko point.

Article 14–Violations

Any of the following actions is determined as a violation, and the offender's opponent, or both robots, will get a yuko point:

- A part (or parts) of the robot that exceeds a weight of 10 grams is separated and dropped from the robot.
- The robot stops moving on the dohyo.
- Both the robots are moving, but don't contact each other.
- The robot emits smoke.

Article 15–Loss by Violation

A contestant who takes any of the following actions will lose the game by violation:

- A contestant does not attend the appointed dohyo when called at the beginning of the game.

- A contestant ruins the game, such as by intentionally breaking, damaging, or defacing the dohyo.

Article 16–Disqualification

A contestant who takes any of the following actions will be disqualified and forced to leave the game:

- A contestant's robot does not meet the robot specifications stated in Article 5.
- A contestant makes a robot using a method restricted in Article 6.
- A contestant displays unsportsmanlike behavior. For example, using violent language or slandering an opponent or a referee.
- A contestant intentionally injures the opponent's operator.

A disqualified person will lose the right to enter any national tournaments.

Section 8: Injury and Accidents

Article 17–Request for Suspension

When a contestant is injured due to the operation of the robot or the robot has an accident, and the game cannot be continued, a suspension can be requested by the contestant.

- A referee must take immediate action necessary to take care of this situation.

Section 9: Objections

Article 18–Objections to the Referee

No objections to the judgment of the referee can be raised.

Article 19

A contestant who has an objection to the operating rules must express dissent to the Tournament Committee before the end of the game.

Section 10: Specifications of Robot Markings

Article 20–Marks on the Robot

The east side contestant must put two red marks of at least 2 cm in diameter on the robot. The west side contestant must put blue marks of the same size and in the same places.

Section 11: Others

Article 21–Modifications and Abolition of the Rules

Modifications or abolition of the rules are made by the decision of a general assembly of the Tournament Committees held according to the Rules of Tournament Committees.

Interpretation of the Rules

Robot sumo is a relatively simple game. Two robots try to push each other out of the ring. There are three matches per game; the winner does best of two out of three.

Game Points

The first robot that has any part of its body touch the ground outside the dohyo (pronounced "doe-yo") loses the match, and its opponent is awarded a yuko (pronounced "you-ko") point. If both robots are still on the dohyo after the three-minute time frame, no points are awarded, and the next match is started. If after all three matches are over and no robots have any points (or both robots have an equal number of points), the referee will decide the winner based on which robot was the most aggressive. If both robots are determined to be equally aggressive, the referee can allow an additional match to be played.

Note that it is possible to win a game without your robot pushing its opponent out of the dohyo or by playing only one match. Penalty points and disqualifications affect the outcome of the game. The rules in Sections 6 and 7 are very clear about how points are awarded. The one rule that can be confusing is how the yusei (pronounced "you-say") points are awarded. There are situations when part of a robot goes off the end of the dohyo, but does not touch the ground outside the dohyo (remember that the dohyo surface is 5 cm above the surrounding ground), and gets stuck. For example, suppose you have a two-wheeled robot, and one wheel goes off the dohyo. Now part of the robot's body will be resting on the surface of the dohyo. If the robot gets stuck or if it can't get back on the sumo ring by itself, the match is stopped and the other robot is awarded a yusei point.

The Weigh-in and Sizing Check

When entering a tournament, your robot must go through a weigh-in and sizing check. Weigh-in is simple—you just place the robot with all of its accessories, except the remote control transistor, on a scale. The robot can have any weight up to the maximum weight for the weight class. If the robot weighs too much, it is disqualified.

The sizing check involves setting the robot on a table in its normal starting position, and then having a square box placed over the robot. The box must be able to slide all the way to the table surface without any interference. Many organizations will use a clear plastic box whose inside dimensions are the maximum dimensions of the robot weight class. Shown here are the sizing boxes used at the Northwest Robot Sumo Tournament (courtesy of Bill Harrison).

The Beginning of the Game

Before the game starts, both contestants will face and bow to each other to show sportsmanship and respect to each other and the game. The referee will indicate when you should do this. After bowing, you place your robot on the dohyo. What happens at this point can vary among different clubs.

The Japanese rules state that the robot must be set down either on or behind the starting line. No part of the robot can be forward of the starting line. The Japanese interpretation of this rule is that some part of the robot must be inside the 20 cm wide box that is behind the 20 cm long starting line. This is a relatively small, rectangular area. Many clubs have adopted the interpretation that behind the starting line is behind an imaginary line that is parallel with the starting line and stretches across the dohyo. Figure 2-4 shows an illustration of this interpretation.

Either interpretation of the starting point rule can be enforced in a local contest, as long as everyone knows which interpretation is being used. It is better to use the more restrictive area interpretation when your contests are open to national and international competitors, since they may be more familiar with the official Japanese rules.

> **NOTE** *To some contestants, the exact location of the robot's position is very important; to other people, the robot's starting position doesn't matter.*

There is no rule that states which robot is set down first. When you set your robot down, you can move it around to position it against your opponent. If the referee feels that you are taking

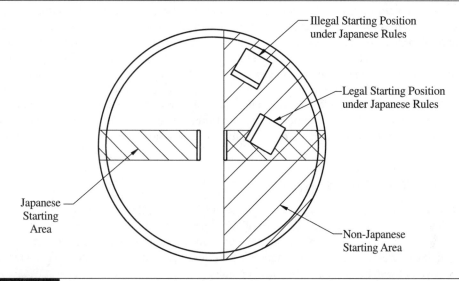

FIGURE 2-4 Robot sumo starting line interpretations

too much time to position your robot, you may receive a warning. The match will start when the referee sees that both robots are ready to begin.

Robot Specifications

The general robot specifications for what is allowed or not allowed are fairly self-explanatory. At the very beginning of a match, the robot must fit inside a 20 cm by 20 cm (7.87 inch by 7.87 inch) square box, and there is no height limit.

What is not stated in the rules is that the robots are allowed to change shape once the match starts. This means that the robot can have arms that deploy sideways, scoops that deploy forward, and a body that expands to increase the width of the robot. All of these actions are allowed, and they are used quite often in many different robots. All the robot must do is fit entirely within the 20 cm by 20 cm square box at the start of the match and not weigh more than 3 kg (6.61 pounds).

In Japan, only commercial radio-control systems are allowed to be used with remote-control systems. The rules that describe the radio-control unit brands and about being able to remove and replace the radio-frequency crystals are intended to prevent radio interference between two robots. There are so many robots that compete in Japan that these rules are necessary.

Some contests allow robots to be controlled by other remote-control methods, such as infrared, sound, and radio modems and tethered connections. All of these methods are fine when there are not very many radio-controlled robots competing. However, when dozens of radio-controlled robots show up, the rules need to be stricter to prevent any chance of radio-control interference. An uncontrolled 3 kg robot flying into a crowd can injure people, so controlling the remote-control method is important for safety and liability reasons.

With autonomous robots, the robot must not move for five seconds after the start button is pressed. This is the only restriction to the control logic inside an autonomous robot. A fundamental part of the Japanese robot strategy is to have several different programs in the robot. The top of the robot will have a set of switches for selecting which program will be used. Each program will use a unique offensive and defensive strategy. This way, challengers can choose the right program based on the competition and the orientation of their robot with respect to their opponent's robot. For some reason, robots outside Japan have not used this capability. Instead, competitors rely on how they orient their robot with respect to the other robot. This is one of the reasons why some people are very sensitive to when their competitor sets down the opposing robot and where it is positioned.

Robot Restrictions

Article 6 of the rules states what the robots cannot do. The first rule relates to your robot not having a feature that will disrupt the ability of your opponent's ability to remotely control their robot. This means that your robot cannot have any radio-frequency-jamming devices or any feature that can prevent the operator from seeing the robot. The strobe light example makes it clear that you cannot blind the radio-control operator.

Your robot cannot shoot, launch, or throw anything. Your robot cannot use anything flammable. Your robot cannot drop, throw, ooze, toss, drip, or spray any liquid, powder, or gas. For example, you cannot use oil slicks, drop metal filings to get caught in your opponent's electric motors, or pour a liquid on your opponent's electronics.

The last rule in this section is the interesting one: "The robot will not include any part that fixes the robot to the dohyo surface and prevents it from moving (such as suckers, glue, and so on)." Many people have interpreted this to mean that your robot cannot use sticky wheels, vacuums, or magnets to help improve traction. This rule does not say you cannot do this.

Your robot must be able to continuously move around. The reason for this rule is to prevent a robot from gluing itself to the ring so that it cannot be pushed out of the ring by its opponent. If your robot has sticky wheels, vacuums, or magnets and can still move, then those components are allowed. As long as your robot continuously moves, you can use anything to help improve traction.

When improving traction, if the device leaves any residue on the dohyo, then you are in violation of the no liquid, powder, or gas rule. If you use a double-sided tape on your wheels, and part of it comes off and sticks to the dohyo, then you are in violation of the last clause of Article 6. Again, as long as nothing attaches or leaves a residue on the dohyo, and your robot continuously moves, you can use anything you want to use to increase traction.

One of the main rules that must be adhered to is that your robot cannot have any feature that can damage or deface the dohyo. An occasional scratch here and there is expected as part of normal wear and tear. But if your robot has some feature that plows into the surface of the dohyo and scrapes off a lot of paint or dents, tears, bends, or gouges the dohyo, your robot will be disqualified—whether the damage was accidental or intentional. Disqualification may occur before, during, or after a match.

This is a subjective evaluation. If the officials feel that a feature might damage the dohyo, your robot will be disqualified. This rule might sound harsh, but many contests have only one dohyo, and the officials do not want to risk the entire tournament because of one robot. When the only dohyo available is damaged, the rest of the tournament must be canceled.

Restarting a Match

From time to time, robots will keep rotating in a circle or stop moving entirely. For example, two robots may get tangled and then just rotate around together in a circle. Another common occurrence is when two robots hit together face to face, and they both stop moving. Their wheels may still be rotating, but the robots are not moving. The referee can stop the match, separate the robots, and then restart the match. The referee will often do this when it looks like the situation will not change. There are other times when both robots fall out of the dohyo at the same time, and so the match is restarted.

If the match is restarted, the time limit is continued from the previous match. No maintenance is allowed when the referee stops the match. The match is restarted immediately.

If the robots keep getting stuck, the referee can designate the starting positions anywhere on the dohyo, as long as the restart positions do not give an advantage to one of the robots (see Figure 2-5).

Modification of the Rules

Article 21 of the rules states that the rules can be modified by the rules committee. Since most contests are privately held, the contest organizers are the local rules committee, and they are responsible for making sure that the rules are posted where all contestants can see the current rules.

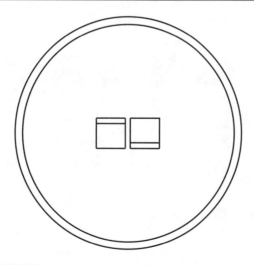

FIGURE 2-5 Possible referee restart position for two robots that keep getting stuck together (both robots are side by side facing opposite directions in this example)

In order for robot sumo to become a true international event, the same set of rules must be followed by all major contests. Many clubs throughout the world are slowly changing their local rules to match the official Japanese rules. The one difference that is not changing is that most of these clubs are keeping the other weight classes (described in the next section) in addition to the Japanese 3 kg weight class. The rest of the rules generally follow the same spirit of the Japanese rules.

Robot Sumo Weight Classes

The Japanese rules have only one weight class, but around the world there are other weight classes that follow the same rules. There are three primary weight classes used around the world in robot sumo: 3 kg, mini sumo, and lightweight sumo. There are many more weight classes, but these are the most popular.

Probably the most popular weight class outside Japan is the mini sumo class. This class was invented by Bill Harrison of Sine Robotics (www.sinerobotics.com). Table 2-1 shows the specifications for the mini sumo robots. The mini sumo robots and mini sumo ring are small, and they are the least expensive to build of all of the classes. Most of the parts that go into a mini sumo robot can come from toys found around the house. There are hundreds, if not thousands, of mini sumo robots competing around the world. Because of their popularity, several companies are selling complete mini sumo kits.

The next popular weight class is the lightweight sumo class. Table 2-1 also shows the specifications for this weight class. This class has all of the same specifications as the 3 kg class, except that the weight is reduced to 1 kg. The lighter weight simplifies some of the materials that can be used to build the robot. LEGO robots are very popular in this weight class. Some clubs have LEGO-only sumo contests using this weight class.

Specification	Mini Sumo	Lightweight	3 kg
Maximum weight	500 g	1000 g	3000 g
Size	10 cm x 10 cm	20 cm x 20 cm	20 cm x 20 cm
Height limit	None	None	None
Dohyo diameter	77 cm	154 cm	154 cm
Dohyo height	2.5 cm	5 cm	5 cm

TABLE 2-1 Robot Sumo Class Specifications

Tournament Styles

There are many different styles of tournaments that can be used in robot sumo, including single-elimination, double-elimination, and round-robin style tournaments.

A round-robin tournament has all of the robots competing against all of the other robots. The robot with the best record wins the tournament. This is a fun style because you get to compete against all of the other robots at the tournament. The only drawback to this tournament style is that it can take a long time to complete if there are a lot of robots.

In Japan, they use a single-elimination style tournament. All of the robots are paired together based on a ranking system. When a robot loses a game, it is eliminated from the tournament. This is a brutal tournament, since is doesn't mean the best robot will make it to the final rounds. A lot of luck comes into play with this tournament style, since pairings can eliminate the best robots early. A small electrical problem that is easy to fix, such as by plugging in the battery, can make the difference between continuing or going home.

In a double-elimination style tournament, you need to lose twice in order to be eliminated. Simple operator error mistakes, like not making sure your batteries are charged, does not eliminate you from the tournament. If the top two robots face each other early in the tournament, they will have a chance for a rematch later in the tournament.

Figure 2-6 shows a double-elimination chart for 16 robots. To use this chart, fill in the robot names in numerical order on the chart. Robot 1 will face robot 9 in the first match. If there are

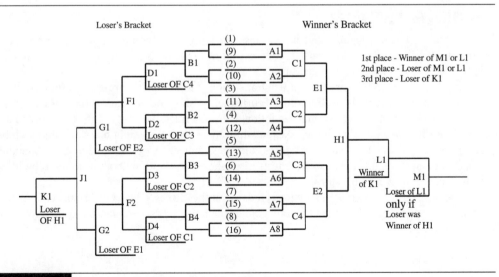

FIGURE 2-6 Double-elimination chart for 16 robots

fewer than 16 robots in the tournament, fill in the remaining spots with a "bye." Then follow the tournament table. All losers will go to the left side of the table. The match order is given by the letters. The "B" matches will play after all of the "A" matches are complete, and so on. The final match will be against the winner of the loser's bracket. If the undefeated robot loses in the final match, there will be a second match against the two robots, since this is a double-elimination tournament.

The drawback to a double-elimination tournament is that it takes about twice as long as a single-elimination tournament. If there are a lot of robots, this can be a long tournament. When planning a tournament, assume that it will take on average three minutes per game (best two out of three matches). Most matches are finished in less than 30 seconds. If there are 16 robots in a single-elimination tournament, it will take a total of 15 games to complete the tournament. At three minutes per game, this will take a total of 45 minutes to complete. If you are holding a double-elimination tournament, it will take 29 games to complete the tournament. At three minutes per game, it will take about an hour and a half to complete all of the games.

As the number of robots increases, the longer it takes to complete a tournament. Large tournaments will require the use of multiple sumo rings to finish in a single day. In Japan, the final tournament has 128 robots competing in one day. When considering the amount of time required to complete a tournament, you can see why a single-elimination style tournament is used in Japan. If you have the time, use the double-elimination style. All of the contestants prefer that style, since they get a second chance to show off their robots.

A Simple Game

Robot sumo is a friendly game with simple rules. The spirit of the competition is to push your opponent out of the ring. As long as everyone is following the rules and the spirit of the game, everyone will have an enjoyable experience.

The rules presented here are the English translations of the Japanese rules, and they are the rules used in the International Robot Sumo Tournaments. Many other major sumo tournaments in the United States and in other countries follow the same rules. The main differences are that many clubs also have other weight classes in addition to the 3 kg weight class and many clubs prohibit the use of vacuum systems, magnets, and sticky wheels. In general, people can take their robots and compete in different robot contests throughout the world, without needing to build a robot specific to each club's rules.

If you have any questions about the interpretation of these rules, contact the events coordinator at your local robot club. The appendix at the end of the book lists several large organizations that organize robot sumo tournaments.

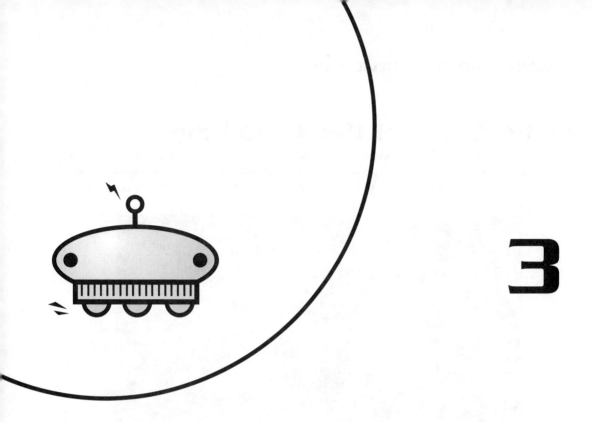

Building the Sumo Ring

The arena where the sumo matches take place is called the *Dohyo* or *sumo ring*. In the human version of sumo wrestling, the Dohyo surface is made out of fine sand that is hardened by salt and water. The perimeter of the Dohyo is marked by tightly rolled straw so that the sumo wrestlers can feel the edge of the Dohyo with their feet.

In robot sumo, the sumo ring surface is a hard, black surface. In order to help the robots "see" the edge of the ring, the outer edge of the ring is painted white, and the ring is also elevated so that robots can "feel" the edge. The sumo ring can be made out of virtually any material, as long as the geometrical specifications are maintained and the proper color schemes are used. This chapter will explain several different techniques for building sumo rings, including mini sumo rings.

Specifications for the Sumo Ring

The following shows an official Japanese robot sumo Dohyo. The Dohyo is mounted on top of an elevated arena, and there is a mini sumo ring on top of a full-size sumo ring. It was built by Fuji Soft Inc. in Japan and donated to Bill Harrison, head of the Northwest Robot Sumo Tournament.

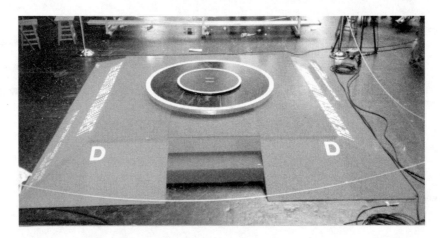

Color and Size Specifications

The sumo ring is a large, flat disk that is 154 centimeters (60-5/8 inches) in diameter, with a height of 5 centimeters (2 inches). The entire surface must be flat and smooth.

The color of the top surface is nonglossy black. The outer 5 centimeters (2 inches) of the ring is called the *border*. The color of the border is glossy white. There should not be any seams that can be felt between the two colors or anywhere else on the surface; this ensures that the robots' front "scoops" do not get caught on the seam and tear the surface material. The surface around the sumo ring can be any color except white. Figure 3-1 shows the specifications for the sumo ring.

Near the center of the sumo ring are two starting lines. The sumo robots must start behind these lines during each match. The color of the starting line is a shade of brown obtained by mixing two parts red with two parts blue and one part yellow. The illuminance of the starting lines must be must be less than 1,000 lux and be halfway between the black area and the white border edge.

The sumo ring can be made out of virtually any material, as long as it meets the specifications for its overall dimensions, the markings have the proper color and orientation, and there are no seams or edges on the competition surface. Most United States sumo rings are made out of wood, but the official Japanese specifications require the sumo ring to be constructed of aluminum.

One of the things to keep in mind about full-sized sumo rings is that they are very heavy. A wooden sumo ring will weigh approximately 100 pounds, and an aluminum one will weigh approximately 150 pounds. A steel surface, as specified by the Japanese rules, will add another 100 pounds to the sumo ring.

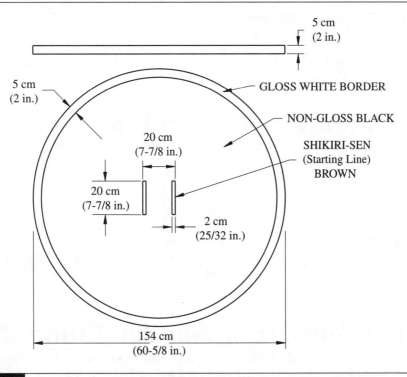

FIGURE 3-1 Size and color specifications for the 154-centimeter (60-5/8-inch) diameter sumo ring

The Surface of the Dohyo

There are two different specifications for the top surface material for the sumo ring: a vinyl surface or a steel surface.

In Japan, the official specification for the vinyl surface material is a long-type vinyl sheet NC, No. R289, made by Toyo Linoleum Inc. Unfortunately, this material is not easily obtainable outside Japan. An equivalent material, called Lonstage, is available from the Lonseal company (www.lonseal.com), located in Carson, California. This vinyl surface material is nearly identical to the R289 material made in Japan.

Lonstage is a flooring material commonly used for theater stage floors. It comes in two different shades of black: glossy black (color number 102) or flat black (color number 101). Either of these colors will work for the sumo ring surface. Lonstage comes in 6-foot wide sheets/rolls. Most other black vinyl material companies sell 3-foot wide sheets. You cannot use the smaller sheets, because they would require a seam between them on the sumo ring surface.

One of the drawbacks to using a vinyl material is that the surface eventually becomes damaged by the robots and storage. Also, this material is relatively expensive to use and replace. Because of these problems, a new Dohyo surface specification was implemented in Japan during the 2001 tournament season. The new specification calls for using a 3.2 millimeter (1/8-inch) thick steel surface instead of the regular vinyl surface. This is not only a more durable surface material, but it also can attract magnets, which gives robot builders more options for improving their robot's traction capabilities.

Building a Practice Sumo Ring

When you start designing sumo robots, it is highly recommended that you build a practice sumo ring on which to test your robots. Having a practice sumo ring is very helpful in testing your sensors and programming strategies.

> **NOTE** *The practice ring can be a simple plywood ring that is painted with the proper color scheme. (The instructions for painting and marking the Dohyo are in the "Painting Your Dohyo" section later in this chapter.) You do not need to use the exact sumo ring dimensions for a practice ring. It can be as simple as a 3/4-inch thick piece of plywood that is 4 feet in diameter. You can use simple wooden blocks to raise the height of the sumo ring.*

Building a Full-Sized Wooden Sumo Ring

Most people find wood the easiest material to use to construct a sumo ring, since it doesn't require any special tools or equipment. However, you will need a large, flat work area, so that you have adequate space to move the ring and materials around during the assembly process. The assembly floor must be flat in order to build a flat ring.

The sumo ring must be designed to allow people to safely walk on it. The easiest approach is to make a solid 5-foot-diameter, 2-inch-thick wood disk. However, even though this method is the simplest, it produces a very heavy Dohyo. An alternative is to use wooden boards to construct the sumo ring. The design is a bit more complicated, but the result is much more portable. This approach is described here.

Wooden Sumo Ring Design

The wooden design presented here uses a wagon-wheel approach, which is much lighter than a solid wood disk. The inner spokes are used to help support the middle of the sumo ring when people stand on it, and the outer wheel edge is used to support the edge of the sumo ring. You can assemble this wooden sumo ring in a few evenings (most of the time is waiting for the glue to dry).

This ring design consists of four different layers. The first two layers are used to construct the wagon wheel, and the other two layers are the plywood surface. The first two layers of the wagon wheel should be as shown in Figure 3-2. (Dimensions are in inches.)

CHAPTER 3: Building the Sumo Ring

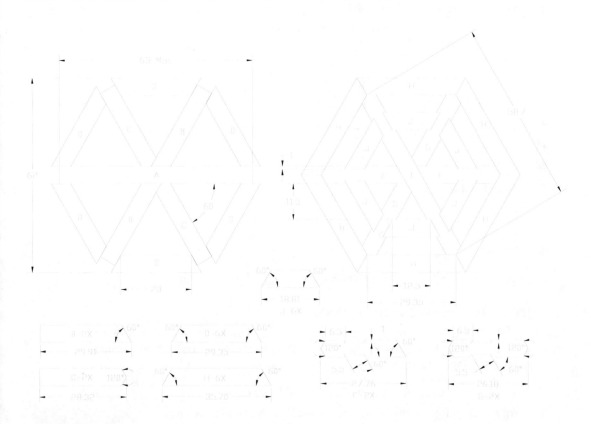

FIGURE 3-2 Schematic of the wagon-wheel design for a full-sized wooden sumo ring

This design uses standard 1 x 6-inch pine boards. Standard 1 x 4-inch boards will also work, but the dimensions of the schematic will be different. The exact dimensions are not critical, because you will trim down this configuration to finish the sumo ring.

Wooden Sumo Ring Construction

There are four basic steps to building the wooden sumo ring:

1. Cut and assemble the first layer.
2. Cut and assemble the second layer.

3. Cut and mount the top wooden surface.

4. Cut the ring to its final diameter and sand the top surface.

You will need the following tools and materials:

- 72 board feet of 1 in. x 6 in. pine boards
- 2 – 4 ft. x 8 ft. x 1/4 in. ACX plywood
- 1 pound of 1-1/4 inch long wood screws
- 1/2 gallon carpenter's wood glue
- 1 pt. general purpose 2 part epoxy
- 120 and 180 grit sand paper
- hand saw
- jigsaw
- trammel

So, if you want to build a full-sized wooden sumo ring, let's get started.

Laying Out the First Layer

The first step is to lay out the first layer of the wagon wheel. This is shown in the left side of the sketch in Figure 3-2. Notice that 63 inches is the minimum dimension; the boards could be longer. You'll lay out 11 boards, which you'll cut in the next steps:

- One board will be longer than the diameter of the sumo ring.
- Two boards will be approximately one half the diameter of the sumo ring and will have a single 60-degree cut.
- Two boards will have two 60-degree cuts on one end of the board.
- Six boards will have a 60-degree cut on both ends, where the smallest side dimension is 23 inches long.

Lay all of these boards on the floor in the arrangement shown on the left side of Figure 3-2. This completes the first step.

Assembling the Second Layer

The second step begins with cutting the six trapezoid boards with the smaller width of 12-1/2 inches and 29-35/100 inches, as shown on the right side of Figure 3-2. Then lay the larger boards on top of the outer perimeter boards from step 1. Don't glue them yet.

CHAPTER 3: Building the Sumo Ring

Next, measure the distance between opposite corners. This measurement should be approximately 58-7/10 inches, as shown on the right side Figure 3-2. Cut a board with the measurement you have just taken, and then make two 30-degree cuts from both ends to create the 120 degree angled cut at the ends of the board, so that this board will fit inside the boards you have just laid down. This long board must be rotated 60 degrees clockwise from the single long board on the first layer.

Now measure and cut to length boards that have the multiple angled cuts at their ends, as shown in Figure 3-2. Don't cut the side notches at this point. Lay all of these boards on top of the boards from the first step. Again, don't glue these boards together yet.

Next, lay the smaller trapezoid boards on top of the boards, as shown in Figure 3-2. Using a pencil, mark all of the edges of these smaller boards on the larger boards. Then cut these notches out. Reassemble this structure until the second layer looks like what is shown on the right side of Figure 3-2. The following shows a photograph of the first and second layers assembled together. The left image shows the boards used to lay out the first layer. The right image shows the second-layer boards laid on top of the first-layer boards.

With the second layer on top of the first layer, you now need to draw a 60-inch diameter circle that is centered on these boards. This circle is used to mark the perimeter boundaries for the wood screws that you will use to assemble this structure. You can use one of the two methods below to draw this circle:

- ■ Use a 3-foot long board, a sharp nail, and a pencil. Drive the nail through the board about 1 inch from the end of the board, so that it protrudes about 1/8 inch from the surface. Measure 30 inches from the nail along the board, and drill a hole just large enough to hold a pencil. Finally, press the pencil through the hole and trace the circle.

- ■ Use a device called a trammel. Two different types of trammels are shown here. With this device, you can make a simple and accurate beam compass. With the trammel, trace a 60-inch diameter circle on the wooden structure. The trammels shown here can be mail ordered from Lee Valley Tools (www.leevalley.com). The trammel on the bottom uses a yard stick, and the trammel on the top mounts onto a 3/4-inch wide board.

Now glue all of the top boards to the bottom boards using a good-quality carpenter's glue. Remember not to move the bottom boards that you positioned in the first layer.

After all of the boards are glued together, use 1-1/4-inch long wood screws to fasten all of the boards together from the top. Make sure you don't place any screws outside the circle you have just drawn. You will be cutting the ring down to the 60-5/8-inch size, and you don't want to run into any screws with the saw. Step 2 is now complete.

Cutting and Mounting the Top Wooden Surface

At this point, the thickness of the two assembled layers is approximately 1-1/2 inches. The next step is to cut and mount the top wooden surface. You will cut and mount two layers of plywood to form a strong sumo ring that can withstand bending without cracking.

You need interior quality (on at least one surface) plywood that is 1/4-inch thick, not 1/2-inch thick. Some lumber stores may carry plywood that is wide enough to be cut into a single 60-5/8-inch diameter circle (such as for Ping-Pong tables or cabinets). However, most hardware stores sell plywood only in sheets of 4 x 8 feet. The directions here are for using 4 x 8-foot plywood sheets, in which you will cut semicircles to join together.

Cut four 31-inch radius semicircles from the 1/4-inch thick plywood. Figure 3-3 shows how to lay out the two semicircles on a sheet of plywood.

The seam for the first two semicircles will be oriented 60 degrees clockwise from the long board on the second layer of wood (built in step 2). Also, this seam must be centered on the boards and be directly over the boards in the two previous layers that do not have a single long board. By comparing Figures 3-4 and 3-2, you should see how the seam is supposed to be oriented with the underlying boards.

Coat the second-layer boards (from step 2) with carpenter's glue, and then place two plywood semicircles on top, to form the circle. Butt the flat ends as close together as possible, and then screw the plywood sheets to the frame. Don't place any screws within 1 inch from the outside diameter of the plywood sheets.

CHAPTER 3: Building the Sumo Ring

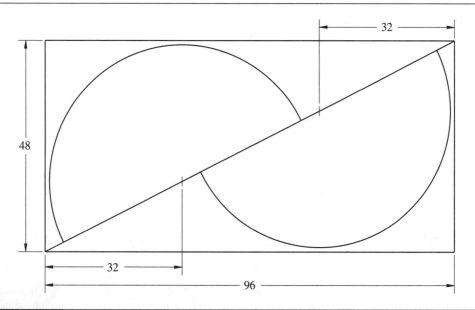

FIGURE 3-3 The semicircle layout on the 4 x 8-foot plywood

Next, glue the other two plywood semicircles on top of the first two. Place the seam orientation parallel and centered over the original long board from step 1. Butt the flat edges as close together as possible, and make sure the outside diameter is centered over the previous plywood layer. Then glue and screw this layer to the first plywood layer. Again, do not place any screws within 1 inch from the outside diameter, or you will end up cutting them with a saw when you are trimming the ring to size. Make sure all of the screw heads are recessed. Finally, place a lot of weight on the top surface, and let it set for a couple days.

Cutting the Ring to Its Final Diameter and Finishing the Top Surface

The last step in assembling the wooden sumo ring structure is to cut the ring to its final diameter. After the glue has dried, lift the ring on top of two sawhorses. Using a trammel, or a 3 foot long board similar to what was described in step 2, mark the 60-5/8 inch diameter circle. Then use a saber saw to cut the ring to the final diameter.

After trimming the ring, recess the wood screw tops below the surface. Then fill in the seams, cracks, screw divots, and wood knots. If you are going to place a vinyl surface on top of the sumo ring, use regular wood filler. If you are going to use a painted wood surface, you should use a two-part epoxy, which is more durable than wood filler. Then sand the top surface so it is smooth and flat.

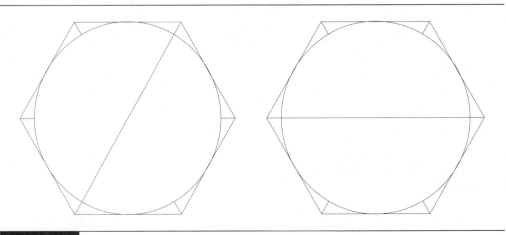

FIGURE 3-4 Top plywood sheet orientation

After you've filled and sanded the top surface, wash off all of the sawdust with soapy water and a damp sponge. Don't pour water on the ring! Use only a damp sponge. Make sure the surface is completely dry before painting or attaching a top surface to the sumo ring. The "Adding the Top Surface to the Dohyo" section later in this chapter describes how to place the final top surface on the ring. The "Painting Your Dohyo" section later in this chapter explains how to paint the sumo ring.

Building a Full-Sized Metal Dohyo

The original Japanese specifications for robot sumo matches require the sumo ring to be made of aluminum, with a 3.2 millimeter (1/8 inch) cold-rolled mild steel disk bonded to the top of the aluminum surface. Aluminum sumo rings are more difficult to fabricate, which is why wooden sumo rings are commonly used. If you plan to build a metal sumo ring, you'll need the proper equipment to weld and flatten large aluminum structures, as well as welding experience. The key is to produce a lightweight, flat structure.

Metal Sumo Ring Design

Figure 3-5 shows a simple schematic for fabricating an aluminum sumo ring. This ring is made from standard 3-inch wide, 0.17-inch-thick web, extruded C-channel aluminum. Extruded C-channel aluminum is specified by its web width and thickness. The flange width for this C-channel is 1-3/4 inches, and the approximate weight is 1-6/10 pounds per linear foot.

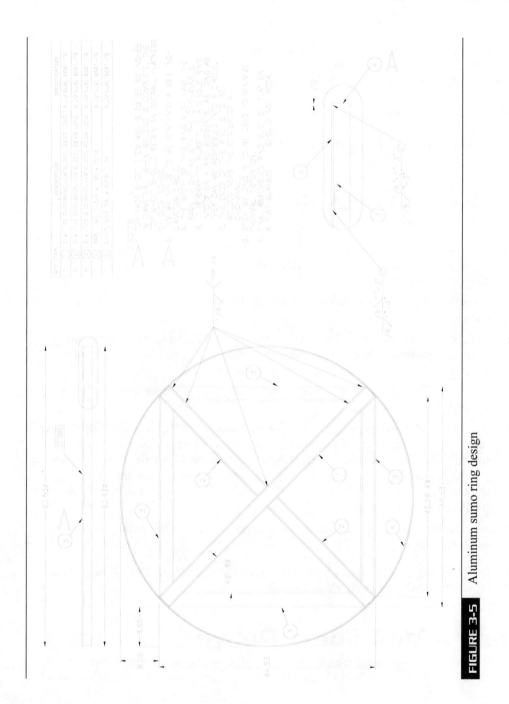

FIGURE 3-5 Aluminum sumo ring design

This configuration uses a simple box shape to minimize the amount of cutting on the extruded channels. Around the perimeter is a 3/16-inch thick by 1-3/4-inch wide band. The purpose of the band is to prevent the outside of the sumo ring from bending when people kneel or stand on the ring.

The top surface should be a single piece. You can get 3/16-inch thick aluminum in 72 x 144-inch long sheets. This is a fairly large piece of metal, so you should have it precut in the 60-5/8 inch diameter disk at a metal fabrication house or by the material supplier. The disk should be transported in a plywood crate to avoid any bending.

Metal Sumo Ring Construction

Here are the steps for building the aluminum sumo ring:

1. Cut to shape and weld together the square box frame. Make sure the frame is flat. If it is not, then manually flatten it. The welds should be quickly made to minimize thermal distortion in the frame structure.
2. Place the frame on top and centered on the aluminum disk. The flanges should be pointing up.
3. Use a stitch-welding process to weld the frame to the disk. Stitch welding minimizes thermal distortions in the structure. Weld individual stitches on alternate sides of the structure to help prevent thermal distortion in the structure.
4. Weld the outside perimeter band to the ring. Again, use a stitch-welding process to attach the band to the ring.
5. After the welding is complete, turn the ring right side up and set it on the floor to make sure it sits flush on the ground. Inspect the top surface to make sure that it is flat and smooth, with no physical distortions at the weld locations. Minor flaws can be grounded smooth. Major flaws may require that you start over and build a new ring. You might want to add a set of threaded foot pads on the bottom of the sumo ring to make sure that it will sit flat on the ground.
6. Sand the top surface with fine sandpaper to clean and prepare the surface for the final surface coating.

See the "Applying the Top Surface to the Dohyo" and "Painting Your Dohyo" sections later in this chapter for instructions on finishing the Dohyo.

Building a Mini Sumo Dohyo

The mini sumo Dohyo is probably the easiest ring to build because its dimensions are exactly half of the 154-centimeter (60-5/8-inch) diameter Dohyo. Figure 3-6 shows the specifications for a mini sumo ring.

CHAPTER 3: Building the Sumo Ring

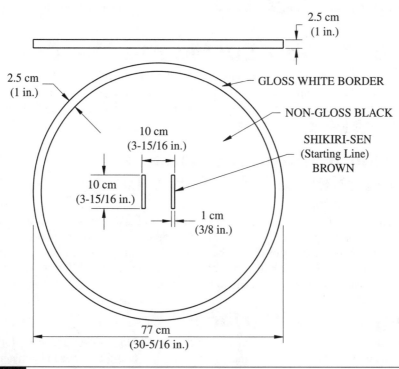

FIGURE 3-6 Specifications for the mini sumo Dohyo

You can construct a wooden mini sumo ring with a wide variety of materials, such as plywood, particleboard, or formica-covered wood. First, you need to decide which type of surface you will place on the sumo ring. This determines the wood's thickness:

- If the top surface will just be painted, use 1-inch thick wood.
- If the top surface will be vinyl, use 7/8-inch thick material.
- If the top surface will be metal, use 15/16-inch thick board.

The next step is to cut the wood into a 30-5/16 inch diameter disk. Use a trammel to mark the diameter on the wood and cut the board using either a saber saw or a bandsaw.

You can build an aluminum mini sumo ring using a design similar to that for the full-sized aluminum Dohyo, as described in the previous section, but you don't need to weld it. You can bolt the structure together. Screw the top surface down to the base frame with flathead screws if you're going to place a vinyl surface on the top. It is best to tack-weld or glue the screws to prevent them from turning when the bolts are being retightened after the sumo ring has been completed.

The next sections discuss how to apply the top surfaces and how to paint the Dohyo.

Applying the Top Surface to the Dohyo

The top surface is what the sumo robots move across. As explained in the "The Surface of the Dohyo" section at the beginning of the chapter, full-sized sumo rings have either a metal or vinyl surface. For practice and mini sumo rings made of wood, you can simply prepare the top wooden layer and then paint it, as described in the next section.

Preparing a Wooden Surface for a Practice or Mini Sumo Ring

Wooden surfaces are usually rough, with small cracks, pits, knots, and seams. As described earlier in this chapter, you need to sand the entire surface flat and smooth, and then clean it with soap and water and a damp sponge. Next, fill all of the cracks, pits, and seams with a wood filler, or coat the surface with a clear epoxy resin that is commonly used to coat tabletops. Finally, sand the surface smooth again and finish it with really fine sandpaper, 240 mesh or finer. Once the top surface has been sanded smooth, it is ready to be painted.

Adding a Metal Surface

For a more durable surface, you can bond a metal disk to the Dohyo's surface. For a full-sized sumo ring, you will need a piece of 1/8 inch thick steel that is 60-5/8 inches in diameter. You will also need some helpers to properly place the metal surface on the ring.

If you're going to put a metal surface on top of your mini sumo ring, the disk needs to be cut to 30-5/16 inches in diameter. Use 1/16-inch thick steel. These thicknesses are minimum guidelines for the lightest weight sumo rings.

The wooden surface must be sanded flat and smooth, but the imperfections in the surface do not need to be filled in. The top metal surface can be painted before you bond it to the wooden surface (see the next section for details on painting the sumo ring). Before attaching the metal surface to the wooden sumo ring, clean both surfaces with soap and water (don't soak the wood with water – use a damp sponge) and make sure it's completely dried before assembling.

You can bond the metal surface to the sumo ring with tape or cement. Double-back foam tape will provide a good bond between the two materials. When using the tape, you do not need to cover the entire surface. Place the tape in small strips around the perimeter and at various random locations in the center of the ring.

One of the best adhesives to bond metal to wood or metal to metal is contact cement. You will use about a quart of this cement for a mini sumo ring and two to three quarts for a full-sized sumo ring. Make sure all of the surfaces are clean and dry before applying the contact cement. Apply the contact cement on both surfaces to be bonded and let it set untilit becomes tacky.

When you're ready to place the top surface, with the help of a couple other people, very carefully apply the top metal surface to the sumo ring. Make sure the two surfaces are concentrically aligned before you let the adhesives come in contact with each other, or you will bend the metal surface when you try to get it off to realign it. After the two materials are bonded together, place some weights on the top surface and let the ring stand for a day.

Adding a Vinyl Surface

To bond a vinyl surface to the top of the sumo ring, you'll need the following tools and materials:

- The manufacturer's recommended adhesive for the particular vinyl material and the material to which it will be bonded. For the Lonstage vinyl, the manufacturer recommends adhesive number 555 for bonding to a wooden surface, and adhesive number 880 to bond to a metal surface. (Their adhesive number 300, a two-part epoxy, is also a good bonding adhesive.)
- A single sheet of vinyl (no seams are allowed on the top surface).
- A 1/16-inch notched trowel for applying the adhesive.
- A three-sectioned, 100-pound roller to obtain good adhesion. You can rent a roller from a local equipment rental store.
- A 2-foot square piece of plywood, to stand on if you need to stand on the sumo ring while using the roller. This way, you will not create any indentations with your feet in the adhesive.
- A sharp utility knife for trimming excess vinyl.

Here are the steps for adding a vinyl surface:

1. Clean the surface of the sumo ring with soap and water and a damp sponge.

2. Cut the vinyl into a square or circle that is larger than the diameter of the sumo ring.

3. Roll up the material backwards, so that the bottom surface of the vinyl is on the outside surface of the roll.

4. Apply the adhesive to the sumo ring's surface using a 1/16-inch notched trowel, starting from the center and working toward the edges. The adhesive should be evenly distributed across the entire surface.

5. Using a roller, gently roll the vinyl from one side of the sumo ring to the other side. Flopping the material on the surface could result in catching some air bubbles. It is best to work from the center of the sumo ring toward the edges. Do not let the roller roll off the edge of the sumo ring. Run the roller parallel to the edge with only a slight portion of the roller overhanging the edge.

6. After the top surface has been attached, let it set for three days to cure.

7. Once the adhesive has cured, trim off the excess vinyl using a sharp utility knife. The vinyl cuts very easy, so be patient and cut slowly. It is best to rough trim the vinyl to within 1/4 inch from the sumo ring edge, then remove the remaining material in a second cutting pass.

Painting Your Dohyo

After you've built your sumo ring and cleaned the entire surface, you're ready to paint it. You'll need the following items:

- Primer paint
- Glossy white paint
- Semigloss black paint (for wooden rings)
- Semigloss or flat brown paint
- Painter's trimming masking tape
- Newspaper
- A sharp knife, such as an X-Acto knife
- Scrap wood to make a marking tool
- A drill with a drill bit that has the same diameter of the pen/pencil diameter that will be used to mark the perimeter edge of the Dohyo

You can use spray paint for the primer and colors.

Painting the Wooden Surface

The first step in painting the wooden surface of a sumo ring is to cover it with a primer paint, which will give it a good bond to the colored paints. Primer spray paint will work for this application.

After the primer is dry, paint the entire surface with the semigloss black paint. Again, spray paint will work here. Hold the paint can about 12 inches away from the surface, and don't let the paint run on the surface. It is best to use multiple thin layers of paint and allow the paint to become semidry between each layer. Let the paint dry for at least two days before painting the starting lines and the border.

Creating a White Border Edge Marking Tool

Prior to painting the border, you will need to make a simple tool to use to mark the inside edge of the white border on the sumo ring. This is a tool you can build to use to mark the white border edge.

Follow these steps to make the border-marking tool:

1. Trace the outside diameter curve of the sumo ring onto a scrap piece of wood.
2. Cut the curve into the piece of wood. You will want to keep the convex side of the curve.
3. Screw another piece of wood on top of the curved piece, and allow it overhang the curved edge by at least 3 inches.
4. Mark a hole center 2 inches, or 1 inch for a mini-sumo ring, from the curved edge.
5. Drill a hole to hold a pen or pencil. You can attach a second piece of wood to the top of this part to make a better pencil support.

Masking the Starting Lines

The next step is to mask the starting lines near the center of the sumo ring. Use masking tape that is designed to be placed on painted surfaces for painting trim work. 3M sells blue masking tape, called Safe-Release Painters Masking Tape (part number 2090), that works well for this application. Don't use masking tape that has a high-adhesion capability, because it will pull the paint off the surface.

With a pencil, lightly mark the edges of the starting lines on the black surface following the dimensions for the particular sumo ring specification. Then lay some masking tape along the marked lines.

Next, lay the masking tape along the outside edge of the entire sumo ring. The tape must cover the area where the inside edge of the white border will be placed. The following shows the sumo ring with the masking tape applied.

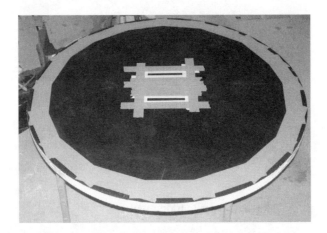

Using the border-marking tool described in the previous section, place a mark on the tape's surface by running the tool along the top surface of the sumo ring. Then place newspaper over the areas in the middle of the ring that are not to be painted. Use the masking tape to cover all of the edges of the newspaper, so that paint cannot get under the newspaper or on the black surface.

Once you've applied all of the paper and tape, you need to trim the outer edge of the masking tape. Using a sharp knife such as an X-Acto knife, lightly cut through the tape along the line that you drew on the tape. Then gently peel the tape from the outside edge of the line. Make sure you don't remove any of the tape that is on the inside edge of the line, or cut into the vinyl surface.

Painting the Border and Starting Lines

Now you have a black border that is 2 inches wide, or 1 inch wide on a mini sumo ring, ready to be painted white. Paint the outer border surface with a glossy white paint. Spray paint will work here. Make sure that the spray paint can is at least 12 inches away from the surface. Use several layers of thin paint to avoid having any paint run on the surface. Here is a Dohyo prior to removing the masking tape and newspaper.

Paint the starting lines with a semigloss or flat brown paint. Spray paint will work. The exact color specification is two parts red, two parts blue, and one part yellow. Use multiple thin layers of paint. The final paint coatings should be very thin and not leave an edge when the tape is removed.

Finishing Up

Let the paint set for a couple days before removing the tape. Then gently remove the tape from the entire surface. The paint should not peel off the surface. If it does, you did not get a good paint bond, and it will scratch off easily when the robots run across the surface.

If you feel an edge between the paint colors with your finger, it must be smoothed out. You can do this by drawing a sharp-edged washer parallel with the paint edge. If you don't smooth out the edge, a robot could catch the edge and start scratching the paint off the surface.

Painting Vinyl and Steel Surfaces

The painting steps described in the previous sections can be applied to all of the surface materials, except the black paint is not needed for vinyl surfaces. Prior to painting the vinyl surface, the anti-paint (a wax) coating must be removed. Use a 00 steel wool pad to sand off this coating. If this coating is not removed, the paint will not adhere to the surface of the sumo ring. The following shows a full-sized wooden sumo ring with a vinyl surface.

The Japanese specification for painting the steel surface is to use a nongloss black melamine paint on the entire surface. The paint must be baked on at 130 degrees Celsius (266 degrees Fahrenheit) for 30 minutes, then the glossy white border and the brown starting lines are painted onto the surface.

Final Comments on Building Sumo Rings

The sumo rings described in this chapter represent a couple different ways to build a competition Dohyo. The main specifications that must be adhered to are for the outside diameter, height, color, and markings. There are no specifications for the actual construction methods used to build the rings. The exact details of the sumo ring construction are up to the builder and the local contest rules. As noted at the beginning of the chapter, when you first get started building sumo robots, it is a good idea to build a simple practice sumo ring for testing your robots.

When you're designing a sumo ring, keep in mind the final weight of the ring, since it can be very heavy. A sumo ring should not be rolled on its edge if it has a vinyl surface. The weight of the sumo ring can cause the edges of the vinyl to become damaged.

Once you have built a sumo ring, you're ready to start building your sumo robots. The next chapter begins your journey in learning what it takes to build a sumo robot. As you read each chapter, remember what you have learned about sumo rings in this chapter. Keep in mind that your robot needs to stay on the surface of the Dohyo throughout the contest, while being able to maneuver and push its opponent out of the ring.

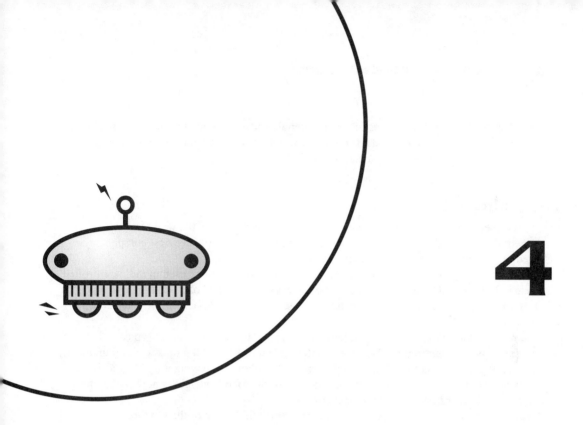

Robot Materials and Construction Techniques

Building a sumo robot is actually a relatively simple and straightforward process. The hardest part is just getting started. The next hardest part is planning what you want your robot to do. As with any project, breaking robot construction down into a series of smaller tasks will make it much easier. This book will help you break down the overall sumo robot construction into a series of smaller steps, so that building a sumo robot becomes an enjoyable process of learning about robots and competing in the exciting world of robot sumo.

Building a sumo robot doesn't require a lot of space. In fact, you can build a sumo robot in your garage or even in your apartment's kitchen. You do not need to use any advanced materials to build the robot, and you can build the robot using common tools. Most robots are made out of plastic or aluminum that you can purchase at your local hardware store.

This chapter will cover the basics of robot materials and construction tools. One of the first things to think about when you build a robot is the weight of the materials that you will use to construct the robot, so that is the topic addressed first.

Robot Weight

Weight is a premium in a sumo robot, so you will want to conserve on the weight of every component that goes into a robot. For most robots, the batteries are the heaviest component, followed by the motors, and then the structural frame.

For example, a typical six-cell nickel-cadmium (NiCd) battery pack used in radio-controlled cars weighs 310 grams. This is about 10 percent of the maximum weight of a 3-kilogram (6.6-pound) robot. If you are planning on using two battery packs, they will be 20 percent of your robot's total weight.

The weight of a small gearmotor can range from 150 grams to 300 grams. Typical robots will use either two or four motors, for a total motor weight of from 300 grams to 1,200 grams. If you use 1,200 grams for motors, that is 40 percent of the robot's total weight. Adding a six-cell battery pack to the four heavy motors means that half of the robot's weight is now spoken for, which leaves the other half for everything else.

Depending on your robot design, the structural frame will be one of the top three heaviest components in the robot. It might be heavier than the motors and batteries combined, or it could be lighter than either one of them.

Knowing the weights of the major components that you will use in your robot will help you to select the right type of materials for robot construction. You want to use the strongest and lightest materials in your robot. It is always easier to add weight to the final robot than it is to drill a bunch of holes in your robot to reduce its weight.

Robot Materials

Wood, plastic, aluminum, and brass are some of the best sumo robot materials in regard to strength, weight, and machinability. There is no point is discussing the virtues of titanium di-boride ceramics for your robot, since only a handful of places in the world can actually machine this material. Machinability is important to consider when selecting a particular material. Remember that you will need to drill, cut, bend, and shape the material into your parts, so keep in mind the types of tools you have available when selecting your robot's materials.

Wood

Wood is a great material for a sumo robot. It is easy to cut and drill, and you can easily glue, bolt, or screw parts together. Making changes to the design doesn't require building new parts, because you can easily move parts around. Wood is also probably the cheapest material you can use to build a sumo robot, and you can obtain it at virtually any hobby store.

Wood is an ideal material for prototyping your robot design. It is easy to cut some parts out of wood and then put them inside your robot to make sure all the parts fit together. You can use a file to shape the parts to make them fit. Then you can use the wooden parts as templates when you cut plastic or aluminum parts.

One of the best wooden materials to use is model airplane plywood. This plywood is very strong, and it is available in thicknesses of 1/32, 1/16, 1/8, and even 1/4 inch.

Plastic

There are hundreds of different types of plastics in common use today. For sumo robot construction, acrylic, polycarbonate, and expanded foam polyvinyl chloride (PVC) are the most common types of plastic.

Acrylic and Polycarbonate Plastic

Acrylic and polycarbonate plastics are hard, strong, rigid, and clear in color. This material is commonly sold as flat sheets with common thicknesses of 1/16, 1/8, 3/16, and 1/4 inch.

Acrylic has a higher tensile strength than polycarbonate. Acrylic can be tapped, so that you can attach threaded fasteners directly to it. A common trade name for acrylic is Plexiglass. To cut acrylic, use sharp tools with low cutting speeds, because acrylic materials have a tendency to chip and crack if the tools are dull. Another form of acrylic is a high-impact acrylic called Implex. This form is easier to machine because it does not have the same cracking tendency.

Polycarbonates have a very high-impact resistance, which is why they are commonly called *bullet-proof glass*. A common trade name for polycarbonates is Lexan. Polycarbonates are much easier to machine than acrylics.

Acrylics and polycarbonates can also be hot-formed or bent by applying heat to the plastic. A common tool that is used to bend plastic is a strip heater. You set the plastic sheet down on the strip heater, align where you want the bend joint to be on the sheet with the heater element, wait a few minutes, and then bend the sheet.

Both acrylics and polycarbonates have a thin, brown paper, protective coating on both sides of the plastic. This material can easily be peeled off. What is nice about the paper coating is that you can mark all of your cut lines without scratching the plastic.

Expanded Foam PVC

Another plastic that is a very good construction material is called Sintra. This material is an expanded foam PVC. The surface of this material is hard, but the inner core is like dense foam. It is not soft like a sponge, but rather hard. If you look at this material closely, you will see tiny air bubbles in the center core. Versacell is another trade name for expanded PVC.

Expanded PVC is the lightest of the three plastics, and it is not as strong as acrylic and polycarbonate. It comes in a wide variety of colors, including black, white, yellow, blue, red, green, and about a dozen other colors. The two common thicknesses for this material are 3 millimeters (about 1/8 inch) and 6 millimeters (about 1/4 inch).

This material is attractive for sumo robots because it is very easy to machine. It can be easily cut and carved with a sharp knife, yet it is strong enough to be tapped (assuming no heavy loads on the fastener/screw). Expanded PVC is ideal for mini and lightweight sumo robots.

> **CAUTION**
>
> *When cutting the PVC with any power tool, make sure that the tool is sharp. If the PVC and saw blade get too warm, the blade will start to push through the plastic instead of cutting the plastic. Then the material behind the saw blade will start to melt back together again. This becomes a problem with scroll saws, since the melted plastic starts to prevent you from backing up.*

Aluminum

Of all of the metals that are available, aluminum is the best one for sumo robots. It is very strong and has the lowest density of all of the common metals available. Magnesium and titanium are also excellent robot materials, but they are very expensive and hard to obtain. Aluminum is relatively cheap, and you can find it at most metal-supply and hardware stores.

Aluminum Alloys

There are many different types of aluminum alloys. Silicon, iron, copper, magnesium, chromium, and zinc are some of the elements that are added to aluminum to make the different types of alloys. The different alloys have unique mechanical properties, such as strength, ductility, hardness, and corrosion resistance.

> **NOTE**
>
> *You may hear someone say that his or her robot is made out of aircraft-grade aluminum. This doesn't mean anything, since aircraft are made out of many different types of aluminum. The type of aluminum that each aircraft part uses is based on the function of that part.*

The most common types of aluminum are 1100, 2024, 5052, 6061, and 7075 series aluminum. These four-digit numbers are specific alloy designations. Do not use 1100 series aluminum for your robot. This material is a commercially pure aluminum that is extremely soft. In fact, you can bend a 1/4-inch diameter rod between your fingers. The 5052 type is a low-strength aluminum that is commonly used in boat bodies and storage tanks and drums.

The 2024, 6061, and 7075 grades of aluminum are the three types that can be used for sumo robots. These types are heat-treatable, which means that their mechanical properties can be altered by different heat-treating processes. The 2024 aluminum is about 20 percent stronger than the 6061 aluminum, and the 7075 aluminum is almost twice as strong as the 6061 aluminum. You can find all three of these aluminum grades at your local metal-supply store. Many hardware stores also sell a small selection of 6061 aluminum as round bars, flat strips, and even 90-degree extruded angles.

CHAPTER 4: Robot Materials and Construction Techniques

NOTE *Usually, the 2024, 6061, and 7075 metals are stamped with an H followed by a number and a T followed by a number. These designations tell you what type of a heat treatment and temper the specific aluminum has undergone. By knowing the particular heat treatment number and the material type, you can look up the mechanical properties for that specific grade of aluminum in a material science book. For sumo robots, the specific heat treatment the aluminum has gone through really doesn't matter.*

Machining Aluminum

What makes aluminum a favorite material for building robots, besides its strength and weight, is that it is one of the easiest metals to machine. Hacksaws, jigsaws, table saws, band saws, and even scroll saws are used to cut aluminum. To cut aluminum, use a sharp, fine-tooth blade with 18 to 32 teeth per inch.

When tapping aluminum, you must use an aluminum-tapping fluid. Aluminum has the tendency to gall. *Galling* means that the metal tries to fuse itself back together. The cuttings may gall in the threads when you try to remove the tap, and the tap may become stuck or break inside the hole, or you may shear off the threads you just made. Aluminum-tapping fluid prevents this from happening. You can purchase aluminum-tapping fluid at any metal-supply store and most hardware stores.

Brass

Brass is a strong metal that is easy to machine, but it is very heavy. Its density is three times greater than aluminum and almost ten times greater than the plastics mentioned earlier in this chapter.

What makes brass an attractive material is the types of shapes you can obtain at hobby and hardware stores. For example, K&S Engineering (www.ksmetals.com) has developed a product line of thin-walled round tubes, square tubes, rectangular tubes, hexagon tubes, 90-degree angles, C-channels, and flat sheets. The wall thickness for the tubes, angles, and channels is 0.014 inch. Their sizes range as small as 1/16 inch in diameter up to 5/8 inch in diameter. The flat sheets have thicknesses of 0.016, 0.025, 1/32, and 1/16 inch.

For a sumo robot, you can solder or braze the brass tubes together to form a strong tubular frame for your robot. You can screw or braze the brass sheets together to form a solid shell around your robot. Thin brass sheets can be glued to a wooden or plastic surface to make a strong armor for your robot.

Material Comparison

Table 4-1 shows a comparison of the different materials discussed in this chapter sorted in increasing strength order. All materials have elastic and permanent deformation zones. When a material is stretched, bent, or twisted, there is a certain amount of stress that is induced into the material. When the stress is less than the yield strength of the material, it will return back to its original shape when the load is removed. But if the stress is greater than the yield strength, the material will be permanently deformed. If the stress is further increased, the material will eventually break. Stress is defined as the applied load on a part divided by the cross-sectional area where the load is being applied. For example, a 1/4-inch diameter bar that is supporting 100 pounds will have an internal stress of 2,038 pounds per square inch (psi), calculated as 2,038 psi = 100 pounds / (3.14 × 0.25 inch × 0.25 inch / 4)).

In Table 4-1, you can see that plastic materials have the lowest strength and 1100 aluminum has second lowest strength. As mentioned earlier, allthough titanium has the highest strength properties, it is the most expensive material and is very hard to obtain and machine. But if you are fortunate enough to get some titanium, you can create some very strong, lightweight robot structures.

When weight and strength are important, use 6061, 2024, or 7075 aluminum. When weight is your major concern and strength is not, consider using either polycarbonate or acrylic materials. For weight-critical applications, you might want to use expanded PVC.

When it comes to sumo robots, don't think that material strength is everything. A mini sumo robot weighs only 500 grams (1.1 pound), and the full-sized sumo robot weighs only 3 kilograms (6.6 pounds). The robot only needs to be able to support its own weight and the impact with another robot of the same weight. Many successful robots in all weight classes are made out of expanded PVC.

Material	Density (lb/in^3)	Yield Strength (lb/in^2)
Expanded PVC (Sintra)	0.025	2,000
Aluminum, 1100	0.098	4,000
Polycarbonate	0.045	9,000
Acrylic	0.043	11,000
Steel, 1018	0.283	32,000
Aluminum, 6061-T6	0.098	40,000
Aluminum, 2024-T4	0.098	43,000
Brass, 260	0.310	52,000
Aluminum, 7075-T6	0.098	78,000
Titanium, 6Al-4V	0.17	145,000

TABLE 4-1 Strength and Density Properties of Various Materials

Fasteners for Robot Assembly

There are four different ways to assemble your robot: screws/bolts, adhesives, rivets, and welding. Both welding and riveting are permanent assembly methods. For smaller robots, you should avoid permanent fastening, because you will need to disassemble your robot for maintenance and the never-ending upgrades you will want to perform.

Adhesives

The two types of adhesives to consider when building a robot are permanent and temporary adhesives. You should use permanent adhesives, such as superglue and epoxies, only on parts that you do not intend to take apart in the future. You should always have temporary adhesives in your collection of tools.

Acrylic, polycarbonate, and expanded foam PVC can be bonded together with two-part epoxies and cyanoacrylates (superglues). For larger robots, or robots that will experience a lot of impacts, you should fasten parts together using screws with metal-support structures, as described in the next section.

One of the best temporary adhesives is the double-sided, sticky, foam tape. This tape is an ideal solution to hold parts together before making permanent fixtures. By temporarily locating a part, you can test the robot to see if that is the correct place for the part.

Screws and Bolts

The best way to assemble parts in your sumo robot is to use screws. Sumo robots are compact, with a lot of components in a small area. You need to be able to assemble and disassemble your robot many times. Using screws is the only practical way to do this. Shown here is a 3-kilogram sumo robot built by Bill Harrison. This robot is packed with internal components, and the assembly screws are clearly visible.

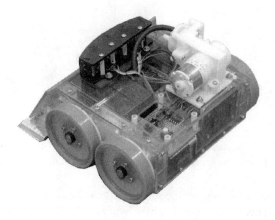

Screws are made from a wide variety of materials, such as high-strength steel, stainless steel, brass, and nylon. There are many different types of screws, including machine screws, wood screws, and sheet-metal screws. The threads on wood and sheet-metal screws are larger and further apart so they can bite into the material and cut their own threads into the work piece.

Machine screws are the preferred type of screw for sumo robots, because these screws are designed for repeated assembly and disassembly. Machine screws require the holes to be predrilled and then tapped. Here is a picture of a small hand tap for cutting threads in a piece of material.

Plastic materials do not have the same strength for holding screws as do aluminum alloys or other metals. When fastening plastic materials, you should use metal nuts or support bars if you anticipate heavy shocks or you need to support high loads. For example, you can place a small, square, aluminum bar stock with threaded holes in the corner between two parts and screw the parts to the bar stock to form a strong joint, as shown here.

TIP *Lynxmotion (www.lynxmotion.com) sells a variety of 1/4-inch square aluminum bars that have pre-tapped holes for #4-40 screws for this type of application.*

Table 4-2 lists some common machine screws that can be used with sumo robots. The Screw Size column shows two numbers, separated by a hyphen. The first number is the screw size, and

CHAPTER 4: Robot Materials and Construction Techniques

Screw Size	Tap Drill Size (in.)	Clearance Hole (in.)
#4-40 (0.112 dia.)	43 (0.089) {3/32}	29 (0.136) {9/64}
#8-32 (0.164 dia.)	29 (0.136) {9/64}	10 (0.1935) {3/16}
#10-24 (0.190 dia.)	25 (0.1495) {5/32}	2 (0.221) {7/32}
1/4-20 (0.250 dia.)	7 (0.201) {13/64}	K (0.281) {9/32}

TABLE 4-2 Tap Drill and Clearance Hole Diameters for Various Machine Screws

the second number is the number of threads per inch. The screw diameter is shown in the parentheses. The Tap Drill Size column shows three sets of numbers. The first number is the specific drill number for making the correct-sized hole, the second number is the actual diameter of the tap drill, and the third number is the closest fractional drill that can be used to tap-drill the holes. The third column shows the same type of information for the clearance-hole diameters that a fastener will pass through when being attached to another part.

One thing to keep in mind is that you need a minimum of four threads to obtain full fastening strength of the thread pattern. This places a constraint on the largest size screw you can use. For example, say you have a 1/8-inch piece of aluminum you want to add some threads to. The minimum thread pitch is 0.031 inches/thread (0.031 inch/thread = 0.125 inch/4 threads). Inverting this number gives you 32 threads per inch. This number tells you that #4-40 and #8-32 screws will work with this part thickness, but larger diameter screws will not provide enough thread engagement for a strong joint.

Tools for Robot Construction

You do not need a computer-controlled numerical (CNC) milling machine, an abrasive waterjet machine, or a laser to make a sumo robot. Although these machines work well for robot construction, they are more cost-effective for high-volume production of robot sumo parts. Besides, most of us do not have one of these in our garage. Most sumo robots are made by using hand and power tools that are commonly found in garages.

Hand Tools

The following are the basic hand tools that will help you build your robots:

Hammer	Screwdrivers
Hack saw with fine-tooth blades	Miter box for the hack saw
Mill and jeweler files	Socket wrenches
Allen wrenches	Pliers

Wire cutters	Clamps
Big vice	Scribe
Adjustable or a Machinists Square	Machinist scale (a 6-inch metal ruler that can measure values less than 1/32 inch)
12-inch ruler	Utility knife
X-Acto knife	Variable-speed hand drill, with a set of drill bits
Thread-tapping set	Safety goggles (which should be worn at all times)

These are the minimum basic tools for mechanically building a sumo robot. Many robots have been built with far fewer tools, and they have done very well in tournaments. Some robots have been built with only a hand drill, an X-Acto knife, and a machinist scale.

A motorized hobby shaping and drilling tool, such as the Dremel MultiPro variable-speed tool, (www.dremel.com) is also useful for building a robot. With its wide assortment of bits, parts can be sanded, drilled, ground, debured, polished, and shaped. If there is any power tool that a hobbyist should obtain, it is a Dremel-style tool.

Power Tools

The following power tools will save you time in making parts:

- Drill press
- Table saw
- Band saw
- Jigsaw
- Scroll saw

Most of these tools are inexpensive and do not take up a lot of space. Figure 4-1 shows a drill press, scroll saw, Dremel tool, and a jigsaw, set on top of a table saw, to show the relative sizes of the power tools. These power tools are not required to build a robot, but they do help simplify some of the various tasks. They are very common tools. If you do not have one of them, there is a good chance that one of your neighbors will have one that you can borrow.

Electronic Tools

The electronic tools you need include the following:

- Soldering iron
- Resin-core solder
- Wire strippers

CHAPTER 4: Robot Materials and Construction Techniques

FIGURE 4-1 Various hobbyist power tools on a table saw

- Desoldering wick
- Several spools of hookup wire
- Multimeter able to measure voltage, current, and resistance (two multimeters are really handy for monitoring two signals—such as current and voltage—at once)
- Solderless breadboard (for testing all of your circuits before you build the final circuits in your robot)
- At least a dozen multicolored test leads (the long wires with either alligator or pigtail clips on both ends of the wires)

An oscilloscope is an optional electronics instrument that can be very useful for analyzing circuits, especially circuits that generate time-dependent signals such as square waves. The oscilloscope is one of those tools that once you have used it, you will never want to work without it. Unfortunately, oscilloscopes are fairly expensive; in fact, they can be more expensive than your sumo robot.

With an oscilloscope, you can measure DC and AC circuits, frequencies, pulse widths, voltage levels, and logic signals. As with a computer, the faster it can operate, the more expensive it becomes. For most robotic applications, a 20 MHz (low-speed) oscilloscope is fast enough for most of your needs. The oscilloscope should have dual-trace capability (to measure two signals at one time—remember the two multimeter suggestion above?).

For those on a tight budget, a very nice, low-cost, PC-based oscilloscope is available. Parallax (www.parallaxinc.com) is one of the distributors of the OPTAscope 81M from Optimum Designs Inc. (www.optascope.com). This is a small, dual-channel, 8-ounce, 5 x 2-1/4 x 1-1/2-inch oscilloscope that connects to your computer via a USB connection. The peak sampling rate for this oscilloscope is only one million samples per second, but this is fast enough for monitoring servo pulses or 40 kHz infrared reflective sensors. With the OPTAscope, you can use measurement cursors to measure the voltage, pulse width, frequency, and differences between the two separate channels. Figure 4-2 shows the OPTAscope window. As you can see, it has all the basic features of a standard oscilloscope.

The OPTAscope does not require any external power, and it can run concurrently with other programs on the computer. Figure 4-3 shows a laptop PC running both the OPTAscope and the Basic Stamp editor software. (The Basic Stamp products will be discussed in detail in Chapter 11) The OPTAscope is being used to measure the output signals from a Basic Stamp on the Board of Education. With this setup, you can use a single PC to write, debug, and electrically test the same circuit. Since the OPTAscope gets all of its power from the PC, the PC can become a true field robotic programming and diagnostic system. This can be very helpful when you need to make some changes to your program and analyze circuit problems while you are at a contest.

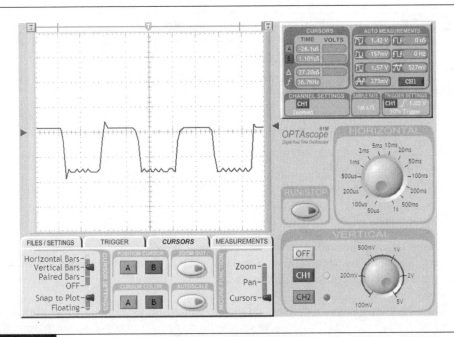

FIGURE 4-2 OPTAscope 81M PC-based oscilloscope

CHAPTER 4: Robot Materials and Construc

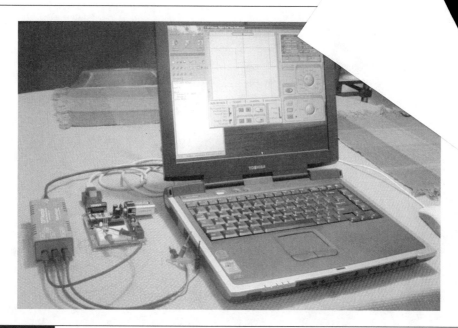

FIGURE 4-3 OPTAscope and a laptop PC being used to analyze and program a Basic Stamp circuit in real time

Final Comments on Robot Materials and Construction Techniques

When you are building a robot, your choice of materials and construction techniques depends on your level of experience, the equipment you have available, and your personal preferences. Some people prefer one method over another, when either method will work just fine. Some people like to custom-build every single part that will go into the robot. Other people like to purchase kits and parts and assemble their robots with off-the-shelf technology. Some people are very mechanical. Others have strong software or electronic skills.

There is no best material, no best construction technique, and no best robot design. If there were, then everyone's robots would look the same. When it comes to building a robot, all of the builder's unique skills are culminated into a unique robot.

All of the materials discussed in this chapter will work in a robot, but whatever material you select, it will have an effect on some of the other parts that will go into the robot. Before building your robot, you should read through this book to learn how all of the various parts that go into a robot affect one another in the final design. When you understand the various relationships, then you can start designing your sumo robot, using the information in each chapter as a guide to selecting the right components, designs, and strategies.

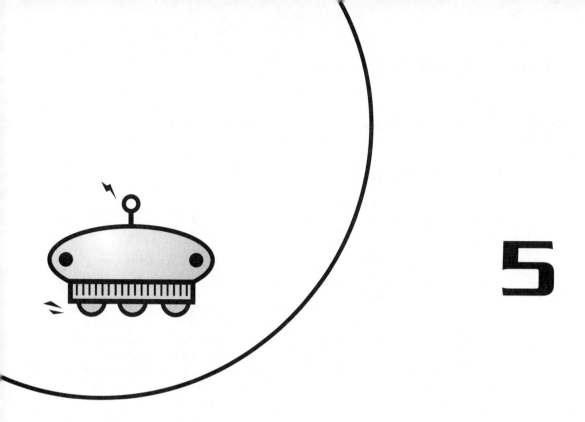

5

Making Your Robot Move

Moving is one of the most important characteristics of a sumo robot. Since the goal of robot sumo is to have your robot push its opponent out of the sumo ring, your robot must be able to move across the sumo ring. The generalized term for describing the mechanics of movement is *locomotion*. This chapter describes the various locomotion styles you can use to make your robot move. Then it provides details on using wheels, which are the most common means of locomotion for sumo robots.

Choosing a Locomotion Style for Your Robot

Usually, the first idea people have for making a robot move is to use wheels, since just about every moving vehicle we see today uses wheels. The next popular choice is tank-style treads, since bulldozers and battle tanks generate a lot of pushing force on soft ground. Though these methods of locomotion are very effective, there are other imaginative methods of locomotion, such as walking on legs or slithering along the floor like a snake.

The Wheeled Robot

The most common sumo robots are the wheeled varieties. The wheel is the oldest, simplest, and most efficient method used to enable a vehicle to move across a surface. Wheels are very popular for robots because they provide good traction, are easy to build or purchase, and are easy to mount to a shaft.

The wheel design and number of wheels for a robot depend on personal preference and on how the wheels will be used. Two- and four-wheel-drive robot configurations are the most common, but other well-built robots have more than four wheels and/or an odd number of wheels. For example, you might see a three-wheeled tricycle robot. For many robot builders, the appearance of the robot is just as important as winning a contest.

Here, we'll look at some examples of wheel design and their advantages and disadvantages. But keep in mind that all of these advantages and disadvantages have not been shown to provide much of a competitive edge in a tournament. Every year, robots with different wheel configurations win tournaments. The fact is that the number of wheels on a robot is more a matter of personal taste than it is a competitive edge.

Two-Wheeled Robots

With any two-wheeled robot, you need some method to stabilize the robot, or it will rock back and forth about its wheels and fall over. A two-wheeled robot will naturally rotate in the opposite direction of motion, because the center of gravity is above the wheel axle. The most common way to solve this problem is to place the wheels towards the rear of the robot and add a pair of casters at the front of the robot, so that the center of gravity is between the front and rear wheels. With the casters, you now have a four-wheeled robot. For sumo robots, casters are more commonly just a ball in a socket, rather than traditional casters such as those used on table legs.

Since the object of sumo is to push the opponent out of the ring, and the surface of the sumo ring is flat and smooth, most robot builders place a scoop at the front of the robot instead of using casters. The following picture of Nemesis, built by Pete Burrows, shows a two-wheeled robot using a front scoop. The front scoop slides on the surface of the sumo ring and provides the balancing point to stabilize the robot. The disadvantage to using a sliding front scoop is that it can get jammed under the other robot and cause your robot to stall.

Four-Wheeled Robots

The other popular robot configuration is the four-wheeled variety. Here's an example of a four-wheel-drive robot, a see-through four-wheel-drive robot, built by Bill Harrison from Sine Robotics.

The main advantage of a four-wheeled robot is that all (or, at least, almost all) of the robot's weight is on the driving wheels; the more weight on the drive wheels, the more traction the robot will have. With two-wheel-drive robots, some of the robot's weight is on nondrive wheels, which is a big disadvantage.

Along with traction, another advantage of four wheels is that the robot will have pushing power available to all of its wheels, which can be helpful if the front or rear wheels are lifted in the air. The disadvantage is that a four-wheeled robot requires more energy to turn than a two-wheeled robot. However, a robot with four drive motors can use smaller motors to get the same pushing capability as a two-wheeled robot.

Another interesting advantage is that if one of the motors should fail during a contest, the robot will still have three motors remaining. A robot with only two motors won't be so lucky if one of the motors fails, since all it will be able to do is move in circles.

Robots with Six or More Wheels

Robots with six or more wheels have all of the same advantages and disadvantages that four-wheeled robots have when compared with two-wheeled robots. The only real advantage to more wheels is that you can use six or more smaller motors to drive the robot. The total pushing capability of these smaller motors can be greater than two larger motors, and sometimes it is easier to mount the smaller motors in the robot. Here is an example of a robot with six wheels and six motors. This is Overkill, built by Jim Frye of Lynxmotion.

The Tank-Treaded Robot

Some of the coolest looking robots are built with tank-style treads. Tank-style robots generally have an intimidating appearance, with their low profile and solid construction. The next illustration shows an example of a robot that uses pure gum rubber treads. It's The Beast, built by Carlo Bertocchini, which uses tank-style treads to beat its opponents. It is one of the top autonomous robots competing in the United States and has been winning contests since 1993.

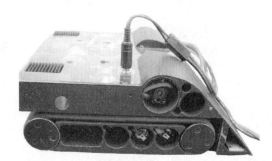

Tank-style treads provide a lot of surface area to be in contact with the sumo ring, improving the pushing capabilities of the robot. During a competition, a robot using tank-style treads can drive almost halfway off the sumo ring without falling off. This effectively gives a tank-style robot a little larger competition ring than a wheeled robot has. Another advantage is that only one motor is needed to drive a single track, whereas a four-wheel-drive robot may require four motors to drive all four wheels. This design does not require any special gearing to drive the front and rear wheels at the same time, since they are coupled together via the tank tread.

There are two disadvantages of using tank-style treads on a robot: keeping the treads on the wheels/cogs and the frictional resistance to turning. Unlike wheeled robots, a treaded robot must use differential steering to turn around. Since the treads are physically laid out on a straight line, most of the tread surface will be sliding sideways when turning. This sideways loading can cause the treads to come off of the wheels/cogs driving them. This is why treads should have teeth in them to keep them in place when turning. It can be very embarrassing to watch your treads fall off during a competition (but not as bad as forgetting to put in your batteries).

The advantage of the larger surface contact area that treads provide is also the main disadvantage. Tracked robots must expend more energy to turn, since they need to break the frictional contact with the ground. This requires the robot to use more powerful motors, and more energy is wasted from the batteries.

Types of Tank Treads

Only a small percentage of sumo robots use tank treads, mainly because it can be difficult to find good-quality tank treads. One popular source for tank treads is toy tanks and bulldozers. You can remove the drive wheels and treads from the toy, or you can convert the entire toy into a sumo robot. Many stores sell small remote-control (R/C) vehicles that can be converted into a competitive sumo robot. Here is an example of how a lightweight sumo robot and a mini sumo robot can be built from the same treaded toy. These are lightweight (top) and a mini sumo (bottom) class robot, made by Bill Harrison of Sine Robotics. They were built from the same type of remote-control, rubber-treaded toy.

Another source of tank treads is LEGO sets. These treads are well suited for the lightweight sumo class. They can be assembled into many configurations and are easily driven by the LEGO gears and motors. This next picture shows two simple chassis made with different styles of LEGO tank treads. If you use tank treads from low-friction hard plastic (as in the robot on the left below), you can improve their traction by gluing large, wide rubber bands to the outside of the tracks.

One of the more common items used for tank treads is a timing belt. Timing belts are also known as *synchronous belts*, since they are used to transmit power with precise positional accuracy. Timing belts are used in factory lines, robotic manipulators, and car engines. A timing belt has teeth on one side that are used to mesh with a set of pulleys. A timing belt without any teeth is called a *flat belt*. You could use a flat belt for tread on a robot, but you would need to add flanges to the sides of all of the wheels to prevent the belt from rolling off the wheels. This is a timing belt configuration for a tank tread. Notice that the pulleys have flanges on their sides to keep the timing belt centered on the pulleys:

Tank Tread Support

In order to take advantage of the traction capability of the treads, the entire surface of the tread that is in contact with the floor must support the weight of the robot. If you are using only two pulleys and a timing belt with nothing between the pulleys, the robot will be nothing more than a four-wheel-drive robot, where the timing belt is being used to drive the front and rear wheels together. This is because the weight of the robot is being supported by only the pulleys.

As with regular tanks, a set of smaller idler wheels on the bottom of tank tread will help support the weight of the robot. Usually, the size of the robot makes it difficult to use multiple idler wheels. Using a low-friction plate, such as Teflon, on the bottom treads becomes a more practical option. Figure 5-1 illustrates how the bottom of the tread must support the weight of the robot.

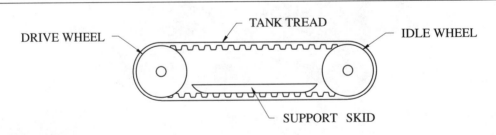

FIGURE 5-1 How to support the bottom of the tank treads

The Walking Robot

The classic robots from movies like *Star Wars* and computer games like *MechWarrior* are walking robots. In their imaginary environments, they engage in combat and defeat all opposing foes. But in real life, walking robots are rarely seen in robot combat events. One of the reasons for this is that walking robots are generally more difficult to build than wheeled robots. However, legged robots are always a crowd favorite during a competition.

The low-profile wheeled or treaded robot with a wedge-shaped scoop at its front has an advantage over the walking robot. Usually, when a robot lifts up a leg to take a step, the wedge-shaped robot gets under the robot's foot and easily knocks it over. But this isn't always the case. This next picture shows a simple, two-legged walking robot (a biped) that was built to compete in a mini sumo event. This robot is fairly unstable, but it has caused many robots to stop in their tracks when they hit the edge of a rubber-coated foot still in contact with the sumo ring surface. Every now and then, it wins a match, and the crowd goes wild.

Six-Legged Robots

Six-legged robots are more stable than bipeds, since they have at least three legs that are always in contact with the ground. A biped balances on one leg when walking and turning.

The next example shows a simple six-legged robot that was built for a mini sumo event. This robot uses the classic three-servo design to manipulate all six legs. Micro R/C servos power the legs to move slowly along the competition ring. Because of the way it walks, it often gets tangled up in the mechanisms of the opposing robots. Many times, this robot looks like it is boxing with its opponent when it gets knocked back onto its haunches. Like the biped, this robot is a great crowd-pleaser when it wins a match, or even when it loses a match.

The Future of Walking Robots in Sumo

As technology improves, robots will become more and more humanlike. Honda has introduced the human-sized P-3 robot and the shorter Asimo robot, which look and walk like humans do. Sony has built a 19.7-inch (50-centimeter) tall SDR-3X robot, which is a very capable biped that can track and kick balls around, and can balance on one leg while standing on teeter-totter.

Very soon, you will start seeing biped robots competing against each other in sumo, just like their human counterparts. When they come up against a wheeled robot, they will simply lean over to flip the power switch and shut it off to win a match. Of course, technically, turning off your opponent's robot is violating the rule about not interfering with its electrical performance, but it sure would be fun to see. Then again, who's to say that the robot didn't shut off due to incidental contact?

Walking Robot Design

When designing a walking robot for sumo applications, balance and traction are very important. Opposing robots will get under your robot's feet and knock it over, so your robot should have a method for getting back on its feet. Getting knocked over doesn't mean you automatically lose; not getting back on your feet or getting pushed out of the ring means you lose. Rubber on the bottom of the feet really helps to improve traction.

Also think about kicking. A walking robot should be able to kick (forcefully push) the other robots out of the sumo ring, unless you build hands that can lift the other robot up and drop it out of the sumo ring.

TIP *If you're interested in a building a walking robot, search the Internet. You'll find many websites that describe how to build both biped and six-legged robots. Use key words "walking robot", "legged robot", "hexapod", and "biped" to name a few.*

Selecting a Steering Method

When designing a robot, one of the first decisions that you need to make is what type of steering the robot will use. The steering method will have a significant effect on virtually every aspect of the robot, thus you need to choose the method at the beginning of the design phase.

With combat robots, either Ackerman steering or differential steering is commonly implemented. Ackerman steering is the type of steering that is used in virtually all automobiles or remote-control cars. Differential steering is found in all tracked vehicles.

Most four-wheeled sumo robots use differential steering, and they are four-wheel-drive robots. If only two wheels are being powered, the Ackerman steering method is preferred, since there will be a lot of wasted energy in trying to slide a set of wheels sideways across the ground.

Although the Ackerman and differential steering methods are the most common, they are not the only methods that can be used to steer a robot. The following sections discuss various steering methods that you can use in a sumo robot.

Ackerman Steering

Figure 5-2 shows a schematic of a simple robot design that uses Ackerman steering. A robot with Ackerman steering needs only one drive motor for going forward and reverse, and a simple servo motor to steer the robot. The three-wheeled tricycle is a simple form of the Ackerman steering method that can be employed in a competition robot.

Ackerman steering is ideal for moving at high speeds and turning in large areas. One of the advantages of this type of steering method is that fewer components are needed for the power train. Also, for small robots, you can use the components commonly used in the R/C car industry, instead of needing to fabricate your own.

The main disadvantage of the Ackerman steering method is that a robot cannot spin around, or "turn on a dime." To turn around, the robot must either do a U-turn or back up and make a three-point turn (a Y-turn). The minimum turning radius is a function of the sharpest turning angle of the front wheels and the distance between the front and rear wheels. But a robot with

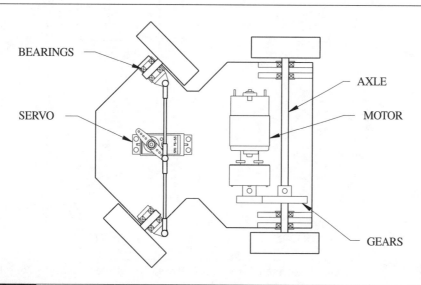

FIGURE 5-2 Schematic of a robot with Ackerman steering

a good driver and good software can overcome the turning problems by avoiding the need to make sharp turns.

Differential Steering

Differential steering, also called tank steering, is probably the most common steering method used in robots today. This type of steering is ideal for maneuvering in small areas. For a sumo robot, this type of steering is very desirable, since the robot needs to be able to turn around quickly to avoid getting stuck on the white border or getting pushed out of the sumo ring.

The term differential steering comes from the fact that the turning radius of the robot is a function of the ratios (or differences) of the wheel or tank-tread speeds on both sides of the robot. Differential steering is one of the easiest steering methods to implement, both from a mechanical and electrical standpoint, and it can be applied to both wheeled and treaded robots.

Figure 5-3 illustrates how differential steering works. When all of the wheels (or treads) on both sides of the robot move in the same direction with the same speed, the robot will move forward in a straight line. When you reverse all of their directions, the robot will move backward in a straight line. When the wheels/treads on both sides of the robot move in opposite directions,

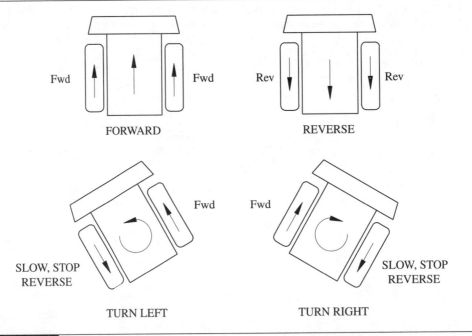

FIGURE 5-3 Differential steering is a result of different wheel speeds causing the robot to change directions.

with the same speed, the robot will spin about its center. With this type of steering, all of the motor-driven wheels on one side of the robot must turn at the same speed.

Moving in straight lines is not always the desired action. Sometimes, you want your robot to move around an obstacle or sneak in behind an opposing robot. Turning on a radius is simple to do with a robot with Ackerman steering: just point the front wheels toward the direction you want to move. With differential steering, turning is accomplished by varying the speeds of the motors. To have the robot turn on a specific radius, the wheels/treads on one side of the robot must move at a faster or slower speed than the wheels/treads on the other side of the robot.

For remote-control robots, turning on a specific radius is relatively easy to do, since the operator of the robot can see the robot turn and can actively compensate for turning errors. For autonomous robots, you will need a variable-speed control for the motors. (Chapter 9 discusses how to control your motor speed.) You will also need to know what the relationships between each of the wheel/tread speeds are and how they affect the turning radius. (Chapter 6 explains these relationships and motor performance in detail.)

The Synchro Drive System

Another, less commonly used steering method is the synchro drive system. With this type of a drive system, all of the wheels have their own independent drive and steering motors. For example, a four-wheeled synchro robot will require eight motors for driving and steering. This automatically makes this type of a robot an all-wheel-drive robot, which can move forward and backward like any other robot. But when the robot needs to turn, all of the wheels will turn in some direction.

The direction in which the wheels rotate has a profound effect on how the robot turns. If the wheels rotate together, always maintaining parallel relationship with respect to each other, the robot will change its direction of motion, but its body will maintain the same orientation as it was before the turn. With this type of a drive system, the front of the robot always faces the same direction, regardless of the direction in which the robot is moving. If the wheels' steering direction is not parallel with respect to each other, then the robot's body will face toward the turning direction as the robot turns.

The advantage of this type of a steering system is that it keeps the front of your robot facing your opponent as your robot circles around your opponent. This type of a design has been successfully implemented in several competitive sumo robots. The drawback to this type of a design is that it requires a more complex drive mechanism for each wheel, and the motion-control electronics and software can be fairly complex.

Holonomic Motion Control

An interesting steering method that can provide some rather unique attack strategies is the holonomic motion control method. A robot using this steering method is a three-wheeled robot that uses an unusual-looking wheel called the omni-directional roller wheel. Although these wheels are not designed for high-traction applications, they do offer some interesting design options for a sumo robot. Here is an omni-directional roller wheel (courtesy of Jim Sincock, Acroname, Inc.).

A robot using holonomic motion control can move in any direction, and you can precisely control the rotational speed of the robot's body. This is similar to the synchro drive system described in the previous section, except that only one motor is needed for each wheel, instead of two. The translational and rotational motion of the robot is a result of the omni-directional roller wheels. These wheels have a set of rollers around the perimeter of the wheel. The rollers spin freely around their center lines, so the wheel will slide freely along the axial direction of the wheel. But when the wheel turns, the rollers will not slide on the surface in the turning direction, so the whole wheel will roll over the surface.

By placing three of these wheels in a triangular pattern, and varying the angular velocities of the wheels, the robot can move in one direction, and the robot's body will either spin or remain in a fixed orientation. The velocity of the robot and its body rotational spin is a function of the angular velocities of the wheels. This shows a robot design that uses the holonomic motion control approach. This robot is called the PPRK. It is from Acroname Inc. (www.acroname.com), courtesy of Steve Richards.

You can use this type of a motion control system in a sumo robot to keep a front scoop always facing the opponent, and it allows the robot to spin away from an opponent when being pushed out of the sumo ring. Like the synchro drive system, holonomic motion control is more complex than traditional differential steering methods.

Selecting the Right Wheel for Your Robot

Though wheeled robots are most commonly used, finding good wheels is often a difficult task for the beginning, and even the expert, robot builder. A big part of the robot-design process is the selection of the wheels, and it should be done early in the construction/design phase. Here, you'll get an idea of where to find wheels and how to make your own.

Once you've found the ideal wheels, you will need to figure out how you will attach them to the axles. This is an often-overlooked detail that has caused robot builders many headaches. Some wheels can mount directly on the drive shaft, but most wheels require a separate mounting adapter, or hub, to attach them to the axles. The "Mounting Wheels to Drive Shafts" section later in this chapter provides details about attaching your wheels.

Wheel Sources

The best places to look for wheels are the hobby stores and hobby robot companies that sell individual parts. When shopping for wheels, you need to consider wheel size and material.

The width and diameter of the wheel are one of the major constraints on the overall robot design. The diameter of the wheel must be large enough to provide ground clearance for the motors and gear trains and narrow enough not to exceed the overall size constraints of the robot. Usually, finding a wheel with the right size is one of the hardest parts in designing a sumo robot.

Another important characteristic of a wheel is the material it is made out of. Since the goal of sumo is to push opponents out of the sumo ring, traction is critical. The wheels must be able to provide enough traction to fully use the power available from the motors; otherwise, the wheels will just spin on the ring surface while your robot is being pushed.

You can use many types of wheels for your robot, including those for model airplanes, remote-control cars, and construction sets.

Model Airplane Wheels

Model airplane wheels are popular choices for all sumo weight classes. The next illustration shows some foam wheels that are designed for model airplanes. These wheels are made out of a dense, lightweight foam with simple plastic hubs that snap together. They come in a wide variety of wheel sizes, with diameters ranging from 1-3/4 inch to 6 inches and widths from 1/2 inch to 1-1/2 inches. These wheels can be mounted side by side to make wider wheels, and they may require a custom adapter to attach them to a drive shaft. You can get them from Dave Brown Products Inc. (www.dbproducts.com).

Radio-Control Car Wheels

Another good source for wheels is the radio-control (R/C) car industry. You can use hollow rubber or foam wheels designed for remote-control vehicles.

Racing Slicks   Most R/C car wheels are hollow rubber tires that mount to a rim, similar to regular automobile tires. However, these tires do not get filled with air to inflate them. Instead, they are filled with foam inserts to increase their internal "pressure" or firmness; the denser the foam insert material, the firmer the tire. The smooth racing slick wheels are a good choice for sumo wheels.

The following shows a set of front and rear wheel F-1 remote-control car rubber racing slicks from Tamiya (www.tamiya.com). The front wheel is 2-2/10 inches in diameter and 1-1/4 inches wide. The rear wheel is 2-1/4 inches in diameter and 1-3/4 inches wide. When the surfaces of the wheels are kept clean, they provide a lot of traction.

Foam Wheels   Some of the best sumo wheels are the foam wheels used in 1/10 and 1/12 scale remote-control racing cars. You can tune the hardness of the wheels for the application. Harder foam wheels have longer life but less traction. Softer wheels have shorter wear life, because they tear more easily, but have better road traction. This shows a collection of different types of foam wheels that can be used for sumo robots.

In the remote-control car industry, foam wheels are given color "Dot" designations that indicate the relative hardness of the foam. Table 5-1 shows the foam color designations along with their relative hardnesses. The table also shows the wheel's Shore A (durometer) rating,

Compound	Durometer Rating	Hardness
Standard Foams		
Green Dot	29-30 Shore A	Medium/Soft
Blue Dot	39-40 Shore A	Medium
Double Blue Dot	50-51 Shore A	Firm
Orange Dot	54-55 Shore A	Very Firm
Exotic Foams		
White Dot	25-26 Shore A	Soft
Grey Dot	29-30 Shore A	Medium/Soft
Black Dot	36-37 Shore A	Medium
Natural Rubber		
Platinum	25-26 Shore A	Soft
Pink	32-33 Shore A	Medium
Magenta	36-37 Shore A	Medium
Purple	44-45 Shore A	Medium/Firm
Red	54-55 Shore A	Very Firm

TABLE 5-1 Foam rubber designations used in remote-control car tires (based on the data from Team Trinity, www.teamtrinity.com)

which is an industry-standard scale for measuring the hardness of rubber materials. The smaller the Shore A number, the softer the material. The formulas used to make the various color Dot compounds are very proprietary to the manufacturer, so there are subtle differences in the foam characteristics between manufacturers. (Though the wheels have different color designations, the actual color of the wheels is black.)

You will need to consider the width of the wheels in the robot design. If a particular foam wheel is too wide for your robot, you can trim it down with a hacksaw. Foam wheels are usually sold as either front wheels or rear wheels. The front wheels are about 1/2 inch to 1-1/2 inches wide. The rear wheels range from 1-1/2 to 2-1/2 inches wide. The wheel diameters can range from 2 to 3 inches, depending on the model number and manufacturer. The rear wheels usually have either a two-, three-, or four-hole bolt pattern to attach to a drive shaft hub. The front wheels usually have only a single hole in the center. You will need to drill additional holes in the front wheels to mount the wheels to a drive shaft adapter.

Lynxmotion (www.lynxmotion.com) sells a wide variety of foam wheels that are specifically designed for sumo applications, as shown here. These wheels are based on the Green Dot foam formula (see Table 5-1). Lynxmotion also sells the appropriate drive shaft adapters to mount their wheels onto a 1/4-inch diameter shaft.

Donuts for Custom Foam Wheels   In the radio-control car racing circuit, many people make their own foam racing wheels. The radio-control car industry has responded by selling a product called *donuts*. These are just the foam tires (made from the foam material described in the previous section) around the hubs. This is a 1/12-scale front wheel Blue Dot foam donut and a 1/10-scale rear wheel Green Dot donut, along with an R/C car style wheel blank for mounting the foam donuts.

Donuts are ideal for custom-made sumo wheels. Custom designing your wheels gives you full control over their outside diameter, width, and mounting method.

You glue the donuts to rims with contact cement or another rubber adhesive. The only constraint with this type of a custom sumo wheel is selecting a rim diameter that has the same size as the internal diameter of the foam donut. The outside diameter of the donut will be approximately 1/2 inch larger than the inside diameter of the donut. Most R/C car manufacturers sell donuts that have internal diameters that range from 1-1/4 to 2 inches, in 1/8-inch increments. See the next section for details on building custom wheels.

Construction Set Wheels

Construction sets, such as LEGO kits, make a great source for wheels. These wheels come in many different types and sizes. They also come with mounting hardware to attach them to frames and gears in a fairly straightforward manner. This shows a wide variety of LEGO wheels that can be used to build robots.

Custom Wheels

If you want full control over the design of your robot's wheels, or you just can't find the right wheels for your robot, you can make the wheels yourself. The general design of a wheel consists of a rim made out of a hard plastic or metal and a soft, high-traction tire material mounted on the outside diameter of the rim. The high-traction material can be foam, rubber, silicon, urethane, or another material.

Custom Wheel Design

Figure 5-4 shows a simple design for a robot sumo wheel that can be used to mount foam donuts or other rubber surface material. The first part of the wheel is the hub mount, which is bolted to the drive shaft via a set screw. The outer wheel shell, the rim, bolts directly to the hub, and a foam donut (or some other soft material) is glued to the outside diameter of the rim. This is a general-purpose wheel design, which you can adjust as necessary for your robot:

- You can change the size of the internal bore inside the hub to accommodate shaft diameters other than 1/4 inch.
- You can change the outside diameter of the wheel to match the inside diameter of a tire or foam donut.
- You can increase or decrease the width of the wheel based on the robot design constraints. The length of the hub generally should not be greater than the width of the wheel.

The general-purpose wheel shown in Figure 5-4 allows the interchangeable use of standard four-bolt remote-control car hubs. If you look carefully at Figure 5-4, you will notice the rim rests on a small shoulder that protrudes from the hub. This is done so that the outside diameter of the rim will run true with respect to the inside bore of the hub. Without the shoulder, the rim will always have different orientation error, since its position is being controlled by a set of bolts. A good design uses shoulders to register the location of rotating components, not bolts.

78 ROBOT SUMO: THE OFFICIAL GUIDE

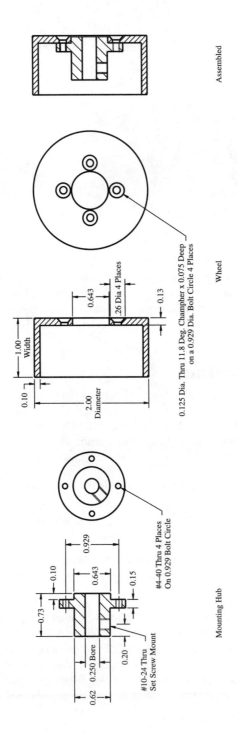

FIGURE 5-4 A custom sumo wheel design. You can make different wheel sizes by changing the diameter and width dimensions.

Custom Wheel Material

An unlikely location for obtaining a "secret" high-traction wheel material is your local grocery store. The secret material, which has been used in many champion robots, is found in the washable lint pickup rollers used to remove lint from your clothing. The reusable lint roller is the one with the colored gel on the surface that gets washed. It is not the ones that use the disposable tape. The lint pickup tubes on these rollers make great wheel materials. All you need to do is glue a wheel/hub inside the lint tube (such as the design shown in Figure 5-4), and then cut the lint tube down to the desired wheel width. Below shows lint tubes that can be used for sumo wheels.

TIP *The rubberlike material that is on the lint pickup tubes is one of the best traction materials you can use, as long as it is kept clean. You should clean the wheels throughout a competition to maintain their traction.*

Another way to make a hub for the lint tubes is to use a rubber tube, as illustrated in Figure 5-5. McMaster-Carr (www.mcmastercarr.com) sells rubber tubes with an outside diameter of 1-1/2 inches (same inside diameter of the lint tubes) and an inside diameter of 1/4 inch. Place a 1/4-inch diameter shaft inside the rubber tube. Use a pair of washers and a nut to compress the rubber onto the shaft and expand it to fill the lint tube. With this type of mounting method, you can easily replace the lint tubes without needing to replace the hubs.

One caveat is that the lint tube material may be considered a "sticky wheel," and thus may be illegal in some contests. Check with your local contest handlers for their competition rules regarding sticky wheels.

Mini Sumo Robot Wheels

Mini sumo robots usually have special requirements for their wheels. Because of their smaller size, finding a good wheel is often a difficult task. The ideal wheels are often too big for mini sumo robots. Thus, many people make their own wheels. One of the simplest mini sumo wheels is the O-ring wheel. This type of wheel gets its name from the rubber O-ring that fits in a groove on the outside diameter of a thin, plastic disk. What makes this type of a wheel attractive is that it's easy to build.

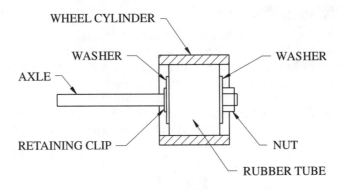

FIGURE 5-5 Compressing a rubber tube to make a wheel/hub

Custom Mini Wheels

Figure 5-6 shows a simple schematic for fabricating an O-ring wheel. The disk can be made out of any hard plastic, wood, or even a metal. The "O-ring groove full radius" label means that the radius of the groove is one half the diameter of the O-ring that is going to fit inside the groove. The table shown in Figure 5-6 lists the appropriate O-ring sizes for three different wheel thicknesses.

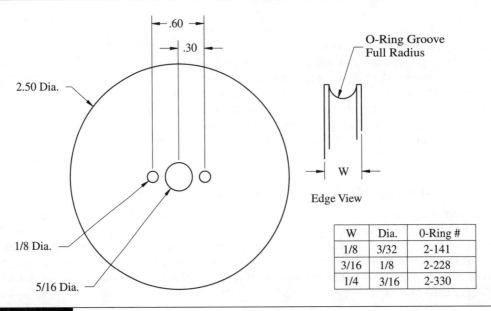

W	Dia.	O-Ring #
1/8	3/32	2-141
3/16	1/8	2-228
1/4	3/16	2-330

FIGURE 5-6 Schematic of an O-ring mini sumo wheel

O-rings are made out of a wide variety of materials, such as Buna-N, neoprene, polyurethane, and silicone. Most O-rings are made out of Buna-N, a black rubber with a Shore-A hardness between 65 to 70. The last three digits in the O-ring number represent its size. The first of these three digits represents the O-ring diameter: 0 equals 1/16 inch, 1 equals 3/32 inch, 2 equals 1/8 inch, 3 equals 3/16 inch, and 4 equals 1/4 inch. The last two digits represent the diameter codes for the O-rings. Here are the steps for creating the mini sumo robot wheel:

1. Mark the center of the 5/16-inch diameter hole in some wheel stock material. Then use a compass and scribe a 2-1/2 diameter circle around the center mark.

2. Drill a 5/16-inch diameter hole in the center, and rough cut the outside diameter close to the scribed line.

3. With a 5/16-inch diameter bolt, sandwich the wheel blank between two washers and thread a nut tightly against the washers.

4. To clean up the outside diameter of the wheel blank, place the bolt in the jaws of a drill and slowly turn the disk. With a mill file, file the outside diameter down to the scribed line.

5. With a small diameter rat-tail file, file the groove into the edge of the disk. This will require a steady hand to start the groove. The maximum depth of the groove should not exceed the radius of the O-ring groove.

The two 1/8-inch diameter holes are added last. The exact spacing should correspond to the drive shaft mounting hub adapter you will be using. Since most people use R/C servos as the motors for their mini sumo robots, the wheel mounting hub is one of the servo horns. There are two choices for the servo horns to use for the hub: one that already has holes or one that does not have holes. If you choose one of the large-diameter servo horns with holes already in it, such as the X-shaped horn, measure the center distances between two holes that are on outermost opposite sides of the servo horn. Use this measured value as the 1/8-inch diameter mounting hole spacing on the wheels. With the servo horn that does not have predrilled holes, use the 6/10-inch spacing shown in Figure 5-6.

6. Drill two holes at the measured hole locations in both the wheel and the servo horn. Drill the 1/8-inch diameter holes in the wheels, and use a #43 drill (0.089-inch diameter; the tap drill size for a #4-40 screw) to drill the holes in the servo horn.

7. Use a #4-40 tap to add some threads to the holes in the servo horn. The next illustration shows a servo horn attached to the wheel with nylon screws.

> **NOTE** *If you do not have a #43 drill and a #4-40 tap, you can use a small (about 1/8-inch diameter) sheet-metal screw to attach the wheel to the servo horn. Since the horn is made out of nylon, you can use this type of fastener without needing to tap the hole. You should predrill the hole with a drill diameter that is about the size of the minor diameter of the sheet-metal screw. If you use a sheet-metal screw, grind/cut off the sharp tip of the screw before using it.*

8. After the wheels are fabricated and the servo horns modified, add the two appropriate O-rings to the outside of the wheels. Bolt the wheels to the servo horns using 3/8 inch long #4-40 screws.

Mini Sumo Wheel Sources

Sine Robotics (www.sinerobotics.com) and Acroname (www.acroname.com) sell O-ring-style wheels with different geometries. Acroname offers a custom O-ring-style wheel-making service, which will build wheel sizes to order. Acroname also offers an aluminum spline option, so that the wheel can mount directly onto a servo shaft instead of needing to use a modified servo horn. Parallax, Inc. (www.parallaxinc.com) sells a similar wheel style that you can use for mini sumo applications. Parallax's wheel has the splines machined into the white plastic, so it can be mounted directly onto the servo's output shaft. The wheels that Acroname and Parallax manufacture can mount directly on Futaba, Tower Hobbies, and GWS servos, but not on Hitec or Aitronics servos.

John Olsen has developed a new custom-injection molded mini sumo wheel that mounts directly on the Futaba servos, just like the wheels from Parallax and Acroname. The main difference is that John's wheels use a soft, blue rubber band (called the broccoli rubber band, since it is commonly used to hold broccoli together at grocery stores) instead of the classic O-ring to improve wheel traction. This new wheel design has become one of the more popular for mini sumo robots, and many of the robot companies are selling them along with their other wheel designs.

All of these wheel designs will work well in a mini sumo. The following shows these three types of wheels: mini sumo wheels from Sine Robotics (right), Parallax (left), and John Olsen's design (center).

Custom-Machined Mini Sumo Wheels

Some very industrious mini sumo robot builders will custom-machine their wheels to meet their design requirements, as opposed to designing the robot around the wheels. For example, take a look at the 1.2-ounce (35-gram) aluminum wheel that Dana and David Weesner built for their mini sumo robot named Predator. Notice the servo horn bolted inside the wheel. This type of craftsmanship shows a true admiration to the sport of robot sumo.

Mounting Wheels to Drive Shafts

How to attach wheels to drive shafts can appear to be a mystery when you're building your robot. For most wheels, you can use a flange-style adapter to attach a wheel to a shaft, as shown next. The flange portion of the adapter consists of a set of holes or lugs for bolting the wheel onto the adapter, and the hub portion of the adapter is used to secure the adapter to the shaft. The flange-style wheel adapters shown are from Lynxmotion (www.lynxmotion.com), Team Associated (www.teamassociated.com), and Solutions Cubed (www.solutions-cubed.com).

You might fabricate adapters to match the specific geometry of the wheel, or you can purchase them. The radio-control car industry is one of the best sources for obtaining prefabricated wheel adapters. You'll want to use these adapters if you are using R/C car wheels.

The adapters can be custom machined from aluminum, brass, steel, nylon, or other types of plastic. Usually, you will design the adapter after obtaining the wheels and drive shafts, so that you can make the proper measurements. If you do not have access to a lathe that you can use to machine a custom adapter, another option is to modify the wheel or the drive shaft to fit the mounting hub.

Here, we'll look at several methods for mounting wheels to shafts.

Set Screw Mounting

Many different types of shaft adapters use set screws to secure the adapter to the shaft. On one side of the adapter's hub there is a threaded hole for the set screw, as shown next. The holding torque capability of the set screw is a function of the coefficient of friction between the set screw and the normal force the set screw is applying onto the surface. If the set screw is overtightened, it will mar the surface of the shaft. The raised edges from the marred surface will make it very difficult to remove or attach the adapter on the shaft.

The proper method for using set screws with shafts is to machine a flat on the side of the shaft. This approach improves the torque transmitting capability and protects the surface that is in contact with the wheel adapter's bore. Since the set screw will be in contact with the flat surface, any marring that might occur will not prevent the adapter from being removed. The load-carrying capability is increased, since the set screw will need to be compressed in order for the shaft to spin free. Without the flat, the set screw cannot apply enough holding force to prevent the adapter from twisting on the shaft under high-torque conditions.

The Pinned Shaft Method

Pinning the adapter to the shaft is another popular method for securing a wheel on a shaft. Using this method, you press a small-diameter pin into a hole that is drilled through the side of the shaft, as shown here. The pin is used to prevent the wheel from twisting on the shaft.

The diameter of the pin should not be greater than one-third the diameter of the shaft, and the length of the pin should be approximately equal to the diameter of the hub that is being mounted on the shaft.

The two most popular types of pins are hardened-steel dowel pins and spring pins. When pressing a dowel pin in a shaft, the diameter of the predrilled hole should be from 0.001 or 0.002 inch smaller than the dowel pin to 0.005 inch larger. If the hole is larger than the diameter of the pin, the pin needs to be glued in place. Spring pins can expand or contract to fit the hole in the shaft.

There are two different approaches to pinning the wheel adapter to the shaft: the semipermanent approach and the quick-release approach.

Semipermanent Pinning

The easiest way to pin an adapter to a shaft is to place the hub on the shaft, drill a hole through both the hub and the shaft at the same time, and then press a pin in the hole. Press-fitting parts together is a semipermanent mounting method; this process is not intended for easy disassembly. However, you can remove the hub by tapping the pin out of the hole.

This type of drilling method is called *match drilling*. The two holes will be perfectly matched up. The drawback to this approach is that matched drilled parts are not interchangeable. If you want the hubs and shafts to be interchangeable, the holes in both the shaft and hub must be drilled separately, and they must be perfectly centered on the diameter. This type of drilling requires a milling machine.

Spring pins are ideal for applications where there is a small misalignment between the holes in the hub and shaft, or when you plan to remove the hub from the shaft multiple times.

Quick-Release Pinning

With the quick-release pinning approach, the pin is pressed only into the shaft. The hub has a slot cut into the end that will slip around the pin when the wheel adapter is attached to the shaft, as shown in the first picture below. The end of the shaft is either threaded so a nut can be used to lock the wheel adapter on the shaft, or a retaining clip is used to keep the wheel adapter on the end of the shaft. The second illustration shows this mounting method used to mount servo style arms to gear motor shafts.

Hex Nut-Style Mounting

Radio-control cars also use the quick-release pinned shaft method for mounting wheels to shafts. The difference is that they use a hexagon-shaped (hex), smooth bore nut, instead of a flange-style wheel adapter. There is a notch machined on the face of the hexagon. The hex nut slips over the end of the shaft and wraps around the pin. Certain radio-control car wheels have the corresponding hexagon shape molded into their rims. The wheels slip over the hex nut, and a regular nut is threaded onto the end of the shaft to lock all of the pieces together.

The next example shows a radio-control car stub axle that uses this mounting approach. The Tamiya wheels shown earlier in this chapter use this method to mount to a shaft. An alternative is to use a set screw in the side of the hex nut and fasten this nut to the shaft, instead of using a pin. Below also shows this type of a hub. These remote-control car stub mount and hex nut style hubs were made by Traxxas Corporation (www.traxxas.com) (right pair), and a similar assembly that uses a set screw in the hex nut, were made by DuraTrax (www.duratrax.com).

Wheel Mounting Alignment

For a high-precision shaft adapter, the flange face must be perpendicular to the shaft mounting bore. The adapter's bore should be 0.001 to 0.003 inch in diameter larger than the shaft, and the length of the bore should be at least twice the bore diameter. The wheel should register directly to the outside diameter of the flange, or on a smaller-diameter shoulder that should protrude on the face of the flange. The mounting bolts should not be used to concentrically position the wheel on the hub adapter or support the weight of the robot. They should be fairly close tolerance fits. This alignment was illustrated with the custom wheel design earlier in this chapter.

Many sumo robots have their wheels mounted directly on the output shafts of gear motors. When the two parts fit perfectly together, they work well. However, when you're trying to fit metric and English parts together, there is often a mismatch. One of the more common mismatches is fitting a 6-millimeter-diameter motor shaft inside a 1/4-inch-diameter bore. Since the exact diameter of the 6 millimeter shaft is 0.236 inches, there is a 0.014 inch gap inside the wheel's hub when assembled. In many applications, this gap will not pose a problem, but it will cause the set screws to loosen. One way to solve this problem is to wrap a small piece of 0.005-inch or 0.006-inch thick brass shim stock around the shaft. This will help fill the gap when assembling the parts together. The ideal solution is to make sure the hub and bore were designed to fit together in the first place.

Quick-Disconnect Coupling

One last method that will be discussed here for mounting wheels to shafts is the use of quick-disconnect shaft couplings. This type of coupling fits over a shaft and the inside bore of a wheel at the same time. These couplings are commonly used to attach pulleys, gears, and sprockets to drive shafts, instead of using keyways, pins, or set screws. You can find them at stores that sell power-transmission parts and equipment, such as McMaster-Carr (www.mcmastercarr.com).

There is a nut on one side of the coupling. As the nut is tightened, a set of wedges inside the coupling slide together, causing the coupling to clamp down onto the shaft, at the same time expanding into the inside bore of the wheel. With one wrench, you can securely attach a wheel to a shaft.

Closing Comments on Locomotion

As you can see, there are many different types of wheels that can be used to build sumo robots, and there are many more types that have not been presented here. At this point, you should have a basic understanding of how to design and build a custom wheel or purchase an off-the-shelf wheel to make a good sumo robot.

Sumo robots are not limited to using wheels for locomotion. You can use tank treads or build walking-style robots. You are only limited by your imagination, so have fun designing your robot.

Near the beginning of the chapter, you learned about differential steering and Ackerman steering methods. Although Ackerman steering is a good steering method for many different styles of robots, differential steering is the preferred method for sumo robots. The rest of the book will focus more on differential steering methods.

One of the topics that is often overlooked when discussing how to build robots is how to attach a wheel to a shaft. After reading this chapter, you should see that attaching wheels to a shaft is a relatively straightforward process, and there is no mystery to this at all. The techniques presented here can be applied to all sizes of robots, and the methods for mounting a wheel to a shaft are also used to attach gears, sprockets, and pulleys to shafts.

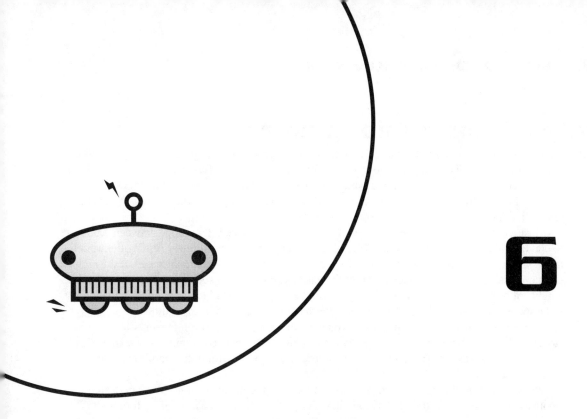

Power Transmission Fundamentals

One of the most important considerations in the design of your robot is the ability to transmit power from the motors to the wheels, which translates into speed and pushing force. Chapter 5 discussed the various types of wheels, treads, and even legs that can be used to propel a robot across the floor. Now you need to know how to transmit the power from the motors to the wheels of your robot. This chapter will present all of the information you need to determine how fast a robot can move across the sumo ring and how much pushing force the robot can generate. In addition, you will learn how power transmissions work, as well as how to build and select them.

Power Transmission Choices

The purpose of the power transmission is to transmit the rotational energy from the motors to the wheels of the robot. This can be done in many different ways. One of the most common methods is to connect the wheel directly on the motor's output shaft. This is known as the direct-drive method of power transmission. In most cases, the motor is actually a gearmotor, which contains a motor and gearbox in a single, self-contained package.

Most electric motors used in sumo robots have rotational speeds that range between 4,000 to 40,000 rpm, depending on the model. This speed is much too fast for directly driving the wheels, unless you want your robot to move at warp speed. To decrease the rotational speed of the motor to a more usable output shaft speed, you use a gearbox. As you will see in this chapter, one of the benefits of a speed reduction is that the torque on the output shaft is increased by the same ratio of the speed reduction. This results in your robot having more pushing capabilities.

At the beginning of the design process, you need to determine the power transmission method for your robot. The type and size of the gearboxes place significant constraints on the motor selection and wheel sizes, since all of these components must fit within the robot's size and weight constraints.

When designing the sumo robot, you have two choices to make up front: You can design the robot around a set of parts that you already have, or you can determine a set of performance specifications and then obtain the appropriate components you need to meet these specifications. In reality, this becomes more of a compromise. You will find a set of components that will be close to an initial set of target specifications, and then design the rest of the robot around these components.

The first two performance specifications you need to determine are the robot's speed and how much weight the robot should push across the sumo ring. You can figure out these specifications as follows:

- The diameters of the wheels and the rotational speed of the wheels determine the robot's speed requirements.
- The wheel's rotational speed determines the speed reduction needed for a particular motor.
- Convert the desired pushing force into a torque value based on the selected wheel diameter.
- With the speed reduction, convert the wheel's torque into a minimum motor torque specification.

With all of this information, you can select a motor to meet these requirements. You have probably noticed by now that all of these elements—robot speed, pushing force, wheel diameter, speed reduction, and motor selection—relate to one another. Changing one value results in changes in all of the other values. This is why most people design their robots around prepackaged gearmotors and live with whatever performance results they can get out of the motor. But this does not need to be the case. With a little upfront planning and flexibility in your selection of power transmission components, wheel diameter, and motor type, you can design the ultimate robot.

Kinematic Fundamentals

Kinematics is the study of motion, and since sumo robots move, you should have a basic understanding of the physics involved before you design your robot. Figure 6-1 shows a simple schematic of a sumo robot moving across the floor.

The speed of a robot is a function of the rotational speed of the wheels and the diameter of the wheels. Equation 6-1 shows that the velocity (v) of the robot is a function of the diameter of the wheel (D) and the rotational speed of the wheel (N) in revolutions per minute (rpm). The units of the wheel diameter can be inches or centimeters, and the corresponding velocity will be inches per minute or centimeters per minute.

Equation 1
$$v = \pi D N$$

Equation 6-2 shows how Equation 6-1 is rearranged to determine the required rotational speed of a wheel to enable a robot to move at some specific speed.

Equation 2
$$N = \frac{v}{\pi D}$$

For example, if your robot has 2-inch diameter wheels, and the rotational speed of the wheel is 300 rpm, then the speed of the robot will be 1,885 inches per minute, or 31.4 inches per second. With this speed, your robot can travel the width of a sumo ring in less than 2 seconds!

Differential Steering

In Chapter 5, you learned about the two most common steering methods: Ackerman steering and differential steering. With Ackerman steering, the turning ability of the robot is primarily due to the geometry of the wheelbase and the turning angle of the front wheels. With differential steering, the turning capability of the robot is only a function of the rotational speeds of the wheels (assuming the wheel diameters are the same).

Controlling the direction of a remote-controlled robot is relatively easy to do, since the operator of the robot can visually compensate for turning errors. With autonomous robots, to

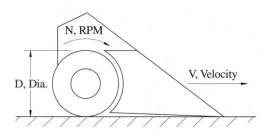

FIGURE 6-1 Kinematic diagram of a moving, wedge-shaped, sumo robot

negotiate a precisely controlled turning attack, you need to know the exact wheel speeds and how to control them. Chapter 9 discusses how set up variable-speed motor control.

Figure 6-2 is a simple schematic showing a two-wheeled robot conducting a turn with a radius (R). This illustration can be applied to four-wheel and six-wheel drive robots or tank-tread style robots by imagining that all of the wheels on one side of the robot are a single, large-diameter wheel.

Once all of the math is worked out, the general relationship for wheel speeds is shown in Equation 6-3.

Equation 3

$$N_1 = \frac{2R + w}{2R - w} N_2$$

Notice how this equation is only a function of the width of the wheelbase (w) and the turning radius (R). The diameters of the wheels do not factor into this equation (as long as they are the same diameter). The equation becomes discontinuous when the turning radius becomes one-half the wheelbase width. When this occurs, one wheel is stopped and the other wheel is moving, and the robot will pivot around one wheel. The wheel speed of N_1 (in revolutions per minute) is a function of the other wheel's speed. This shows that the actual speed of a motor is not important for the turn; what is important is the ratio of the wheels' speeds.

Equation 6-4 shows what the turning radius will be based on the relative differences between the wheel velocities and wheelbase width. Again, this equation is only a function of the rotational speeds of the wheels and the wheelbase width (w).

Equation 4

$$R = \frac{w}{2} \left(\frac{N_1 + N_2}{N_1 - N_2} \right)$$

These equations are useful in calculating turning radii and wheels speeds for turning. For example, say you want a robot to turn with a 12-inch radius, and the inside wheel of your 6-inch

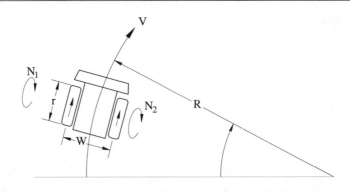

FIGURE 6-2   Schematic of a two-wheeled robot making a smooth, large-radius turn

wide robot is turning at 100 rpm. From Equation 6-3, you can calculate that the outside wheel will need to turn at 167 rpm, or the outside wheel must turn 67 percent faster than the inside wheel.

These two equations assume that there are no frictional effects from multiple wheels or tank treads slipping on the surface, and that all of the wheels have the same diameter. Although most people will only have their robot spin about its center ($R = 0$ and $N_1 = -N_2$), these equations are presented here for robot builders who want their robot to execute advanced turning strategies. These algorithms can be programmed into a microcontroller so that a robot can drive in circles around another robot. Equations 6-3 and 6-4 help illustrate why this is called differential steering.

Pushing Force

Since the objective of robot sumo is to push your opponent out of the sumo ring, it is important to maximize your robot's pushing capability. Pushing force is a function of both torque and traction. Torque comes from the motors, and traction comes from the wheel and ground materials.

In order to push an object across the floor, you need to overcome the friction that is keeping the object in one place. The amount of force that is required to overcome the frictional force is a function of the force that is normal (F_N), or perpendicular, to the surface, which is a function of the object's weight (F_W), and the coefficient of friction (μ) between the object and ground. This is illustrated in Figure 6-3.

When the surface that the object is placed on is parallel to the ground, the normal force (F_N) is equal to the object's weight (F_W). This reaction force is known as the frictional force (F_f) and is shown in Equation 6-5. In order to move the object, the applied force (F) must exceed the frictional force.

Equation 5 $$F_f = \mu F_N$$

Friction

One of the hardest things to do in estimating how much pushing force your robot is going to need is determining what the coefficient of friction is between your opponent's robot and the sumo ring.

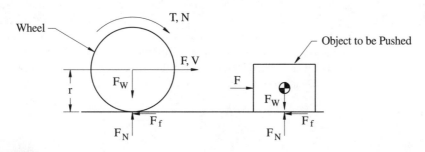

FIGURE 6-3   Free-body diagram of the various forces acting on a wheel and a block

This is because the value of the coefficient of friction is a function of the surface materials and the cleanliness of the surfaces.

In classical physics, the coefficient of friction has a value that is never greater than one and it is not a function of the contact surface area. This is generally true when considering inelastic materials sliding on inelastic surfaces, such as steel on steel. But when it comes to elastic materials such as rubber, the coefficient of friction can vary greatly. For example, the coefficient of friction between rubber and dry metal surfaces can range from 0.5 to 3.0, although it is hard to find coefficients of friction greater than 2.0. For all practical purposes, when considering firm rubber wheels, it is best to assume that the coefficient of friction is between 1.0 and 1.5.

For estimating purposes, you can use a coefficient of friction of 1.5 and assume the surfaces are clean. Then size all of your components so you will not stall the motors in these conditions. This should give your robot a small safety margin. For example, if your opponent's robot weighs 6.6 pounds, with a friction coefficient of 1.5, the pushing force needs to exceed 8.1 pounds.

Torque

By definition, torque (T) is equal force to the force (F) applied to an object multiplied by the distance (r) between where the force is acting on the object and the point where the object will rotate. Equation 6-6 shows this relationship.

Equation 6
$$T = Fr$$

One way to think about how torque works is to hold the end of a yardstick parallel to the ground in one hand. If you tie one end of a string to a 2-pound weight and attach the other end of the string 1 inch away from your hand, your hand must exert 2 inch-pounds of torque to hold the yardstick parallel to the ground. Now if you slide the string 30 inches away from your hand, your hand will need to exert 60 inch-pounds of torque to hold the yardstick parallel to the ground, which is a lot of torque. If the string were to suddenly break, the yardstick will smack you in the head. This is a result of the other part of the torque definition. If the applied torque on an object exceeds the reaction forces, the object will accelerate around the point of rotation. This is why automobiles can accelerate on the road.

When torque is applied to a wheel, it will roll forward until all of the reaction forces stop the wheel from moving forward. In order for a wheel to roll, there must be some friction between the wheel and the ground, or the wheel will spin in one place. The torque from the drive shaft is transferred from the axle to the point that is in contact with the ground. This results in a force that is pointing in the same direction of the torque (as illustrated earlier in Figure 6-3). From Equation 6-6, this force can be calculated by solving for the force when the torque and the wheel radius are known. Equation 6-7 shows this relationship.

Equation 7
$$F = \frac{T}{r}$$

Using this relationship, you would think that the more torque your motors can develop, the more pushing force your robot can generate. Well, this is not exactly true. The friction between the wheels and the ground come into the play here. Remember that if the applied forces on an

CHAPTER 6: Power Transmission Fundamentals

object exceed the frictional forces, the object will begin to slide across the floor. This is just as true for wheels as it is for solid blocks.

By plugging Equation 6-5 into Equation 6-6, you obtain a new relationship that states the minimum torque required to cause the wheel to spin. This new relationship is shown in Equation 6-8.

Equation 8
$$T_f = \mu F_N \, r$$

By plugging this result into Equation 6-7, you obtain the maximum pushing force (F_{max}) this wheel can generate. Equation 6-9 shows an interesting relationship: The maximum pushing force is only a function of the weight on the wheel trying to do all of the pushing and the coefficient of friction. Applying more torque to the motors will only result in the wheels spinning, not in an increase in the pushing force.

Equation 9
$$F_{max} = \mu F_N$$

To obtain the total force the robot can generate, sum up all of the individual forces from each wheel. Thus, the sum of all of the normal forces becomes the total weight of the robot. The maximum pushing force of the robot is shown in Equation 6-10.

Equation 10
$$F_{max} = \mu W_{robot}$$

This equation tells you three things about how to maximize your robot's pushing capabilities:

- Your robot must weigh as much as it can without exceeding weight limits.
- Your robot must maximize the coefficient of friction between its wheels and the sumo ring.
- The motors must be strong enough to generate this force.

As stated earlier, if the motors generate a torque that can spin the wheels, the robot has reached its maximum pushing capability. Any additional torque is just wasted energy. In fact, spinning the wheels can actually reduce the maximum pushing force, since the coefficient of friction is generally reduced between sliding surfaces. If the maximum torque of the motors is less than the spinning torque shown in Equation 6-8, the motors will stall if the robot comes up against a stronger robot (or a wall). As you can guess, stalling your motors is not a good idea.

Speed Reduction

Once you have an idea as to what the rotational speeds of your robot's wheels need to be (from Equation 6-2), the next step is to determine the amount of speed reduction between the motor and wheels your robot is going to need. In order to do this, you need to have an idea of what the motor shaft speed is.

Speed reduction is accomplished by coupling two wheels together through physical contact on their outside diameters. Gears are common speed reducers; the teeth on their outside diameters mesh together to provide the coupling point between the two gears. Another type of speed reducer is a belt or chain drive system. Although the wheels (pulleys) do not physically touch each other, the belt between them provides the coupling medium.

You can think of speed reduction as being just a combination of the different gear diameter ratios. Equation 6-1 shows that the velocity of a point on the outside diameter of a wheel is a function of the wheel diameter and the rotational speed of the wheel. When two wheels are in contact with each other, the point of contact will be moving at the same speed (assuming the wheels do not slip). Equation 6-2 shows how the rotational speed of a wheel is a function of the diameter and the surface speed of the wheel. If you add subscripts to the first wheel in Equation 6-1 and subscripts for the second wheel in Equation 6-2, and then plug Equation 6-1 into Equation 6-2, you obtain the formula for a two-wheel speed reduction. Figure 6-4 illustrates a simple two-gear speed reducer.

Equation 6-11 shows the results of how the speed of the output gear relates to the speed of the input gear. When the driving gear's (gear 1) pitch diameter (D_1) is greater than the driven (output) gear's (gear 2) pitch diameter (D_2), the output gear will spin faster (N_2) than the driving gear spins (N_1). When D_1 is less than D_2, the output gear will spin slower (gear reduction) than the driving gear. D_1/D_2 is called the *gear ratio*, and N_2/N_1 is called the *speed ratio*.

Equation 11
$$N_2 = \frac{D_1}{D_2} N_1$$

For example, if you had a 5,000 rpm motor and wanted a wheel speed of 250 rpm, you would need to reduce the speed of the motor by a factor of 20. Equation 6-11 tells you that the output gear (D_2) must be 20 times larger than the input gear (D_1). This is a pretty big gear reduction with only two gears. Also, finding a pair of gears that will give a 20-to-1 gear reduction will be difficult. In this example, if you were using a 1/4-inch diameter pinion as the driving gear, the output gear

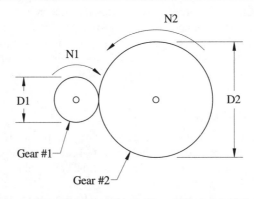

FIGURE 6-4 A simple two-gear speed-reduction schematic

CHAPTER 6: Power Transmission Fundamentals

would need to be 5 inches in diameter. For sumo robots, this gear would be larger in diameter than the wheels of the robot, which would cause many difficulties in the designing process.

To avoid the need for large gears, you can handle the overall speed reduction by breaking it up into two or more intermediate speed reductions. As a general rule, the greater the gear reduction is, the more gears you will need to use to achieve the gear reduction.

Figure 6-5 shows a more complex speed-reduction system, which is actually only two separate two-gear systems. The first speed reduction occurs between gear 1 and gear 2, and the second speed reduction occurs between gear 3 and gear 4. There is no speed reduction between gear 2 and gear 3, since they both spin on the same shaft. In fact, in order for the system to work, gears 2 and 3 must be permanently connected together.

Equation 6-11 defines the speed of gear 2. Since gear 3 drives gear 4, the speed of gear 4 is defined in Equation 6-12. Note how Equations 6-11 and 6-12 have the same mathematical form. Since gear 2 and gear 3 are physically attached to the same shaft, they both will spin at the same speed. Because of this, we can substitute Equation 6-11 into Equation 6-12 to develop an overall speed reduction of this system, which is shown in Equation 6-14. D_1/D_2 is the first gear-reduction ratio, and D_3/D_4 is the second gear-reduction ratio.

Equation 12
$$N_4 = \frac{D_3}{D_4} N_3$$

Equation 13
$$N_3 = N_2$$

Equation 14
$$N_4 = \frac{D_3}{D_4} \frac{D_1}{D_2} N_1$$

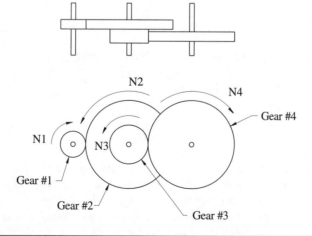

FIGURE 6-5 Double-speed reduction system using four gears

This approach can be expanded to produce very high gear reductions. The total gear reduction in the gearbox is the result of all of the individual gear reductions multiplied together. With this type of speed-reduction system, you have a lot of options for individual gear reductions. For example, for a speed reduction of 20 to 1, all that is required is that the product of the first and second gear reduction stages in the overall speed reducer be 20. You could choose the first gear reduction to be 4 and the second gear reduction to be 5. The total gear reduction would be 20 ($20 = 4 * 5$). In this case, you could use two 1/2-inch diameter gears for gears 1 and 3. Gear 2 must be four times larger than gear 1; thus, it will have a diameter of 2 inches. Gear 4 will be 2-1/2 inches in diameter (five times bigger than gear 3). These gear sizes are much more manageable than the 5-inch diameter gear you would need to do the entire speed reduction in one step.

Often, you will need to make a compromise based on available gear diameters and desired gear reductions. In the real world, you may not find the exact gear, pulley, or sprocket diameters you want. This may be because the actual sizes do not exist, or you may be limited by physical size constraints of the robot. In the example using four gears, the last gear was 2-1/2 inches in diameter. This places a constraint on the wheel. Remember that the wheel diameter must be larger than the gear that is attached to the wheel shaft. You will usually end up selecting gear diameters that are close to the values you want. Thus, the overall speed reduction will be a little lower or a little higher than what you originally planned.

Increasing the Wheel Torque

An added bonus to using a speed reducer is that there will be an equal increase in output torque from the motors. Equations 6-6 and 6-7, shown in the "Torque" section earlier in this chapter, show how torque and force are calculated. Equation 6-15 shows how torque is increased for the gear reduction shown in Figure 6-4. The directions in which the torque is being applied are identical to the rotational directions of each of the gears.

Equation 15
$$T_2 = \frac{D_2}{D_1} T_1$$

T_1 and D_1 are the torque and the diameter of gear 1, and T_2 and D_2 are the torque and diameter of gear 2. If D_2 is greater than D_1, the output torque is increased. The output torque for the system illustrated in Figure 6-5 is shown in Equation 6-16.

Equation 16
$$T_4 = \frac{D_4}{D_3} \frac{D_1}{D_2} T_1$$

The simple example uses a 20-to-1 speed reduction from the motor. This will translate into increasing the output torque by a factor of 20. By looking at Equations 6-11 and 6-15, you should see that torque and speed are inversely related to each other.

One thing to keep in mind about speed reducers in power transmissions is that the power going into the transmission is equal to the power coming out of the transmission (assuming no frictional losses). Shaft power is defined as the product of the shaft torque and shaft speed, which is shown in Equation 6-17. Since power is conserved, decreasing the speed results in the same proportional increase in torque. Conversely, an increase in the output speed results in a proportional decrease in torque.

Equation 17
$$P = 2\pi TN$$

Power Transmission System Components

There are basically two ways of transmitting power from the motor to the wheels: the direct-drive method and the indirect-drive method.

With the direct-drive method, the wheel in mounted directly on the motor's or the gearbox's output shaft. The following illustrates a direct-drive mounting of a wheel to two different types of gearboxes. It shows a direct-drive system using a gearmotor and a foam wheel from Lynxmotion (top) and a Dave Brown foam wheel mounted directly to a High Powered Gear Box, made by Tamiya, which is then directly connected to a motor (bottom).

The indirect-drive method uses chains, belts, and gearboxes between the wheels and motors. The next picture shows an indirect drive using a timing belt. This configuration has the motor inline with the wheel, which saves space on the inside of the robot for other components. It is an unusual configuration, but it will work in a robot.

You can also design a transmission that includes a combination of direct-drive and indirect-drive components. The following sections will describe some of the components that make up power transmission systems.

Chain-Drive Systems

Chain-drive systems are one of the more popular methods of transmitting power from the motors, along with providing speed reductions, to the wheels in robots. Chain drives are predominately used with larger robots, such as the ones that compete in BattleBots, BotBash, and Robot Wars, but they are occasionally used in sumo robots. When considering their versatility, it is surprising that they are not used more often in sumo robots.

One of the big advantages of chain-drive systems is their ability to take up "slop" in the system without requiring precise spacing between the motors, wheels, axles, sprockets, and gears. Also, they have the capability to transfer a significant amount of power to the wheels. Their main disadvantage is weight. Depending on the materials the chains and sprockets are made from, they could become a significant percentage of the total weight of the robot. You need to use chains and sprockets that are strong enough for your needs, and bigger components are heavier components.

Types of Chains

There are many different types of chains that can be used in robots. The chain types that are applicable to small-scale robots are roller, ladder, and cable chains, which are illustrated below. Ladder and roller chains are the most commonly used.

As you can see from the above, chains come in different shapes and sizes; a #25 steel roller chain (top), 0.187-inch pitch ladder chain (middle), and 0.1475-inch pitch cable chain (bottom). Table 6-1 lists the chain sizes that are suitable for sumo robots. All chains are specified by their pitch. The *pitch* is the centerline distance between each link. Figure 6-6 shows a sketch of a typical chain and sprocket in a speed-reduction configuration. All three of these chain types use the same nomenclature and have a similar style sprocket.

CHAPTER 6: Power Transmission Fundamentals

> **NOTE:** *With roller chains, there is an ANSI number that is more commonly used to specify the chain. An ANSI #25 specification is for a 1/4-inch pitch chain, and #40 chain has a 1/2-inch pitch.*

Table 6-1 also lists the working and ultimate tensile load specifications for these chains. You can calculate the chain tensile loads from the force-torque equation shown in Equation 6-7, earlier in the chapter. If the load is kept below the working load, the chain will not break under normal loading. (If your robot is constantly slamming in forward and reverse, cut the working load rating in half.) If the tensile forces on the chains are between the working and ultimate tensile load ratings, the chain will have a short life. (A short life could still mean many years of competition, since the robot runs only a couple of minutes at a time.) If the loading on the chain exceeds the ultimate tensile strength, it is guaranteed that the chain will break. Ladder chains come in cheap steel, high-strength steel, stainless steel, and brass. The load ratings shown in Table 6-1 are for the high-strength steel. Do not use the low-grade steel. The strength is the same for both the stainless steel and high-strength steel chains. Brass ladder chains have one-third of the strength that is shown in Table 6-1.

The two cable chain sizes have the same load ratings, since they both use the same 1/32-inch diameter steel cables inside them. The steel roller chains are by far the strongest chains, but the cable chains offer the best strength-to-weight ratio. The plastic subminiature chains are ideal for mini sumo robots.

Chain Drive Center Distance Calculations

When implementing a sprocket-and-chain system, all of the sprockets must have the same pitch as the chain to which they are connected. Equations 6-18 through 6-20 show how to calculate the

Chain Type	Pitch (in.)	Weight (oz/foot)	Chain Width (in.)	Max. Working Load (lb)	Ultimate Tensile Strength (lb)
Subminiature plastic roller	0.1227	0.067	0.082	2	7
Roller	0.1475	0.56	0.210	25	180
Roller	0.250	1.44	0.340	120	875
Ladder	0.167	0.23	0.222	7	35
Ladder	0.185	0.40	0.297	10	55
Ladder	0.250	0.51	0.342	20	90
Cable	0.1475	0.15	0.230	20	100
Cable	0.250	0.30	0.340	20	100

TABLE 6-1 Standard Chain Size and Load Specifications

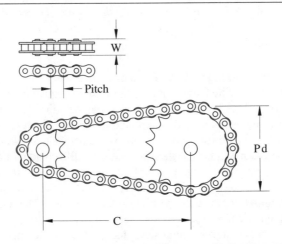

FIGURE 6-6 Typical speed-reduction setup using a chain and sprocket

chain length (L) and the center distance (C) based on the pitch diameters of the larger sprocket (D) and the smaller sprocket (d). The pitch diameter of a sprocket is the diameter of where the center of the chain will be. It is not the outside diameter of the sprocket.

Equation 18
$$L = 2C + \frac{\pi(D + d)}{2} + \frac{(D - d)^2}{4C}$$

Equation 19
$$C = \frac{b + \sqrt{b^2 - 2(D - d)^2}}{4}$$

where:

Equation 20
$$b = L - \frac{\pi(D + d)}{2}$$

These equations will provide physical lengths. To determine the correct chain length and center distance, use this procedure:

1. Estimate what the center distance (C) should be.
2. Chain lengths are based on an even number of pitches. To determine the number of pitches, divide the chain length by the pitch length. If the number of pitches of the chain is not an even number, round up or down for an even number.

3. Multiply this result by the pitch length to get the final physical length of the chain.
4. Plug the pitch length into Equations 6-19 and 6-20 to calculate the center distances between the sprockets.

Many catalogs will list the cable lengths only by the total number of pitches, and all catalogs will list the sprockets by both pitch diameter and number of teeth. The center distances can also be calculated in a generic form of the number of pitches. To do this, replace the pitch diameters with Equation 6-21, where N represents the number of teeth on the large diameter (D) sprocket and n represents the number of teeth on the small diameter (d) sprocket. All of these calculations will be in terms of pitch numbers. The only number that will not be an integer will be the center distance (C).

Equation 21
$$D = \frac{N}{\pi}$$

Cable chains are fixed-length chains and should not be cut and spliced to make a smaller chain. You should size the centerline spacing based on available cable chain lengths.

NOTE *Although cable chains can be cut and spliced back together, you should not do this. The splicing process is a short-term repair operation, rather than a permanent solution.*

Roller and Ladder Chain Assembly

Both roller and ladder chains come in random lengths, which you will need to cut down to their final length.

With a roller chain, you will need a "master link" to connect the chain into a continuous loop. The master link consists of a side piece of the link with two pins that will fit through the roller parts of the ends of the chain. There will be a second flat, figure 8 piece that will fit over the pins and a clip that snaps over the slotted ends of the pins, locking the master link in place. You cut a roller chain to length by simply punching one of the pins out of the chain with a small-diameter pin punch. Sometimes, this can be simplified if the ends of the pins are ground off. The length of the roller chain, including the master link, must be a even number of links. If there is an odd number of links, the master link will not connect the chain together. You also need to take this into account when calculating center distances.

With a ladder chain, you can bend the wires open with a pair of needle-nose pliers, and reassemble it by bending the wires back. Do not bend the wires back more than 90 degrees or induce a sharp right-angle bend, since this will weaken the wires.

Sprocket-to-Shaft Attachment

The sprockets used with roller chains look a little like gears but have larger, rounded teeth and are not meant to mesh with each other as in a standard gear. The sprockets used with the chains listed in Table 6-1 have bore diameters that range from 1/8 to 1/4 inch in diameter, and they will usually have set screws in the hubs. For most light-duty applications, the set screw will be enough to secure the sprocket in place on a shaft with a flat on one side.

Whenever you use set screws in rotating components, especially components that change directions, you should glue the set screws in place. The purpose of the glue is to prevent the set screw from loosening and causing the drive system to fail. The glue should be a temporary type, not a permanent type—something that will hold the set screw in place and require some force to remove. One of the best adhesives for this application is Loctite number 222 (www.loctite.com), for fasteners less than 1/4 inch in diameter.

Another common method that you can use is to pin the sprocket to the shaft. Pinning a sprocket to a shaft will provide a stronger attachment joint than using a set screw. The details of these two mounting methods are described in the "Mounting Wheels to Drive Shafts" section at the end of Chapter 5.

A third method is to machine a keyway into both the shaft and the sprocket. Keyways provide one of the strongest methods for attaching a sprocket to a shaft. Sprocket bores greater than 1/4 inch in diameter usually already have a keyway machined into them. Using keyways on shafts 1/4 inch and smaller is a nonstandard method for attaching sprockets to a shaft (they usually use set screws), so you will need to machine a keyway into these smaller sprockets. Figure 6-7 shows a keyway in a sprocket. A set screw is used to keep the sprocket from sliding lengthwise along the shaft.

One other method for attaching a sprocket to a shaft is to use a cylindrical adhesive. Loctite number 609 or 680 is an excellent adhesive for permanently gluing a sprocket to a shaft. The gap between the shaft and the sprocket bore should not exceed 0.005 inch for the Loctite number 609 and 0.015 inch for the Loctite number 680, or a strong bond will not occur.

When all of the parts are assembled, the chain should be fairly tight with not a lot of deflection when you push it with your finger. If the chain deflects too much, it could come off the sprocket when in use. If the chain is too loose, you can tighten it by removing a link, moving the sprockets further apart, or adding an idler sprocket. An idler sprocket is used to put tension on the chain, so it should have an adjustable mount to vary the tension.

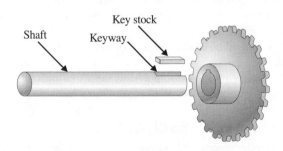

FIGURE 6-7 Using a keyway to attach a sprocket to a shaft

Belt-Drive Systems

Another method that can be used to transmit power from motors to the wheels are belt drive systems. Virtually every industry uses belt-drive systems in some form or another, and most products that use motors to move or spin something will have at least one type of belt-drive system in them. Belt-drive systems must be kept under tension in order to transmit power from one pulley to another. Because of this, a belt-tensioning device is required.

Belt-drive systems do offer one unique advantage. Since these belts do slip, they can be used to protect the motors from burning out if the wheels become stalled, because the motor will be allowed to continue spinning, thus burning out the belt. Sometimes, it is cheaper to replace a belt than it is to replace a motor.

The different types of belts commonly used are flat belts, synchronous belts, V-belts, and O-ring belts. Although V-belts and O-ring belts are very common in industry, and even robots, they are generally not used in sumo robots, primarily because they cannot transmit enough power from the motors to the wheels. Synchronous belts, more commonly known as timing belts, can be used to power sumo robots.

NOTE *The name "timing belt" is derived from the belt's very popular use to make sure that the timing of the cam shafts and piston motion are in sync in an automobile engine. Otherwise, the piston will strike the valves inside the engine and destroy it in a heartbeat. Timing belts have no backlash (slop when changing directions), so they are ideal in precise positioning systems.*

For sumo robots, timing belts are typically used to make four-wheel or six-wheel drive robots using only two drive motors, and they can also be used in the speed-reduction transmission and used as tank treads. Timing belts can transmit significantly more torque than regular flat belts, and they provide a much quieter operation than chain-drive systems. The drawbacks to timing belts are that the costs for the belts and pulleys are fairly high as compared to the other types of belt systems and chain-drive systems, and they require the pulleys to be parallel and aligned with each other.

Timing belts are similar to flat belts, except that they have teeth on one or both sides of the belt that fit into grooves in the pulleys. This allows precise synchronized motion among all of the pulleys that are being driven by the belt. In fact, timing belt tension requirements are much less than the requirements for regular flat belts. Timing belt operation is more like a chain-drive system than the other belt systems.

Types of Timing Belts

Table 6-2 lists the timing belts that have the appropriate size for sumo robots. The MXL and XL belts are commonly referenced by their inch pitch size. The HDT and GT belts are always referenced by their metric pitch lengths. The working tension load is the manufacturer-recommended maximum working tension for near-infinite life. The belts will break with approximately ten times the recommended working load. As with the chains, the maximum belt load occurs on the smallest diameter pulley and can be calculated by Equation 6-7 (shown earlier in the chapter).

The working tension ratings shown in Table 6-2 are related to the width of the belt. The tension rating is for each inch in width. The actual working load is computed by multiplying the working tension load by the width of the belt. For example, a 6 millimeter wide, 3 millimeter pitch GT belt has a working tension of 114 pounds per inch. The actual working load for this belt is 26.9 pounds (26.9 lb = 114 lb./in. * 6 mm/(25.4 mm/in.)). As you can see, the strongest belts are the GT belts.

If the tension in the belt is not high enough, the belt might slip on the pulley. If this were to happen, the belt life would shorten significantly. Also, exceeding the recommended working load will greatly shorten the life of the belts. Unless you exceed ten times the recommended working load, belt failure will be in the form of shearing off the teeth of the belt.

Belt-Drive Center Distance Calculations and Shaft Attachment

You can calculate the timing belt pulley centerline distances by using the same equations that work for calculating chain center distances, Equations 6-18 through 6-21 (shown earlier in the chapter). Timing belts are purchased in fixed pitch lengths. Thus, you will need to determine the final center distances after selecting the final timing belt size.

You attach the timing belt pulleys to drive shafts using the same methods to attach a sprocket or a wheel adapter to a drive shaft. See the "Mounting Wheels to Drive Shafts" section at the end of Chapter 5 for details.

Belt Type	Pitch (mm)	Pitch (in.)	Standard Widths	Working Tension (lb/in.)
MXL	2.032	0.080	1/8, 3/16, 1/4, 5/16, 3/8 in.	32
XL	5.08	0.200	1/4, 5/16, 3/8 in.	41
2 mm GT	2	0.079	3, 6 mm	25
3 mm GT	3	0.118	6, 9 mm	114
5 mm GT	5	0.197	9, 15 mm	160
3 mm HTD	3	0.118	6, 9 mm	60
5 mm HTD	5	0.197	6, 9, 15 mm	100

TABLE 6-2 Synchronized Belt Size Designations and Performance Ratings

Gears

Most speed reducers and power transmissions use gears. Gears come in a wide variety of materials, types, and sizes. Also, as you learned earlier in the chapter, you can assemble gears into many different configurations to accomplish speed reduction.

Meshed Gears

In order for gears to mesh correctly, they must have the same pitch and pressure angle. Coarse-pitch gears have a pitch of 8, 12, and 16; fine-pitch gears are 24, 32, 48, and 64; and extra-fine-pitch gears are 72, 80, 96, and 120.

Gears are generally specified by diametrical pitch size, pitch diameter, and total number of teeth. Diametrical pitch size is the number of teeth per inch located on the pitch circle. The higher the pitch number, the greater the number of teeth, and the smaller the size of the teeth on the gear. On a properly meshed gear, the point where the two gears actually touch is on a diameter that is smaller than the outside diameter of the gear. This is called the pitch diameter. The pitch diameter is also a function of the number of teeth on the gear and the diametrical pitch. Equation 6-22 shows the relationship between the pitch diameter (D_p), the number of teeth on the gear, (N), and the diametrical pitch (P_d).

Equation 22
$$D_P = \frac{N}{P_d}$$

The pressure angle is part of the geometry that is used to design the gear shape. Most gears have a 20-degree pressure angle, but some gears have a 14-1/2-degree pressure angle. You should mesh only gears that have the same pressure angle.

Gears for Parallel-Shaft Applications

Spur and helical gears are used for parallel-shaft applications. The spur gear is the most common type of gear. The face of the teeth on a spur gear is parallel with its bore. Below shows nylon spur gears with brass bore inserts for increased strength.

Helical gears have their gear faces at some nonparallel angle with respect to the bore of the gear. Helical gears are quieter and can transmit greater loads than spur gears, because the gear face is longer than the face on a spur gear for the same gear thickness. One of the drawbacks to helical gears is that they add an axial load to the bearings. Thus, the type of bearings that are used will need to support axial loads in addition to radial loads.

Gears for Perpendicular-Shaft Applications

For applications where the output shaft is perpendicular to the input shaft, miter, bevel, and worm gears are used. When you combine two miter gears, they must have the same pitch and number of teeth. Thus, they can be used only to change the direction of gearing. By mixing different miter gears together, you can obtain speed reductions down 5 to 1.

You can also use a worm gear to change the direction of an output shaft. A worm gear looks similar to a screw or auger screw, as shown next. It is placed on top of a special type of a gear that looks like a spur gear. The axes of these gears are at 90 degrees to each other. The big advantage to using a worm gear is that you can get very large speed reductions in a small area, such as 100-to-1 speed reduction with only two gears.

The speed reduction of a worm gear is a little different than it is with regular gears. Equation 6-23 shows how the output speed of the gear (ω_{gear}) is a ratio of the number of teeth on the spur gear (N_{gear}), the number of threads (N_{thread}) on the worm gear, and the speed of the worm gear (ω_{worm}). The total number of threads on the worm gear, rather than the number of threads per inch on the worm gear, is used in the speed-reduction calculation. Worm gears have one, two, or four threads (leads).

Equation 23

$$\omega_{gear} = \frac{N_{threads}}{N_{gear}} \omega_{worm}$$

One of the advantages of worm-gear systems is that they generally cannot be backdriven. (Spur, helical, miter, and bevel gear systems can be backdriven with enough force.) *Backdriving* means that applying a torque on the output shaft will cause the input shaft to turn. In other words, this type of a gear system can hold its position when power is removed from the motor, and the overall system will immediately stop when power is removed from the motors. The main disadvantages are that worm-gear systems are very inefficient and generate high-axial loads on the worm shaft due to the sliding friction between the two gears.

Gear-to-Shaft Attachment

Chapter 5 described how to attach a wheel adapter to a shaft. That entire discussion is applicable to attaching gears to shafts.

Part of the design process is to make sure that the gear will not move axially along the drive shaft. When this happens, the contact area on the teeth decreases, thus increasing the teeth wear and the bending stresses in the teeth, which will result in premature failure.

Bearings

Bearings are one of the most important parts of any power transmission system. Bearings are used to allow rolling or sliding motion between two objects, while minimizing wear and frictional losses. Bearings can be classified in two broad categories: sliding (or plain) bearings and rolling-element (roller) bearings. With a sliding bearing, the load-carrying members directly slide on its supports. A rolling-element bearing uses balls or rollers between the two sliding surfaces.

In an automobile engine, plain and rolling-element bearings are used in the same engine. Each bearing is selected based on its loading capacity and speed. For sumo robots, you should use bearings on all rotating shafts in gearboxes and drive shafts. Each shaft should have two bearings to support all of the loads on that shaft. Usually, bearings are on both sides of the gears/pulleys/sprockets. In most cases, both sliding (bronze sleeve) bearings and rolling-element bearings will be able to support all of the loads that will be applied on a sumo robot. However, you should always calculate the capacities your robot design requires before you select the bearing to make sure it will work. For combat robots, you should always assume worst-case scenarios.

Sliding (Plain) Bearings

Sliding bearings come in two different types: a sleeve or journal bearing and a thrust bearing. The sleeve bearing is used to support radial loads (loads that are perpendicular to the axis of the shaft), and thrust bearings are used to support loads that are parallel to the shaft. Plain sleeve bearings are typically made out of bronze, Teflon, nylon, Vespel, or Rulon. Each of these materials has its own pressure and velocity specifications. For most plain bearing applications, the bronze material will work well. The following shows two shafts with bronze sleeve bearings and a smaller shaft made of a composite material.

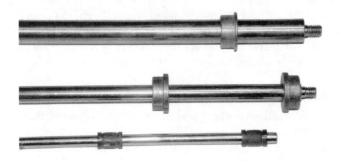

NOTE *Sliding bearings require periodic lubrication.*

The main advantages that plain bearings have over roller bearings is that they occupy less space and are usually less expensive. This is why most small gearboxes use plain bearings instead of roller bearings.

Sliding bearings are rated by three numbers:

- P_{max} is the maximum pressure, in pounds per square inch (psi), that the bearing can support at 0 rpm.
- V_{max} is the maximum shaft surface speed, in feet per minute.
- PV_{max} is the combined maximum dynamic load-carrying capability of the bearing.

When selecting a plain sleeve bearing, first you need to calculate the maximum load and maximum shaft speeds based on the P_{max} and V_{max} ratings for the bearing. Equations 6-24 and 6-25 show these calculations.

Equation 24
$$Max\ Load = P_{max} \times Bearing\ Length \times Shaft\ Diameter$$

Equation 25
$$Max\ Shaft\ RPM = \frac{V_{max}}{\pi \times Shaft\ Diameter}$$

Next, compare these numbers with the actual load on the shaft and the actual rotational speed. If either the load or the shaft speed is greater than the calculated maximum values, you need to use a different bearing.

Finally, calculate the combined pressure velocity product (PV), as shown in Equation 6-26, and compare it to the PV_{max} rating of the bearing. If the result is less than the PV_{max} rating, this bearing should work fine.

Equation 26
$$PV = \frac{Actual\ Load}{Bearing\ Length \times Shaft\ Diameter} \times \pi \times Shaft\ Diameter \times Actual\ RPM$$

Here is a simple example for determining if a plain sleeve bearing will work for a sumo robot—a really fast robot that has a shaft speed of 250 rpm. For this example, assume that the entire weight of the robot, six pounds, is being supported by one bronze sleeve bearing on a 1/4-inch diameter drive shaft. The length of the bearing is 1/4 inch. Assuming that the entire weight of the robot will be supported by only one bearing is a worst-case scenario. In reality, several bearings will be sharing the load.

The SAE 841 bronze sleeve bearing has the following specifications (SAE 841 is a specification code for a particular grade of bronze): P_{max} of 2,000 psi, V_{max} of 1,200 ft/min, and PV_{max} of 50,000 psi-ft/min. From Equation 6-24, the maximum load will be 125 pounds (125 lb = 2,000 psi x 1/4 in. x 1/4 in.). The actual weight of the robot is less than the 125 pounds maximum load. From Equation 6-25, the maximum shaft speed is 18,335 rpm (18,335 rpm = 1,200 ft/min / (π x 1/4 in. / 12 in./ft)). The actual shaft speed is much less than the calculated maximum shaft speed. The actual PV value is 18,850 psi-ft/min (18,850 psi-ft/min = (6 lb/(1/4 in. x 1/4 in.)) x π x 1/4 in. x 250 rpm). Since all three of the calculated values are less than the maximum values for the selected bronze material, this sleeve bearing will work well with this robot application.

Rolling-Element Bearings

Rolling-element bearings come in two different classifications: the ball bearing and the roller bearing. Ball bearings are best for higher-speed applications, and roller bearings are best for higher-load applications. The following shows a shaft with ball bearings. Roller bearings are broken down into three smaller categories: radial-load bearings, thrust bearings, and angular contact bearings (a thrust and radial-load-carrying bearing). Rolling-element bearings are rated by the maximum revolutions per minute, maximum radial loads, and maximum axial loads.

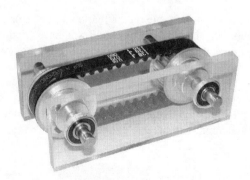

Rolling-element bearings can support significantly higher loads and speeds than plain bearings can support, and they have less friction loss than sleeve bearings do. This is why ball and roller bearings are predominantly used in precision mechanisms.

Gearboxes

A gearbox is typically a stand-alone assembly that contains a set of gears that is used to change the speed of an output shaft relative to the input shaft. Also, you can use a gearbox to change the orientation of the output shaft relative to the input shaft. For example, in a right-angle gearbox, the output shaft is oriented 90 degrees from the input shaft.

A gearbox may use only one type of gear, or it could use a combination of spur gears, helical gears, worm gears, bevel gears, sprockets, and even timing belts in a single assembly. Here are some gearboxes made by Tamiya (www.tamiya.com). These gearboxes can be assembled with different gear ratios, which gives them a lot of flexibility for different speed and torque requirements. These motors are often used in lightweight and mini sumo robots.

Of all of the gearboxes that are available on the market, only a small percentage of them can be used for sumo robots. This is due to the size and weight restrictions of the robot. In most cases, you will need to use a gearmotor or modify an existing gearbox or gearmotor. You also might build your own gearbox.

The most common off-the-shelf gearbox that can handle high speeds and torques for sumo robots are right/left-angle worm gearboxes. These gearboxes have short input and output shafts, which require a shaft coupling adapter to connect two shafts end to end. The following shows a helical shaft coupling that is commonly used to connect motor shafts to gearbox shafts.

CHAPTER 6: Power Transmission Fundamentals

This discussion of gearboxes focuses on those that use gears for the speed reduction. In reality, you can use timing belts and chains instead of using gears, or use a combination of all three. The goal of the gearbox is to increase the output torque and reduce the motor speed. This can be accomplished in any manner you want to use. One of the main advantages of using timing belts and chains is that they don't require precise location of the pulley/sprocket centerlines. These systems can tolerate a significant amount of misalignment when compared to gear systems.

Gearmotors

A *gearmotor* is an electric motor that has a gearbox mounted to it. The following shows some gearmotors that can be used for sumo robots. The various gearmotors shown for sumo robots come from Maxon, Jameco, Lynxmotion, and Servo Systems.

Typical gearmotor manufacturers include Hsiang Neng, Pitman Motors, Maxon, Micro Mo, and Tamiya. (See this book's appendix for details about these companies.) Some of these motors can be very expensive, so they are most often purchased from surplus companies such as Servo Systems (www.servosystems.com), Marlin P. Jones (www.mpja.com), and The Electronic Goldmine (www.goldmine-elec.com). Hsiang Neng sells a wide variety of gearmotors through Jameco Electronics (www.jameco.com).

Some companies, such as Maxon (www.maxon.com), sell high-quality electric motors and gearboxes. Maxon allows you to select the gearbox and the electric motor separately and then assemble them together as a gearmotor. This shows two different versions of these electric motors.

Modified Gearboxes

There are many different types of gearboxes that will make great sumo robot gearboxes, even though they were designed for completely different applications. Some sources of gearboxes that you can modify for a sumo robot are cordless screwdrivers and drill motors and remote-control servo motors. Also, you can replace the electric motor that comes with a gearbox with a more powerful motor.

Cordless Screwdriver and Drill Motors   One really good source for high-powered gearboxes are cordless screwdrivers and cordless drills. Inside these devices is a powerful planetary gearbox. Cordless drill and screwdriver motors offer the best performance for their price, compared with any other gearmotor you can get for your sumo robot. You will need to make some special mounts to use them, but they are very good gearmotors.

The following shows a 3-volt Black & Decker screwdriver, a 7.2-volt Dewalt screwdriver, and a 9.6-volt Ryobi cordless drill. All three of these motors offer very high-output torque capabilities and fast-output shaft rotational speeds. For example, the Dewalt cordless screwdriver is rated for 500 rpm, no load speed, and a stall torque of 75 inch-pounds. It offers a very powerful gearmotor in an 11-ounce package.

The drawback to using these motors is finding a way to mount them in your robot. These motors were not designed to be placed into anything other than their original housings. You need to design a special mount to hold both the electric motor and the gearbox together and inside the robot frame. This is not always an easy task, since these motors do not have convenient mounting attachment points. The other drawback to using these motors is that the output shaft does not have any convenient shape for mounting wheels or shaft adapters. To use these motors in a robot, you'll need to be creative.

NOTE *If you use a cordless screwdriver or drill motor, you will need to take it completely apart to remove a set of cylindrical pins that seem to do nothing. These pins prevent the output shaft from being backdriven. These pins need to be removed in order to be able to reliably reverse motor directions in a rapid manner.*

CHAPTER 6: Power Transmission Fundamentals

Figure 6-8 shows an example of how to combine two Black & Decker screwdriver motors. It consists of the following construction:

1. Place an approximately 3.8-inch long plastic or aluminum tube with an internal diameter of approximately 1-1/8 inch to 1-1/4 inch diameter between two gearboxes. Machine down the outside diameter at the ends of the tube so that it will fit inside the gearbox housing.

2. Match-drill a set of four holes into the tube so that you can use the same retaining clip that was used to hold the original screwdriver together. (*Match drilling* is when the holes are drilled and reamed together in one setup.) Drill a 1/2-inch hole at the center of the tube, to allow the motor wires to exit the motor mount.

3. Add some shims around the motor body to keep the motor centered inside the tube. You will also need to press a pin through the wall of the tube to keep the motor from sliding inside the tube. The pin only needs to protrude inside the tube by about 1/8 inch. You could also use a set of set screws to actually clamp down on the motor body.

If you use this configuration, you will need to cut most of the rotating shaft off, since it will be too long to use in a robot sumo contest. You will need to use either a thin wheel or a custom wheel, where this motor mount will be partially inside the wheel.

There are many other types of configurations that can be used. Instead of making a doubled-ended motor mount, you could make two single-ended motor mounts and place them side by side facing opposite directions, and use them to make a two-motor four-wheel drive robot. Some people have mounted these motors vertically and used a right-angle gearbox (using bevel or worm gears) to drive the motors. Remember that there are no vertical-size constraints.

Motor Replacement Another option for modifying motors is to take the gearbox from one gearmotor and replace the electric motor that originally came with it with a more powerful

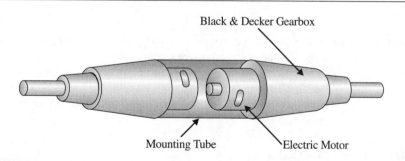

FIGURE 6-8 A pair of Black & Decker gearboxes connected together by a tube to produce a two-wheel-drive system

motor. The following shows two gearmotors where the gearboxes are separated from the electric motor. Both of these gearboxes are attached to the electric motor with two screws. The bottom of the Hsiang Neng gearbox is attached directly to the electric motor. The gearbox on the Maxon motor is attached to an adapter plate that is between the motor and gearbox. To change motors and gearboxes, you will need an adapter plate.

The following shows a close-up view of the adapter plate used on the Maxon motor. By making a special adapter plate that has an identical set of mounting holes for the gearbox and a set of mounting holes for the motor, you can create different gearboxes and motor combinations.

Radio-Control Servo Motors   One of the best methods for obtaining small and powerful gearmotors is to modify radio-control servo motors. There are a wide variety of sizes to choose from, and most hobby shops sell these motors and parts for them. Radio-control servo motors are the preferred motors for mini sumos, they are popular for lightweight sumo robots, and some 6-pound sumo robots use them. Chapter 8 describes how to modify radio-control servo motors for sumo robot applications.

Custom Gearboxes

Before taking on building a custom gearbox, it's best to see if you can find a gearbox that has already been made, and then adapt it for your robot. When considering all of the work that goes

CHAPTER 6: Power Transmission Fundamentals

into designing and fabricating gearboxes, spending a couple hundred dollars for a top-quality gearbox from a gearbox manufacturer can be far cheaper than contracting someone to build the gearbox for you.

The advantage of custom building your own gearbox is that you have control over the speed reduction, mounting locations, and physical size. Most robot builders compromise on the speed reduction and physical size of the gearbox just to find something that will fit inside their robot.

Only experienced machinists should build custom gearboxes. They require working with values on the order of a few thousands of an inch. The gear shafts must be very precisely located. If the gears are too close together, they will bind up and not move. If they are too far apart, they will wear out very quickly and start slipping. The finer the gear pitch is, the tighter the gear locational tolerances become. The total tolerance error includes a combination of the positional error in machining the shafts relative to each other, the gap between the bearing and the gearbox housing, the gap between the bearing and gear shaft, the gap between the gear and the shaft, and the parallelism between the two gear shafts. All of these errors can add up.

To make a good-quality gearbox, the gear shaft bearings should be press-fit into the frame and the gears press-fit onto precision ground shafts. The holes that are drilled for both sides of the gearbox should be drilled and reamed together in one setup (match drilled).

Equation 6-27 shows the gear center distance (C) equation. The center distance is a function of the pitch diameters (D) or the number of teeth (N), divided by the diametrical pitch (P_d), for both gears.

Equation 27

$$C = \frac{D_1 + D_2}{2} = \frac{N_1 + N_2}{2P_d}$$

Though gearboxes are complicated machines to build, many robot builders have been successful in custom fabricating them. The following shows a custom-machined gearbox used in one of the top United States competing sumo robots, named the Beast.

Closing Comments on Power Transmission

During the robot design process, you need to consider the power transmission at the same time you are selecting the motors. The number of gears, sprockets, and pulleys and their sizes, along with the wheel diameter, can have a significant impact on the overall structural design of the robot. To simplify the overall power transmission design, choose a motor that has the lowest revolutions per minute, to minimize the number of components in the power transmission (or speed reducer).

The actual selection of the motor, wheel diameters, and speed reduction is often an iteration process. In the beginning of the selection process, have a set of wheel diameters in mind along with a set of motor types. Then determine a set of speed-reduction combinations with the different motor combinations to obtain the optimal set of parts.

This part of the design process is often a compromise. You may not be able to find certain components, you may want to use components that you already have, and there may be cost and size constraints. One robot builder may spend a lot of time trying to select to ultimate set of components to make the perfect robot. Another robot builder will just grab a gearmotor and build the robot around the motor. In the end, they both have working robots. Winning a contest is the only way to tell which robot is a better robot.

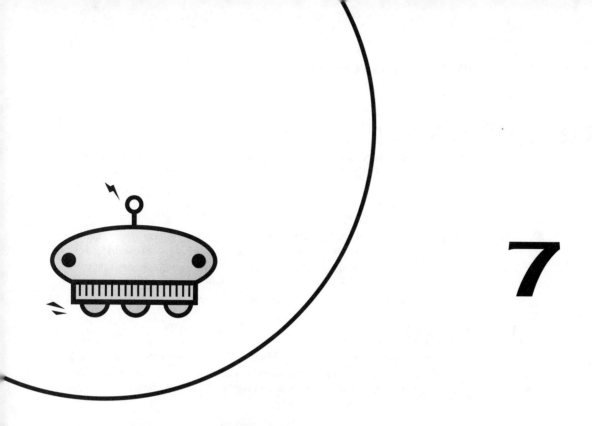

Batteries: The Heart of a Robot

The heart of any robot is its batteries. The batteries must be able to provide all of the power that the robot needs, when the robot needs it, and for how long the robot needs it. The power requirements for competition robots really push battery technologies to their limits.

When it comes to selecting batteries for sumo robots, you will find that you need to start thinking like a NASA engineer. You will want a battery that is rechargeable, has high power capacities, is lightweight, and can fit in a very small area. Often, you will find that you need to design your robot around the batteries, since they are usually the largest single component in your robot.

This chapter will explain how to estimate the performance characteristics of the different battery technologies, along with their advantages and disadvantages. A properly selected battery could make the difference between winning or losing a tournament.

Battery Types

There are many different types of battery technologies to choose from—alkaline, nickel-cadmium (NiCd), nickel-metal-hydride (NiMH), lithium, hydrogen fuel cells, and nuclear to name a few. Here, we will cover the battery types that are more commonly used for sumo robots. Figure 7-1 shows some different types of batteries.

> **NOTE** *Batteries have been around for more than 100 years. Some of us old-timers still remember using the classic carbon-zinc batteries in our electrical gadgets. We were excited when alkaline batteries became commonly available because of their significant improvement in longevity. Today, carbon-zinc batteries have become obsolete (although you can still purchase them), and alkaline batteries have become the low-end batteries.*

Alkaline Batteries

The alkaline (actually, alkaline-manganese) battery is probably the most commonly used battery in the United States. Just about every store has a large selection of alkaline batteries on its shelves. They are used to power many of the portable products we use today, from calculators to personal

FIGURE 7-1 Various types of NiCd, NiMH, and SLA batteries

data assistants, from flashlights to compact disc players, and from radios to most electronic toys. Many robots use alkaline batteries as their only power source.

The biggest advantages of alkaline batteries are their wide availability and their relative low startup cost. They have the least maintenance requirements since they do not require recharging, and they have the longest shelf life compared to the other batteries discussed in this chapter.

The disadvantage of alkaline batteries is that they have the highest internal resistance compared to the other battery types. Their discharge curve is not flat. The output voltage continually decreases as the battery capacity is used. (See the "Battery Basics" section later in this chapter for details on battery capacity and internal resistance.)

Although alkaline batteries have a very high energy density, they perform very poorly under high-current-draw applications. Thus, they should be used for only low-current applications, such as powering the robot's microcontroller.

Also, using alkaline batteries in applications where they must be repeatedly replaced can become expensive. In these applications, using rechargeable batteries is more cost-effective.

Another battery type to mention here is the rechargeable alkaline, developed by Rayovac. These batteries have been around since the early 1990s and have energy densities that are comparable to nickel-metal-hydride batteries. They have the best shelf life of all of the rechargeable batteries. They lose only 0.3 percent of their capacity per month. Their only real drawback is that they also have a relatively high internal resistance, which limits the maximum current draw.

Nickel-Cadmium Batteries

Nickel-cadmium (NiCd) batteries have been commercially available since 1950 and have become a very mature technology. NiCd batteries are used in high-discharge-rate applications. They operate very well in abusive environments. One of the main advantages of NiCd batteries is that they can tolerate very high discharge currents, upwards to 20 times their rated capacity, without suffering any damage to performance. They can handle high-current fast charging, which makes them well-suited for robot applications. NiCd batteries can be cycled (charged/discharged) more than 1,000 times, which makes them the most cost-effective rechargeable solution.

Although they are very forgiving if abused, NiCd batteries do need to be taken care of. They need to be fully discharged to prevent the *memory effect* from taking place. NiCd batteries do not like partial discharging between charging cycles. When this happens, crystals build up on the cell plates, reducing the overall performance of the battery. To keep this from happening, always fully discharge the batteries before recharging them.

Another drawback is that NiCd batteries have a relatively fast self-discharge rate. They can lose up to 20 percent of their charge every month in storage, which makes them a poor choice for long-term power-storage solutions. Also, since cadmium is considered a toxic material, NiCd batteries should not be disposed of in the regular trash. Because of this, other battery technologies are becoming more popular.

Nickel-Metal-Hydride Batteries

Nickel-metal-hydride (NiMH) batteries are a relative newcomer to the battery market. NiMH development efforts started in the 1970s as a method of storing hydrogen gas in a nickel-metal matrix for satellite and automotive power applications. Since then, the technology has been refined and improved. NiMH batteries became commercially available in the early 1990s.

NiMH batteries have energy densities that are 30 to 40 percent higher than NiCd batteries, and they do not exhibit any of the memory-effect problems that NiCd batteries have. Thus, they do not require complete discharging between charging cycles, and they are environmentally safe.

The drawbacks to NiMH batteries are that they require special battery chargers, their total cycle life is about one-third that of NiCd batteries, and their self-discharge rate is the fastest of all the rechargeable batteries. When stored, they can lose up to 30 percent of their charge every month. And because of their lower number of lifetime recharge cycles, they are one of the more expensive types of rechargeable batteries.

Sealed-Lead-Acid Batteries

In 1859, French physician Gaston Planté invented the lead-acid battery, which became the first rechargeable battery. Since then, the sealed-lead acid (SLA) battery, also commonly called a *gel-cell*, has evolved from the original lead-acid batteries.

SLA batteries have very high amp-hour capacities, but they are very heavy compared to other battery types. SLA batteries can tolerate high discharge currents, up to ten times their rated capacities for short durations. They do not have any memory-effect problems, and they do not require any electrolytes to be refilled. They have an extremely low self-discharge rate, approximately 5 percent per month, which makes these batteries ideal for long-term power-storage devices for devices such as uninterruptible power supplies (UPSs).

One of the biggest drawbacks to using SLA batteries is that they are fairly large and heavy, which makes them difficult to use in sumo robots. Although some very successful sumo robots use SLA batteries, a lot of careful planning is required to handle their size and weight constraints. SLA batteries have a low-charge cycle life of around 300, and battery performance can be permanently damaged if the batteries are deeply discharged. Since these batteries use an electrolyte that contains lead, they are considered environmentally unfriendly and must be disposed of at proper disposal facilities.

Lithium-Ion Batteries

Lithium-ion (Li-ion) and lithium-ion-polymer batteries are relatively new to the commercial battery market. Actually, the initial work in lithium battery technology began in 1912, and the first nonrechargeable lithium batteries were commercialized in the early 1970s. But it wasn't until 1991 when Sony Corporation commercialized the first rechargeable Li-ion battery.

Li-ion batteries are gaining a lot of popularity in the portable telephone and laptop computer markets. The single-cell voltage for a Li-ion battery is three times greater than a NiCd or NiMH battery cell, and the energy density is approximately twice that of a NiCd battery. These batteries are ideal for applications where weight and size constraints are important. They do not have any memory-effect problems and do not require any scheduled charge cycling to maintain battery life. Also, Li-ion batteries have a low self-discharge rate of only 10 percent per month.

A drawback to Li-ion batteries is that they have a finite life that is independent of how the battery is used. Battery performance starts to degrade after one year. They are considered to have only a two-year lifespan, after which they should be replaced. The other drawback to these batteries is that they have very specific charging requirements and discharging limitations. Abusing these requirements will result in metallic lithium building up inside the batteries. Metallic lithium is a very dangerous material to handle. Because of this, Li-ion battery packs have their own built-in circuits to control charging and discharging of the batteries, and individual battery cells are not sold to the general public.

 Some robot builders will hack a Li-ion battery from a cell phone to use in their robots. This is very dangerous and should be avoided due to risk of explosions. If you decide to do this anyway, make sure that the charge/discharge circuit that came with the battery remains with the battery and is used with the battery at all times.

Other Types of Batteries

There are many other types of batteries that can be used in robots such as regular lithium, mercury, silver, and zinc air.

Nonrechargeable lithium batteries output 3 volts, and a pair of them can be used to supply the power for your microcontroller. Lithium batteries are commonly used in cameras. They are expensive when compared to alkaline batteries, but two of them take up less space than four alkaline batteries. They can be very advantageous in space-constrained robots such as mini sumo robots.

Mercury and silver batteries are commonly in the form of coin-shaped batteries found in calculators, watches, and memory-backup devices in computers. Carbon-zinc battery chemistry is still commonly used in the higher voltage 6- and 12-volt lantern-style batteries. Carbon-zinc batteries can be found in many different types of specialty batteries that have voltages ranging from 1.5 volts to more than 60 volts.

Zinc-air batteries have the highest energy density of all of the batteries described here, but they have not been fully commercialized yet. Their common use is as tiny hearing-aid batteries.

For most robots, the common battery technologies that are used are alkaline, NiCd, NiMH, and SLA. The remainder of the chapter will focus on these types of batteries.

Battery Basics

Batteries are used in just about every modern portable device, yet they are often misunderstood. Batteries are actually very complicated electrochemical devices, which convert chemical energy into electrical energy. For average battery users, the only thing that really matters is that the battery will work in their equipment. But when it comes to extracting every bit of energy out of the battery, the subtle differences in the battery technologies become a significant factor. Here, we will look at some battery characteristics that play a role in battery selection.

Battery Capacity

Most batteries advertise their *amp-hour* (Ahr) rating as a measure of capacity to distinguish them from other batteries. In simple terms, amp-hour capacity shows how many amps can be drawn out of the battery in one hour. In reality, this is only true under certain ideal conditions.

In small print next to the amp-hour number, you will see something like C/5, C/10, or C/20. The *C* is the rated capacity number (amp-hour) you see in the advertisements. The 5, 10, or 20 represents the number of hours it takes to discharge the battery at a lower current draw. For example, an 1,800 milliamp-hour (mAhr) battery with a C/5 rating actually can deliver 360 milliamps for five hours. If you try to draw 1,800 milliamps of current out of this battery, the battery will last less than one hour.

The true capacity of a battery is a function of the energy density of the battery. Table 7-1 shows typical energy densities for several different types of batteries. The energy density of a battery is in units of watt-hours per kilogram (Whr/kg). The actual capacity of the battery is the product of the energy density of the battery and its mass. Thus, the true units of the battery capacity are watt-hours.

If the battery were 100 percent efficient, watts could be viewed as the product of current and voltage. Since a particular battery chemistry has a unique cell voltage, a battery's capacity can be viewed as a function of an amp-hour rating as described above, although this is not a technically correct method of rating a battery.

In reality, battery capacity is very dependent on the current draw, internal resistance, battery temperature, battery chemistry, the efficiency in converting chemical energy into electrical energy, and the age of the battery. All of these have nonlinear relationships, which make predicting how long a battery will last a difficult task. Many times, you will need to look at the technical specification charts for a particular battery to estimate how long a battery will last in operation. Figure 7-2 shows a typical plot of how the battery voltage changes based on different current draws from the battery. The data in this plot came from the published data for a Sanyo 1,900-mAhr NiCd battery.

In Figure 7-2, notice that the initial voltage for this battery is approximately 1.35 volts. This is typical for a freshly charged NiCd battery. The voltage will quickly drop to its nominal values after it has been used for a short while. This plot shows that the voltage remains fairly constant for most of the life of the battery for current draws of 2, 4, and 8 amps (1C, 2C, and 4C), and then rapidly drops off. The 16-amp draw case has a relatively short life. Also notice that as the current draw increases, the battery voltage curves shift downward. This is a result of the voltage loss due to the internal resistance of the battery (described in the next section).

Chemistry	Cell Voltage	Energy Density (Whr/kg)	Internal Resistance (ohms)
Alkaline	1.5	130	0.100
Lead acid	2.0	40	0.006
NiCd	1.25	50	0.004
NiMH	1.25	90	0.004
Li-ion	3.6	150	0.300

TABLE 7-1 Energy Densities of Various Battery Chemistries

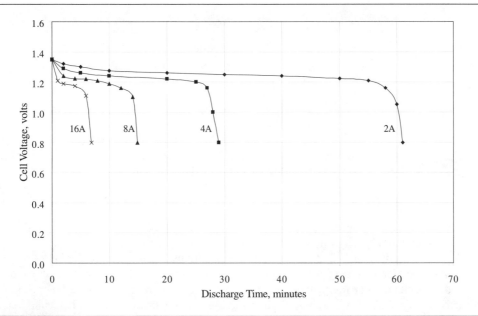

FIGURE 7-2 Voltage drop in a Sanyo 1,900-mAhr battery under different current draws

For most people, the amp-hour rating is the only way to estimate battery life. To estimate battery life in hours, divide the battery amp-hour rating by the average current that the battery will need to supply, as shown in Equation 7-1.

Equation 1
$$time = \frac{amp\text{-}hour\ capacity}{average\ current\ draw}$$

To estimate the average current-delivery capacity of the battery over the discharge period, divide the amp-hour capacity by the amount of time (in hours) in which you expect to drain the batteries, as shown in Equation 7-2. Note that this is average current draw. Peak current draws can be 10 to 20 times this for NiCd, NiMH, and SLA batteries.

Equation 2
$$average\ current = \frac{amp\text{-}hour\ capacity}{discharge\ time}$$

This method can be used to estimate how much current your batteries can deliver and how long your batteries will last under a certain average current draw.

Keep in mind that the amp-hour capacity of a battery is not constant. It decreases based on how fast you are pulling the current out of the battery. In other words, higher current draws result in lower effective amp-hour capacities of the battery. NiCd and NiMH batteries are not affected by this as much as SLA batteries are. Table 7-2 shows a list of amp-hour capacity derating values

Current Draw	NiCd	NiMH	SLA
C/20	1.00	1.00	1.00
C/10	1.00	1.00	0.94
C/5	1.00	1.00	0.88
C	1.00	1.00	0.68
10C	0.90	0.92	0.33

TABLE 7-2 Battery Capacity Derating Factors Based on Current Draw

based on current draw. A current draw of C/5 means that the current draw is one-fifth of the advertised amp-hour capacity for the battery.

As you can see in Table 7-2, the current draw capabilities of NiCd and NiMH batteries are similar, and they decrease only under high-loading conditions. Unlike the SLA batteries, the amp-hour capacity of the battery starts to drop off when the current draw exceeds 5 percent of the rated capacity. With SLA batteries, it is better to use a battery that has a much higher amp-hour capacity than is needed for the environment in which it will be used.

When estimating the battery life and average current draw from Equations 7-1 and 7-2, multiply the battery's rated capacity by the derating values shown in Table 7-2. These values are only guidelines; different battery chemistries and battery manufacturers will affect the actual values.

Internal Resistance

The internal resistance of a battery is an important number to understand, especially if you are planning on drawing high currents. In reality, all batteries have some internal resistance. Here is a simple battery model.

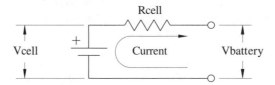

You will notice that there is a resistor in series with the battery's internal voltage. The output voltage of the battery is defined in Equation 7-3.

Equation 3
$$V_{battery} = V_{cell} - IR_{cell}$$

From Equation 7-3, you can see that the output voltage decreases as the current demand increases or as the internal resistance of the battery increases. Figure 7-3 illustrates this for three different

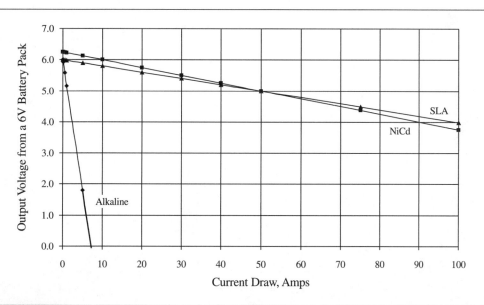

FIGURE 7-3 Voltage losses due to the battery's internal resistance under high current draws

6-volt battery packs. This plot shows a Duracell C-size alkaline battery pack, a Panasonic P200SCP sub-C NiCd battery pack, and a Panasonic 7.2 amp-hour 6-volt SLA battery.

It is obvious that the alkaline battery pack is a poor choice for high-current demands, but it is shown here for comparison. What is interesting is the voltage drop for the NiCd and the SLA batteries. In this plot, the internal resistance of a single NiCd battery cell is 5 milliohms, and five cells are needed to produce 6 volts. Thus, the total internal resistance of the NiCd battery pack is 25 milliohms. The total internal resistance for the 6-volt SLA battery is 22 milliohms (6.7 milli-ohms per cell). There is not much difference between the NiCd and the SLA batteries in this case, so the voltage drop is approximately the same. The internal resistance for the C-cell alkaline battery is 210 milliohms, so a four-cell battery pack will have a total internal resistance of 840 milliohms. This is why alkaline batteries should not be used for high current draw applications. Remember that the internal resistance inside a battery pack adds up just like resistors in series.

One thing to keep in mind is that the internal resistance of all batteries increases as the battery is drained, and the internal voltage loss increases, so the output voltage decreases. This is in agreement with what is observed in Figure 7-2. Figure 7-4 shows a generalized plot of how the internal resistance of a battery changes as the remaining capacity is drained. The actual values of the resistance depend on the battery chemistry, temperature, and life of the battery.

The only way to truly know if your battery has a good charge remaining in it is to test it under some external load. If you test a battery with a multimeter, it might show that the battery is perfectly fine. But that battery may run out of juice very quickly as soon as you put it in use. A multimeter will measure only the potential difference across the battery terminals. It does not measure the internal resistance of the battery. Since no current is flowing through the battery, the multimeter will show a higher voltage reading than what the battery can deliver.

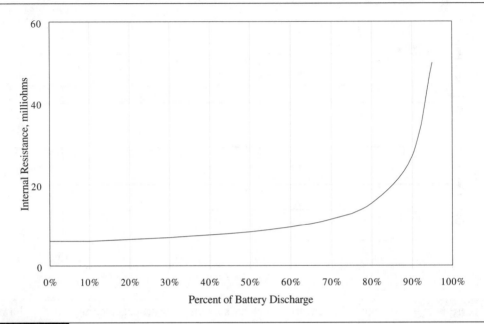

FIGURE 7-4 Internal resistance of a battery as a function of the remaining capacity in the battery

With all batteries, as the current draw increases, the battery gets hotter, its internal resistance increases, and the efficiency of the chemical-to-electrical energy decreases. In other words, as the current demand increases, the voltage drop in the battery increases at a greater rate.

So why is this important? Well, the electronics that use this voltage may be adversely affected by receiving a lower voltage than expected. If the voltage is too low, microcontrollers will stop functioning properly, and motors will run slower than expected.

Battery Summary

Table 7-3 shows the size, weight, and amp-hour capacity ratings for several different types of batteries that are commonly used in sumo class robots. The alkaline battery capacity ratings are based on a resistance load placed on the batteries. Except for the SLA batteries, you will need to combine multiple batteries together to increase the overall voltage of the battery, as discussed in the next section.

Battery Packs

Quite often, a single battery will not meet all of your power requirements. Thus, you will need to build a battery pack that will have both the voltage and current requirements that you need.

To increase the voltage of a battery pack, connect the batteries together in series. To increase the current capacity of the batteries, connect the batteries together in parallel. To increase both the voltage and current capacities, wire together multiple serial battery packs together in parallel.

CHAPTER 7: Batteries: The Heart of a Robot

Chemistry	Size	Volt	Dimensions (dia. x height)	Weight (oz)	Capacity (mAhr)	Discharge Rate
Alkaline	AAA	1.2	0.41 x 1.75	0.42	1150	75 ohms
	AA	1.2	0.57 x 1.99	0.85	2870	75 ohms
	C	1.2	1.03 x 1.97	2.29	7800	39 ohms
	D	1.2	1.35 x 2.42	4.76	17000	39 ohms
	9V	9	1.91 x 1.04 x 069	1.65	570	620 ohms
NiCd	AAA	1.2	0.41 x 1.75	0.35	250	C/5
	AA	1.2	0.57 x 1.90	0.78	700	C/5
	SC	1.2	0.91 x 1.69	1.73	1800	C/5
	C	1.2	1.02 x 1.97	2.65	2400	C/5
	D	1.2	1.3 x 2.4	4.90	4000	C/5
	9V	7.2	1.04 x 1.95 x 0.67	1.54	120	C/5
NiMH	AAA	1.2	0.41 x 1.75	0.42	650	C/5
	AA	1.2	0.57 x 1.97	0.92	1500	C/5
	SC	1.2	0.91 x 1.69	1.94	3000	C/5
	D	1.2	1.30 x 2.39	6.00	6500	C/5
SLA		6	3.82 x 0.94 x 1.97	10.56	1300	C/20
		6	2.76 x 1.89 x 4.02	27.52	4200	C/20
		6	5.95 x 1.34 x 3.70	44.48	7200	C/20
		12	3.82 x 1.87 x 1.97	20.80	1300	C/20
		12	6.97 x 1.34 x 2.36	28.16	2200	C/20
		12	5.28 x 2.64 x 2.36	42.40	3400	C/20
Li-ion		3.6	0.72 x 2.55	1.48	1800	C/5
		3.6	0.41 x 1.34 x 1.97	1.38	1800	C/5

TABLE 7-3 Typical Battery Size and Capacities

Figure 7-5 illustrates this type of wiring. When connecting batteries together in series and in parallel, all of the batteries must be the same type.

The radio-control car and airplane industry sells a wide variety of battery packs with different voltages and current capacities that can easily be used for sumo robots. Figure 7-6 shows battery packs that are used for radio-control racing cars and for the transmitters and receivers, from

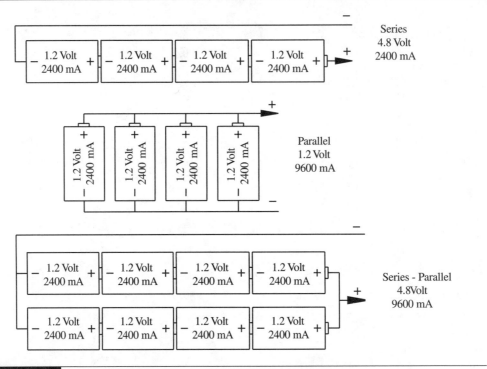

FIGURE 7-5 Connecting batteries together in series, parallel, and combined series and parallel to increase the overall voltage and current capacity of the battery pack

Duratrax (www.duratrax.com, Futaba (www.futaba.com), Radio Shack (www.radioshack.com), and Tower Hobbies (www.towerhobbies.com).

Battery Recharging

NiCd, NiMH, and Li-ion batteries have their own charging requirements. You should select a battery charger that was specifically designed for charging the batteries that you are using. There is no general-purpose charger that can be used to charge all of the different types and capacities of batteries. Use one of the microprocessor-controlled chargers that monitors the charging process and automatically shuts off when the charging is complete. Otherwise, you risk overcharging and damaging your batteries. Figure 7-8 shows a four- to eight-cell NiCd and NiMH battery charger from Duratrax, and a 6-volt SLA battery charger from Power-Sonic (www.power-sonic.com).

Li-ion batteries cannot be charged on NiCd and NiMH battery chargers. They have their own very specific and complicated charging requirements. Because of the complexity of their requirements, Li-ion battery packs come with their own charging circuits.

NiCd and NiMH batteries can be charged using a slow process, called *trickle-charging*, or a fast-charging process. Trickle-charging is the common low-cost method for recharging these batteries. Trickle-charging means to charge the batteries with a charge that is less than C/20 of

CHAPTER 7: Batteries: The Heart of a Robot **131**

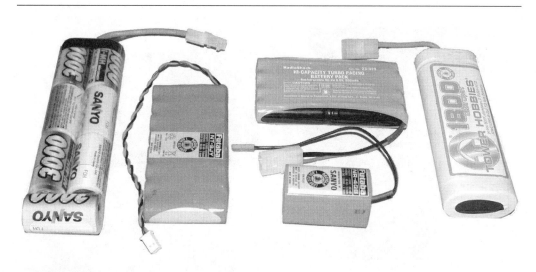

FIGURE 7-6   Battery packs from radio control cars and airplanes

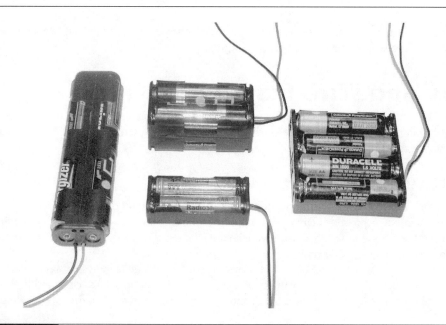

FIGURE 7-7   Battery packs using standard batteries and battery holders

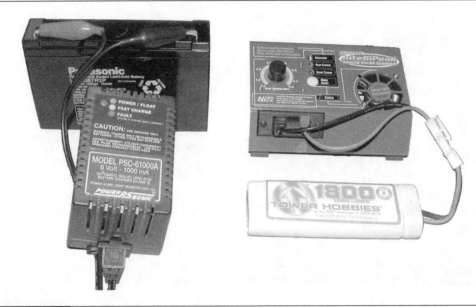

FIGURE 7-8 Four- to eight-cell NiCd and NiMH battery charger from Duratrax (right) and a 6-volt SLA battery charger from Power-Sonic (left)

the battery capacity rating. NiCd batteries can be trickle-charged indefinitely, but NiMH batteries can become overcharged when using trickle-charging.

Battery and Wire Installation

Along with selecting the correct type of battery for your robot, you also need to consider how you will install it. Battery placement, electrical wire requirements, terminal blocks, and multiple power supplies are important issues.

Battery Placement

Batteries are often the heaviest component of a robot, and they should be located as close to the bottom of the robot as possible. This is to keep the robot's center of gravity as low as possible. Another consideration for battery placement is that the batteries need to be in a location where they can be easily recharged or replaced.

During a competition, there really isn't that much time to recharge batteries between matches, so the batteries must have the capacity to run all day long, or the batteries must be replaced. When replacing the batteries, there must be a quick and easy access port to the batteries so you do not need to totally disassemble the robot to replace the batteries. Illustrated here is how the battery is located in the Viper sumo robot from Lynxmotion. By removing the sideplate, you can quickly get to the battery without dealing with any other part of the robot.

If the batteries are going to be permanently mounted inside the robot, the robot must have a port for recharging the batteries. The batteries should be securely mounted inside the robot so they do not slide around inside the robot. If a battery is overcharged, it may vent some excess gas. This could result in building up internal pressure inside the battery box. Batteries should always be placed in a well-ventilated area of the robot, but they should also be secured so that they do not move around in the robot.

Electrical Wire Requirements

In every robot, different amounts of current will flow through different sections of the robot. In order for the current to flow properly, the wires must be sized correctly to allow the current to flow without overheating.

Wires are like resistors, where the internal resistance is proportional to the length of the wire and inversely proportional to the cross-sectional area of the wire. Long wires have more resistance than short wires, and larger diameter wires have less internal resistance than small diameter wires. This is important because there is a voltage drop across any length of wire; the more current going though the wire, the greater the voltage drop. This energy loss results in a temperature increase inside the wire. If the wire gets too hot, the insulation around the wire melts off, which exposes bare wires that can short-circuit with other wires. If the current draw is too high, the wire will melt.

Table 7-4 shows the maximum safe current for different American Wire Gauge (AWG) sizes. Wires with high-temperature insulation, such as silicone, can handle higher current draws than

AWG	Max. Current	AWG	Max. Current
#20	13	#18	18
#16	20	#14	28
#12	38	#10	53
#8	78	#6	105

TABLE 7-4 AWG Maximum Current Ratings for Copper Wire

traditional vinyl-coated wires. However, for safety purposes, the values in this table should be used as a guideline for selecting the appropriate wire sizes for different current draws.

Terminal Blocks

When building a robot, one of the ideal ways to wire the robot, especially the heavy-gauge wires, is to use terminal blocks. A *terminal block* is a small nonconductive strip that contains a set of screws on both sides of the strip. The screws opposite each other are electrically connected, so that when wires are attached to both sides of the strip, they are electrically connected. Terminal blocks make the wiring more manageable and simplify maintenance requirements. Shown here is a robot that used a terminal block to attach the motor, battery, and motor controller wires together.

A terminal block allows you to easily change locations of parts in your robot without needing to remake new connectors. Connectors are another popular choice for connecting components together in a robot. Because your robot will be assembled and disassembled many times, using connectors and terminal blocks will save you time and a lot of headaches.

 When using terminal blocks, make sure the power is removed from the system before making wiring changes. It is easy to create a short circuit if a bare wire touches one of the closely spaced exposed terminal block screws.

Multiple Power Supplies

Motors draw a lot of current from the batteries. The amount of current that is drawn will not be constant. When a motor first starts or suddenly changes directions, there will be a large current spike that is drawn from the batteries. When the robot is pushing its opponent, there will be a large current draw from the batteries. As you learned earlier in this chapter, as the current draw increases, the internal resistance from the batteries results in a drop in voltage from the batteries. During motor startup, instantaneous current draws could be twice the nominal current draw from the motor. This will momentarily drop the battery voltage. If the voltage drops below 4.5 volts, this could result in the microcontroller resetting, which can create a lot of problems when you are testing the robot or competing with it.

Because of the current demand from the motors, many robot designers will use one battery pack for the drive motors and another battery pack for the electronics. The battery pack for powering the electronics can use smaller batteries, since the microcontrollers and sensor electronics do not require the same level of power the motors do. When using multiple battery packs, make sure the grounds for all of the power sources are connected together, or you will experience odd behaviors from your microcontrollers and sensors, since they will not have a common reference in the power circuit.

Closing Comments on Batteries

When it comes to selecting the batteries for your robot, you should have an understanding of all of the power, weight, and size requirements for the batteries. You should also know the advantages and disadvantages of the different battery technologies.

Because of the size and weight constraints of a sumo robot, you may need to make compromises in selecting its battery. Many times, a small battery is used in the robot, and the power requirements from the motors and electronics exceed the capabilities of the battery. This results in a poorly performing robot and a frustrated robot builder. By understanding the subtleties of how batteries work, you can select the correct battery so that the robot will perform exactly as required.

Sometimes, several smaller battery packs located in different parts of the robot are a better solution than having one battery pack for all of the power. Rechargeable batteries should be used in motor power circuits, and either rechargeable or nonrechargeable batteries should be used for the electronic control circuits.

At this point, you should have all the information you need to evaluate the different battery technologies and to determine how they will best fit in your robot design. Before selecting the batteries for your robot, you should have a good idea as to what type of motors your robot will use (see Chapter 8) and the space constraints in your robot.

Remember that batteries are the heart of a robot; they provide the lifeblood for all of the motors and electronic circuits. The heart needs to be strong enough to supply all of the power that is needed for however long it is needed.

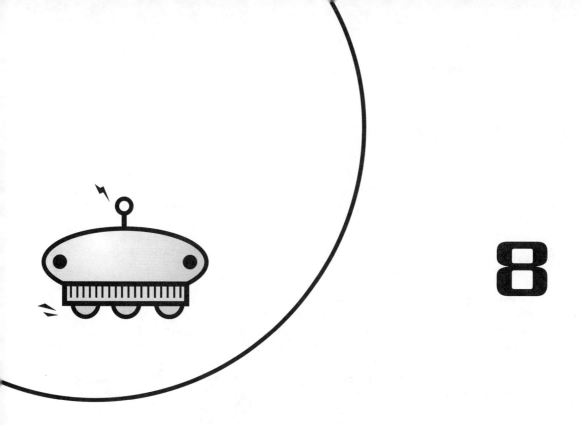

It's All About Power: Selecting the Right Motor

It's all about power when it comes to selecting motors for sumo robots. Sumo robots must be quick, nimble, and able to push their opponents out of the sumo ring. When the head-to-head pushing starts, the robot that has the better motors, geartrains, and traction will usually win the match. Faster and higher torque motors don't guarantee that you will win a sumo match. To be the winner, you need a good balance of speed, torque, strategy, and luck.

There are many types of motors that can be used in a robot, but this book will focus on traditional ferrite and rare-earth magnet permanent magnet direct current (PMDC) motors. This chapter will explain the basics of how motors work and how they are specified. Then you will learn how to use this information to determine the right motor for your robot. At the end of this chapter, you will see how traditional radio-control (R/C) style servo motors make excellent motors for mini, lightweight, and even some 3-kilogram sumo robots.

Motor Types

Most sumo robots use PMDC motors, which are fairly low in cost and relatively easy to control. These motors are found in many electrical devices—hobby models, medical instruments, cordless toothbrushes, high-powered cordless drills, and other tools. They typically use ferrite magnets and have running speeds from 5,000 to 20,000 rpm. Figure 8-1 shows several different PMDC motors.

NOTE *Mabuchi Motor Company (www.mabuchi-motor.co.jp) and Johnson Electric (www.johnsonmotor.com) are two of the major manufacturers of high-volume production motors.*

Another option is to use radio-controlled (R/C) car motors. These ultra-high-velocity motors offer a lot of power in a small package. Many of these motors can draw over 30 amps in normal nonstalled operation. R/C car motors can make a great choice for a robot sumo when you are not worried about high current draws. Since these motors typically run at speeds up to 20,000 to 40,000 rpm, fairly large gear reductions will be required to use them. If you use an R/C car motor in your robot, it is highly recommended that you use an R/C car electronic speed controller to control the motors. This is because the R/C car industry uses a nonstandard method for specifying motor performance. Unfortunately, there are no conversion factors to convert their performance numbers into understandable numbers. Chapter 9 will provide more information on R/C car motor controllers.

Rare-Earth and Ferrite Magnet Motors

High-performance motors use rare-earth magnets and have efficiencies up to 80 to 90 percent. Typical ferrite magnet motors have efficiencies between 50 to 70 percent. Motors with rare-earth magnets generally run much cooler than ferrite motors do. When running at maximum efficiency, a

FIGURE 8-1 Different styles of PMDC motors

rare-earth magnet motor will convert only about 10 to 20 percent of the input energy into heat. A ferrite magnet motor will convert about one-third of the input energy into heat.

Rare-earth magnets are made from either cobalt or neodymium alloys. Cobalt alloy magnets are virtually immune to demagnetization, no matter how much voltage you pump into them or how hot they get. The major drawback of using rare-earth magnet high-performance motors is that they are significantly more expensive than regular motors.

Brushless PMDC Motors

Another class of high-performance motors are the brushless PMDC motors. The brushes in an ordinary motor can be the source of several problems: They spark and cause radio interference, they are a source of friction, and they wear out.

Brushless motors have sensors that detect the position of the rotor relative to the windings, and this information is sent to a special motor controller that energizes the windings at the optimum moment on each revolution of the motor. In a brushless motor, the windings are stationary and the magnets spin, which is the opposite of how a conventional motor works.

Brushless motors can generate more torque and rotational speed in a smaller package than a conventional PMDC motor. Like rare-earth magnet motors, brushless motors are more expensive than regular motors.

Electric Motor Basics

Knowing how your motors will perform is important for your robot's design. Your robot's electric motors directly affect its speed and pushing capability. You need motor performance information to select the wheel diameters and how much gear reduction is required in the robot. The current requirements from the motor will dictate what type and size batteries you will need, and they are also a factor in determining the minimum current requirements for your motor speed controllers.

All PMDC motors have two simple characteristics:

- The motor speed is proportional to the applied voltage across the motor.
- The motor's output torque is proportional to the amount of current the motor is drawing from the batteries.

In other words, the more voltage you supply to the motor, the faster it will turn; and the more current the motor is drawing, the greater the output torque the motor is supplying. These relationships are shown in Equations 8-1 and 8-2. The motor-speed constant (K_v) is in units of revolutions per minute per volt. The motor-torque constant (K_t) is in units of ounce-inches (or inch-pounds) per amp.

Equation 1
$$RPM = K_v \, Volts$$

Equation 2
$$Torque = K_t \, Amps$$

Motor Speed

Equation 8-1 shows that the motor speed is only a function of the applied voltage, but the relationships shown in the equations are true only for the "ideal" motor. We know from experience that as we apply torque on the motor shaft, the motor will slow down. With enough torque, we can even stop the motor.

In reality, a motor slows down because of the internal voltage losses due to the current going through the motor. All motors have a certain amount of internal resistance; thus, the net voltage the motor uses to turn the rotor is proportionally reduced by the current flowing through the motor. The effective voltage that the motor actually uses (V_{motor}) is shown in Equation 8-3. V_{in} is the voltage across the motor terminals, I_{in} is the current draw from the motor in amps, and R is the internal resistance of the motor in ohms.

Equation 3
$$V_{Motor} = V_{in} - I_{in} R$$

Replacing *Volts* in Equation 8-1 with V_{motor} gives the actual equation for motor speed as a function of the applied voltage. Equation 8-4 shows that the motor speed is a function of both the voltage across the motor terminals and the current going through the motor.

Equation 4
$$RPM = K_v V_{motor} = K_v (V_{in} - I_{in} R)$$

As you can see in this relationship, as the current going through the motor increases, the motor speed decreases; and as the applied voltage increases, the motor speed increases. This relationship is in agreement with what is observed in real motors.

Motor Torque

The motor torque and current are proportionally related, and Equation 8-2 is a fairly accurate representation of this relationship. However, the equation needs a small modification to account for the energy required to kickstart the motor.

When there is no load on the motor (the motor is allowed to spin freely, which is called the *no-load speed*), you will notice that the motor is drawing a small amount of current from the batteries. This current not only includes the current due to internal resistance of the motor, but it also includes the current required to overcome the internal "frictional and inertial" losses inside the motor. The current that is required to overcome these losses is known as the *no-load current*. Equation 8-5 shows the actual torque as a function of the current draw, where I_o is the no-load current in amps.

Equation 5
$$Torque = K_t (I_{in} - I_o)$$

PMDC motors are very interesting machines. Speed is a function of how much voltage you apply to the motor and how much current is going through the motor. A motor will try to maintain its speed as best as it can. When an externally applied torque on the motor shaft is increased, the motor will automatically draw more current to counter the applied torque in an effort to maintain

CHAPTER 8: It's All About Power: Selecting the Right Motor

the original rotational speed. (Explaining the physics behind how this happens is beyond the scope of this book.) But from Equation 8-4, you can see that the motor speed will decrease as the current going through the motor increases. Since current and torque are directly related to each other, the motor speed is a function of the applied torque on the motor. Again, this is in agreement with what we have experienced with direct-current (DC) motors.

Since PMDC motors will draw only the amount of current that is needed to counteract the applied torque on the motors, Equation 8-5 can be rewritten to determine the amount of current that is needed for a particular torque requirement. Equation 8-6 shows the current as a function of the applied torque relationship. This equation will tell you how much current your motors will need to overcome external forces, so that you can select the correct sized batteries and motor controllers for your robot.

Equation 6

$$I_{in} = I_o + \frac{Torque}{K_t}$$

By combining Equation 8-6 into Equation 8-4, you will obtain the motor speed as a function of the applied voltage and torque on the motor. Equation 8-7 shows how the motor speed and the applied torque are directly coupled together.

Equation 7

$$rpm = K_v \left(V_{in} - R \left(I_o + \frac{Torque}{K_t} \right) \right)$$

These equations will tell you everything you need to know to determine how well your motors will perform under different loading conditions and applied voltages. With this information, you can determine how fast your robot will move and how much it can push when you apply the speed reduction relationships that were described in Chapter 6.

Stall Current

The motors inside a sumo robot will always be pushed to their limits. This is because a sumo robot is trying to push another robot that does not want to be pushed. When two robots contend, wheel traction, motor gear reductions, and motor torque are believed to be the deciding factors that determine which robot wins. In theory this is true, but it is not always the case. There are a couple other factors that must be considered when pushing your motors at their extreme conditions. Many times, a robot with less traction and pushing capabilities has beaten a stronger robot because the robot builder ignored the effects of stall current and internal heat buildup inside the robot's motors.

As the applied torque on a motor increases, the motor will draw more current to counter the applied torque, which results in slowing down the motor. When the externally applied torque is high enough, the motor will eventually stall. At this point, the motor will be drawing the maximum current from the batteries, which is called the *stall current* (I_{stall}). Depending on the motor, the stall current can be over 100 amps! A motor's stall current can be calculated by dividing the applied voltage by the motor's internal resistance. Equation 8-8 shows this relationship.

Equation 8

$$I_{stall} = \frac{V_{in}}{R}$$

When the motor is stalled, it will continue to draw the stall current from the batteries until one of the following three things occur: the batteries go dead, the motor burns out, or the motor controllers burn out. All of this generally happens within a few seconds. Hence, you do not want to stall the motors on your robot.

Knowing your motor's stall current will help you select a correctly sized motor controller and determine your battery's peak current capacity requirements. Many times, the motor controller will cost more than the motor, so you will want to make sure that the motor controller can handle more than the motor's stall current. If the motor controller's peak current handling capacity is less than the motor's stall current, you should place a fuse that has a lower current rating between the motor and motor controller. It is always better to burn out a fuse than to damage a motor controller. It is easier to replace a blown fuse between sumo matches than it is to replace a motor controller.

Power and Efficiency

The other set of relationships that should be considered is power and motor efficiency. Mechanically speaking, motor power is the product of the output torque and shaft speed (revolutions per minute). Electrically speaking, power is product of voltage and current. Motor efficiency is the ratio of the output and input power. In other words, motor efficiency tells you how well the motor converts the electrical input power (P_{in}) into useful output power (P_{out}). Equations 8-9 through 8-11 show the electrical relationships for the motor input power, output power, and efficiency. The standard unit for power is watts.

Equation 9
$$P_{in} = V_{in} I_{in}$$

Equation 10
$$P_{out} = (I_{in} - I_o)(V_{in} - I_{in} R)$$

Equation 11
$$Efficiency = \frac{P_{out}}{P_{in}} = \frac{(I_{in} - I_o)(V_{in} - I_{in} R)}{V_{in} I_{in}}$$

The output power can never be greater than the input power. The difference between the input power and the output power is due to the electrical, frictional, and dynamic losses in the motor, and this changes as the motor loading changes. This energy loss is converted into heat, which is why motors get hot when they are running. It is best to design and operate your robot's motors at their highest efficiency range to minimize motor heating.

Electric motors react to their environments. You cannot push current into a motor, nor can you force the motor to generate a specific output torque. All you can do to a motor is increase or decrease the voltage to the motor. The torque a motor generates is to counter the applied torque on the motor. The amount of torque the motor is generating is always changing to match the demands placed on the motor. The current going through the motor is directly related to the applied torque on the motor, and the motor will draw as much current es it needs to respond to the external torque. As the current going through the motor increases and decreases, the motor speed will increase and decrease because of the internal voltage losses due to the internal resistance of the motor. The equations presented here show you how all of the motor characteristics relate to each other, so that you can use them to estimate how well the motor will perform based on applied voltages and external factors. For a more detailed explanation on how electric motors work, read the book *Electric Motors Handbook* by Robert J. Boucher. A copy of the book can be obtained at Astroflight Inc. (www.astroflight.com).

CHAPTER 8: It's All About Power: Selecting the Right Motor   **143**

Determining the Motor Constants

To use the equations, the motor constants K_v, K_t, I_0, and R must be known. The best way to determine the motor constants is to obtain them directly from the motor manufacturer. Sometimes, when you buy a motor from a store (say a hobby store or a surplus store), the motor doesn't come with any performance data or the data is incomplete. Also many times, we get our motors from some motorized contraptions we have lying around the house. In these cases, the motor constant information is not available. When this happens, you will need to either contact the motor manufacturer or measure the values yourself. Fortunately, all of these values can be easily measured through a few experiments.

Using a Motor Performance Chart

Many motor manufacturers provide a motor performance chart to graphically show how all of the motor parameters relate to one another. Figure 8-2 shows a typical motor performance chart for a Mabuchi RK-370SD motor. Table 8-1 shows the motor constants for this motor.

In Figure 8-2, you can see how the motor's shaft speed (revolutions per minute) decrease and how the current draw increases as the applied torque on the motor increases. This plot also shows the motor's output power and operating efficiency. You can see that the output power increases until the speed drops to 50 percent of the no-load speed; then it decreases. Maximum power is always at the 50 percent no-load speed and 50 percent stall current condition. The motor efficiency rapidly increases, and then slowly decreases as the applied torque increases. The shape of the efficiency plot is fairly typical of DC motors. Ideally, motors should be operated in their maximum efficiency range.

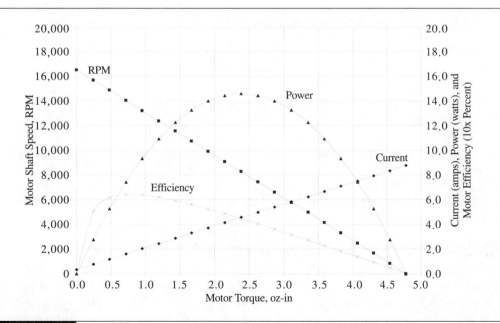

FIGURE 8-2   Motor performance chart for a Mabuchi RK-370SD motor at 7.2 volts

Constant	Rating	Constant	Rating
I_o	0.340 amp	K_v	2385 rpm/volt
No-load speed	16,000 rpm	K_t	0.567 oz-in./amp
R	0.821 ohm	Maximum efficiency	64.5%

TABLE 8-1 Motor Constants for a Mabuchi RK-370SD Motor at 7.2 Volts

If you know the various motor constants, you can create a different motor performance chart for different operating voltages. The motor performance charts change based on the different input voltages.

Measuring the Motor Speed Constant

There are two methods you can use to measure the motor speed constant. Both methods require a voltmeter.

One method is to attach the motor shaft to another motor with a known speed. A drill press is a good source for known rotational speeds. Chuck up the motor shaft in the drill press. Set the speed of the drill press to its lowest speed, and clamp the motor body so that it will not spin. Turn on the drill press, and measure the voltage coming from the motor. Record the voltage and the revolutions per minute on the drill press. Then shut everything off, and repeat the test with a higher speed on the drill press. Finally, use Equation 8-12 to calculate the motor speed constant (K_v). The subscripts 1 and 2 represent the measured results from test 1 and test 2.

Equation 12

$$K_v = \frac{RPM_2 - RPM_1}{Volts_2 - Volts_1}$$

Do not hold the motor with your hand to prevent it from spinning. Holding a motor while applying speed could result in injury.

The other method for measuring the speed constant requires a tachometer. Apply a known voltage to the motor, and use the tachometer to measure the motor shaft's revolutions per minute. Use a voltmeter to measure the voltage at the same time, and record these values. Again, repeat the test with different voltage level and record the second set of values. Then use Equation 8-12 to calculate the motor speed constant.

Calculating the Motor Torque Constant

All PMDC motors have the same physical property, wherein the product of the motor speed constant and the motor torque constant is 1,352. The units for this constant is revolutions per minute divided by volts times ounce-inches divided by amps: ($RPM / Volts$) × ($oz.$-$in. / amps$). The 1,352 value will change if you are using different units of measurements, such as metric numbers. With this knowledge, you can calculate the motor torque constant without needing to use an expensive piece of equipment (called a dynamometer). This relationship is shown in Equation 8-13.

Equation 13

$$K_t = \frac{1352}{K_v}$$

Measuring the No-Load Current

The next step is to measure the no-load current. First make sure nothing is attached to the motor shaft, and then apply the normal operating voltage to the motor. The motor should be spinning at its maximum speed. With a current meter (an *ammeter*), measure the current draw from the motor, and record this value as the no-load current.

Measuring the Motor's Internal Resistance

The last step is to measure the motor's internal resistance. This cannot be measured with an ohmmeter; it must be calculated. Clamp the motor and output shaft so that they will not spin. Apply a very low DC voltage to the motor, such as 1.5 or 3 volts. This voltage must also be less than the motor's nominal voltage rating and high enough to cause the shaft to spin. The voltage needs to be constant throughout the entire test. If you do not have a variable-regulated DC power supply, one or two D-cell NiCd batteries should work. The test should be done quickly so that the motor doesn't heat up. A warm or hot motor will distort the measurements, because the internal resistance changes with temperature.

NOTE *Remember that large motors can generate a lot of torque and draw a lot of current, so you need to make sure your clamps are strong enough to hold the output shaft still. Also, your ammeter must be able to measure at least 10 amps.*

Now measure both the voltage and current going through the motor at the same time. You'll get the most accurate results when you are measuring several hundred milliamps to several amps. The internal resistance (R) can be calculated using Equation 8-14 by dividing the measured voltage (V_{in}) by the measured current (I_{in}). You should take at least three measurements and average the results.

Equation 14

$$R = \frac{V_{in}}{I_{in}}$$

After conducting these experiments, you should now have all of the motor constant parameters to calculate how your motor will perform under different loading conditions.

Calculating Motor Constants from Data Sheets

You may find that the motor specifications are not presented in an easy-to-understand table. For example, the motor speed and torque constants may not be shown, but the motor current draws are given for two different operating speeds. Stall torque may not be given, but a starting current is listed. A nominal voltage and a voltage range for the motors should also be given. Sometimes, all of the information is given, but it will require some calculations to convert it into a useful form. In some cases, the terminology is different. For example, starting current and stall current has the same numerical value. They have different meanings, but they have the same numerical current draw value. You can calculate the motor's internal resistance by using Equation 8-14 and dividing the motor's nominal voltage by the starting current rating.

Many motors will list either the no-load speed and no-load current, or the motor speed and current draw at maximum efficiency. Sometimes both of them will be listed. If you know the motor's internal resistance, you can use either one of these equations (Equations 8-4, 8-5, and 8-11) to calculate the motor speed constant. If you do not know the motor's internal resistance, you will need two know two different speeds at two different current draws. Then, using a bit of linear algebra, you can apply Equation 8-4 to solve for both the motor's internal resistance and motor speed constant. Once you have the motor speed constant, you can calculate the motor torque constant from Equation 8-13.

Power and Heat

Power and heat are important factors when working with electric motors. Electric motors are energy converters. They convert the input electrical energy into mechanical energy with some heat. The heat that is generated is the difference between the input energy and the output energy. A 100 percent efficient electric motor will not generate any heat, but all motors have some inefficiencies, so they will generate some heat. The amount of heat that is generated depends on how the motor is used.

The generated heat values can tell you how hot your motor will get when it is operating at different input voltages and different loading conditions. Equation 8-15 shows a simplified relationship for the heat generated from an electric motor. The units of the heat are watts.

Equation 15 $\qquad H = P_{in} - P_{out} = V_{in} \, I_{in} - (V_{in} - I_{in} R)(I_{in} - I_o)$

Figure 8-3 shows a plot of heat generated in the same Mabuchi motor used to generate the plot in Figure 8-2. In the heat plot, you can see that the maximum amount of heat is generated when the motor is stalled. When the motor is running near its maximum efficiency, the heat generated is near its minimum, but then it exponentially increases as the motor torque increases. Motors generally cannot tolerate the heat that is generated above the 50 percent stall torque rating for long periods of time without failing.

The heat generated in the armature windings and brushes is inside the motor. Because air is a poor conductor of heat, the internal components of the motor cool through conduction through their bearings. This is not a very efficient way to cool a motor, so the internal components are always much hotter than the outside case of the motor (and the outside case can get very hot!).

So why is this important? The heat that is generated results in a temperature rise in the motor. The temperature rise can be high enough to destroy a motor in several ways:

- ■ Most low-cost PMDC motors use ferrite magnets, which can become permanently demagnetized if they are overheated.)

- ■ The flexible braided copper leads that feed current to the brushes (called *shunts*) can melt after just a few seconds of severe overcurrent demands.

CHAPTER 8: It's All About Power: Selecting the Right Motor

- The insulation on the copper windings can melt which can short out or melt the windings.
- Depending on the motor brush mounting technique used, the springs used to keep the brushes on the commutator can heat up and lose their strength, thus causing the brushes to press less tightly against the commutator. When this happens, the brushes can arc more, heat up, and finally disintegrate.
- Most motors have plastic parts that hold the motor together. The plastic parts usually melt long before the metals parts do.

When the temperature of the motor increases, the resistance of the wires inside the motor increases, which results in a reduction in the output shaft speed. This reduces the pushing force and speed your robot has. When you notice that your robot is not pushing or moving as fast, you give it more voltage, which further increases the heat output from the motors, which further increases the motor temperature, which increases the internal resistance of the wires, which reduces the motor speed. Eventually, one of the wires inside the motor will melt or break, and then the motor stops. This almost always happens when the motor is stalled for a short period of time. The failure time is very dependent on the motor and the applied voltage. The higher the motor voltage, the shorter the motor failure time. The failure time can be as short as a second to as long as a few minutes.

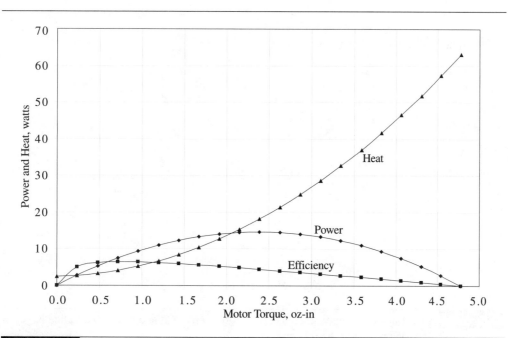

FIGURE 8-3 Heat generated in a 7.2-volt Mabuchi electric motor

One of the most impressive sounds you can hear at a sumo tournament is when two high-powered robots stall against each other. Their heads are locked together, and their motors are stalled. You may be able to hear a high-pitched whine from the stalled motors. When this happens, the length of the contest comes down to whose motor will burn out first. In robot sumo, stalling the motors is a common occurrence, but the stall time should be minimized, or you should use some sort of active cooling on the motors to keep their temperatures down. Ideally, you should never stall a motor, since it will eventually fail when stalled.

Battery Effects

The discussions about motor performance thus far have assumed that the input voltage (V_{in}) is constant throughout the operation of the motor. In reality, this is not the case. There are voltage losses in the batteries, wires, and the electronic speed controller (ESC). All of these losses add up, so that the actual voltage the motor receives is less than the theoretical voltage from the battery. Equation 8-16 shows how the input voltage is adjusted based on the internal resistances of the various components in the power circuit. Figure 8-4 illustrates how the motor shaft speeds are affected by using NiCd and alkaline batteries.

Equation 16
$$V_{in} = V_{battery} - (R_{battery} + R_{ESC} + R_{wiring}) I_{in}$$

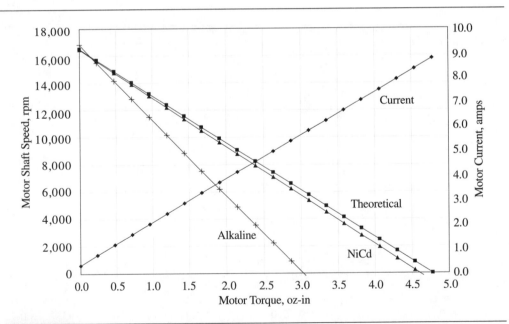

FIGURE 8-4 Motor speed adjustments based on six-cell NiCd and five-cell alkaline batteries used to power the robot

The NiCd battery curve is nearly identical to the theoretical speed curve, but the alkaline battery line shows a large drop in motor speed for a given applied torque on the motor. This is because the internal resistance of a NiCd battery is approximately 0.005 ohm, and the internal resistance of an alkaline battery is approximately 0.100 ohm. The total resistance of a six-cell (7.2-volt) NiCd battery is 0.030 ohm. The internal resistance of the five-cell (7.5 volt) alkaline battery pack is 0.500 ohm—about 16 times greater than the NiCd battery pack. The current line is not affected, because the current draw is a function of the applied torque. Since the motors are receiving a lower voltage from the batteries, they will also have a lower stall torque.

As discussed in Chapter 7 and illustrated in Figure 8-4, alkaline batteries should not be used in high-current applications. The internal resistance for NiCd, NiMH batteries, the ESCs, and wiring is relatively small and generally does not have any significant effects on the motor speed performance. But with high-performance motors, the internal resistances of the various components start to have a significant impact on the motor performance.

Pushing the Limits of Your Motors

In competition events, it's common for people to push their equipment past its design performance limits. Robot sumo is no different. All motors are rated to operate continuously in a certain voltage range. Many times, we want more speed and torque from the motors, so we increase the voltage to the motors.

Effects of Increasing Voltage

Running a motor at twice the rated voltage is common, and there are some robots that run their motors at three and four times the rated voltage. The robot builders that do this know that this will shorten the life of their motors, but they are willing to risk destroying a motor to get that extra speed and torque.

Figure 8-5 shows how the motor speed and torque change when doubling the input voltage of the 7.2-volt Mabuchi motor (used in Figures 8-2 and 8-3) to 14.4 volts. In this plot, you can see that the motor speed is doubled for the same applied motor torques, which means the motors will run twice as fast as the same motor running at 7.2 volts. The motor current lines for both voltage ratings lie on top of one another (since they are the same motor). This is because current draw is only a function of the applied torque. The 14.4-volt motor case has a higher stall current, since it requires more applied torque to slow down the motor and stall it.

By doubling the motor's voltage, you can double the top end speed of your robot, and you can even double the stall torque of the motor. But all of these improvements do not come without a price. Figure 8-6 shows the power and heat generated in this motor for the two different input voltages. Merely doubling the voltage results in quadrupling the maximum output power and increases the heat generated at stall conditions by a factor of four! Stalling a motor operating at twice its operating voltage will cause the motors to overheat and be seriously damaged in a very short period of time. Nothing is free in the world of physics.

Motor heating is proportional to the *current^2 × resistance*. You've seen that doubling the input voltage increases the heating by a factor of four. Tripling the input voltage can increase the

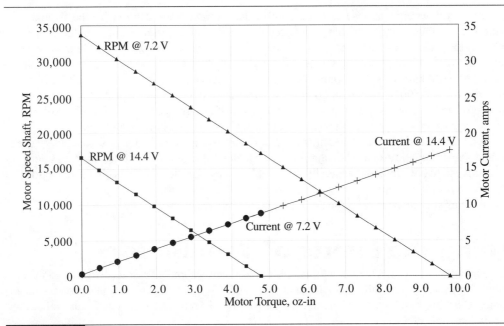

FIGURE 8-5 Motor speed and current changes caused by operating a 7.2-volt motor at 14.4 volts

heating by a factor of nine or almost ten times the normal heat generation from the motor. Doubling the voltage to a high-performance R/C car motor and then stalling that motor can generate more heat output than a typical home electric space heater. Now imagine all that power coming from a lump of metal that can fit inside your cup of coffee. You should be able to see why motor survival time is very limited, and stalling should be avoided.

Cooling Motors

Okay, so you want to use a greater than recommended voltage on your motor to get more power out of it, but you are worried about damaging it. What should you do? First, you must realize that you always run the risk of destroying your motor if you choose to boost its performance past the manufacturer's specifications. That said, here are some steps you can take to reduce the risk:

- At the very least, you should have a can of compressed air—the type that is used to clean cameras—at a competition. Between each match, blow the high-velocity air on the motor to rapidly cool it off before starting the next match.
- Use active cooling fans in your robot. The miniature fans that are used to cool the central processing unit (CPU) inside your computer will help keep the motor cooler.
- Use encoders on the wheels to monitor if they are actually moving. If the software monitoring the wheel encoders detects that the wheels are not moving when they should be, the software can initiate an evasive action that will protect the motors.

CHAPTER 8: It's All About Power: Selecting the Right Motor 　　**151**

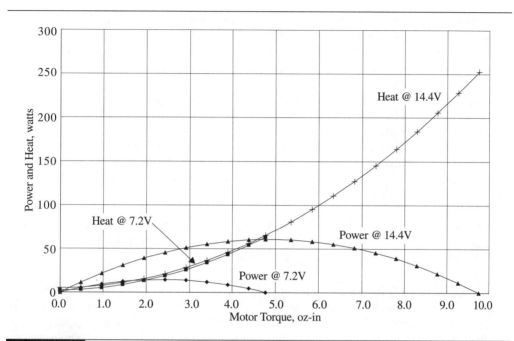

FIGURE 8-6　Heat generated by doubling the applied voltage

- Use a motor controller that has built-in current limiting capabilities that can monitor the current draw from the motors and change the motor speed controls to limit the maximum current that goes to the motor.

How to Select a Motor

When it comes to selecting a motor for your robot, your first decision is whether you will build a gearbox for your robot or you will use a gearmotor. This will determine the types of motors that are available to you. In this section, we will work through an example of picking a motor for a particular robot. For this example, we will use gearmotors for our robot, taking into consideration that machining a gearbox requires special tools that most of us do not have.

Defining Your Robot's Performance Specifications

To figure out what you need to look for in a motor, you first need to define how you want your robot to perform. The first step is to pick the speed you want your robot to move and the weight your robot must be able to push. For this example, we want our robot to move from one end of the sumo ring to the other in less than two seconds. The diameter of the sumo ring is

154 centimeters, so this equates to a speed of 77 centimeters per second (154 cm/sec / 2 = 77 cm/sec), or 1,819 inches per minute (77 cm/sec × 60 sec/min / 2.54 cm/in. = 1,819 in/min).

Next, we want the robot to be able to push at least twice the maximum weight of a sumo robot. Since the maximum weight of the robot is 3 kilograms, the maximum pushing force should be 6 kilograms, or 13.2 pounds.

With the speed and pushing force information, you now have a performance specification for a sumo robot. You can use these values to figure out what type of gear reduction you are going to need and how much torque the motors will need to be able to provide.

Before you can determine a gear reduction, you need to decide on a wheel diameter. For this example, we will use the 2-7/8 inch diameter foam wheels from Lynxmotion. In order to move at 1,819 inches per minute, the wheel must spin at 201.4 RPM (1,819 in./min / (3.14159 × 2.875 in) = 201.4, from Equation 2 in Chapter 6).

NOTE *We selected the 2-7/8 inch foam wheels from Lynxmotion for this example because they are made from a high-traction Green Dot foam material, and you can get the mounting hubs from the same company. Also, they are low in cost.*

For simplicity, we are going to assume that the coefficient of friction between the wheels and the ground is 1.0, and all of the wheels on the robot have the same diameter. Thus, the torque the motors must generate is 19 inch-pounds (13.2 lb × 2.875 in. / 2 = 19 in.-lb, from Equation 6 in Chapter 6) or 304 ounce-inches (16 oz/lb). This is the total torque all of the wheels combined must be able to produce. The robot in this example will be a four-wheel-drive type, so the individual wheel torques will be one-fourth of this, or 76 ounce-inches.

At this point, we have determined that the gearmotors must have a speed of 201 RPM and a torque of 76 ounce-inches. These are the minimum specifications for this robot. Since motor performance is a function of the applied voltage, we will add the constraint that the motors must meet these requirements using a standard 7.2-volt NiCd battery pack.

Finding the Motor

Determining the performance specifications is the easy part. Finding a motor that can meet these performance goals is the hard part. The first thing to do is to look through different catalogs and search the Internet to collect a list of different motor types that appear to be suitable for the robot. During this selection process, consider the physical size, weight, and cost of the motor.

Then make a table that lists the motor, its nominal operating voltage (the nominal voltage is usually 6, 12, or 24 volts), the no-load speed, the stall torque, and the stall current. This information provides the extreme performance data that you will use to select the motor. Table 8-2 shows a sample list of five gearmotors with their performance data. (Two part numbers are shown for the Maxon motor: the first number is for a 9-volt motor, and the second number is for an 18:1 gearbox for the motor.)

Since these motors have different nominal voltage ratings, it's difficult to make an "apples-to-apples" comparison. So, let's look at the same values adjusted to the 7.2-volt battery pack that will be used in the robot. Since revolutions per minute are proportional to voltage, we can linearly scale the no-load speed, stall torque, and stall current values to 7.2 volts. For

CHAPTER 8: It's All About Power: Selecting the Right Motor

Manufacturer	Part No.	Volts	No-Load (RPM)	Stall Torque (oz-in.)	Stall Current (amps)
Tamiya	3633k36	7.2	196	173	5.4
Jameco	161381	12	200	50	1.4
Lynxmotion	GHM-01	6	186	75	2.1
Servo Systems	BC-101	12	180	144	1.4
Maxon	118742/ 233150	9	218	227	7.3

TABLE 8-2 Sample Selection of Gearmotors for Sumo Robots

example, the Maxon motor speed can be scaled from 218 rpm to 174 RPM using the calculation 218 RPM × 7.2 volts / 9 volts = 174. Table 8-3 shows the results of the value adjustments.

NOTE *The linear scaling used to adjust the motor values is not 100 percent accurate, but it is good enough for calculating motor differences.*

The goal is to find a motor that has a minimum speed and torque of 201 RPM and 76 ounce-inches. Table 8-3 shows only one motor that meets both of these specifications, but there are two other motors that are close to these requirements. A third table converts these motor specifications into robot speed and pushing force. Table 8-4 shows how the different motors will perform in the robot.

From Table 8-4, it is obvious that we need to eliminate the Jameco motor from the list, since it is not strong enough to push an opponent out of the ring. You can assume that all robots will weigh the maximum weight of 6.6 pounds. The Lynxmotion motor is the fastest of the four remaining motors. The Servo Systems motor has approximately the same pushing torque, but it has half the speed of the Lynxmotion motor, so we can eliminate it, too.

The Tamiya motor and Maxon motors are good alternatives. Although they are slower than the Lynxmotion motor, they are very close to the target speed of the robot. Their big advantage is

Manufacturer	Part No.	Volts	No-Load (RPM)	Stall Torque (oz-in.)	Stall Current (amps)
Tamiya	3633k36	7.2	196	173	5.4
Jameco	161381	7.2	120	30	0.8
Lynxmotion	GHM-01	7.2	223	90	2.5
Servo Systems	BC-101	7.2	108	86	0.8
Maxon	118742/ 233150	7.2	174	181	5.8

TABLE 8-3 Motor Performance Specifications from Table 8-2 Normalized to 7.2 Volts

Manufacturer	Part No.	Maximum Speed (in./min)	Time to Move Across Ring (sec)	Maximum Pushing (lb)
Tamiya	3633k36	1770	2.06	30.1
Jameco	161381	1084	3.36	5.2
Lynxmotion	GHM-01	2014	1.81	15.6
Servo Systems	BC-101	975	3.73	15.0
Maxon	118742/233150	1572	2.30	31.5

TABLE 8-4 Robot Speed and Pushing Performance Using Different Motors

that their pushing force is twice that of the Lynxmotion motor. For the same pushing force of 15 pounds, you can use two Tamiya or Maxon motors instead of four, which can save both cost and space inside the robot. The other advantage is that when pushing a 15-pound object across the floor, the Lynxmotion motors are near their stall torque and not moving very fast, which greatly increases the chance of burning the motors out in a heavy pushing match. The Tamiya and Maxon motors will be pushing this weight across the sumo surface at about half their maximum current. These motors are less likely to burn out during a competition.

Because of the higher pushing capabilities of the Maxon and Tamiya motors, they should be considered the better motor choices. Although these motors are great choices for sumo robots, they have their drawbacks. The Tamiya motors cost about the same as the Lynxmotion motors, but they are only sold in Japan. Getting this motor outside Japan is extremely difficult, so we'll need to toss it off the list. This leaves the Maxon motor. The drawback to this motor is its cost—more than ten times the cost of a Lynxmotion motor. If you have the extra funds to invest into your sumo robot, the Maxon motor (which is precision made) is the preferred choice. If you are on a tight budget, the Lynxmotion motor is the one to choose.

At this point, you should have an idea of how to select a motor using a decision-based process. Performance and costs are major deciding points when it comes to selecting motors. The motors shown in this example represent only a fraction of all of the motors you can use. The more motors you include in the list, the more options you have.

This same type of an analysis can be conducted with regular DC motors when you are building your own gearbox. On the list, you will add a gearbox option that lists the gear reduction, and type of gear reduction. It is a lot easier to reduce a 5,000 rpm motor to 200 rpm than it is to reduce a 40,000 rpm to 200 rpm. Also, the gearbox costs needs to be considered in the selection process.

R/C Servos for Gearmotors

R/C servos make a great choice for gearmotors. They are small and lightweight, provide a lot of torque, and have their own built-in motor controller. The servos are specified by their torque-holding ability and their *transit time*. Some manufacturers call the transit time the *servo speed*.

CHAPTER 8: It's All About Power: Selecting the Right Motor

Transit time is given as the amount of time, in seconds, that is required for the servo arm to rotate 60 degrees. Equation 8-17 shows how the transit time can be converted into revolutions per minute. The units of transit time must be seconds/60 degrees.

Equation 17
$$RPM = \frac{10}{Transit\ Time}$$

There are many different companies that make these servo motors, such as Futaba, Hitec, Airtronics, FMA Direct, Cirrus, and JR. They are sold in many different sizes with different torque and transit times. Table 8-5 lists some of the different types of servos that are available and their equivalent RPM speeds (if they are modified for continuous rotation). The Tower servos are from Tower Hobbies. Figure 8-7 shows these servos for relative size comparisons.

Using Servos as Sumo Robot Motors

In order to use an R/C servo as a motor, you must modify it so that it can make a 360-degree continuous motion. Not all servos can be modified into continuous rotating motors, because on some servos, the final output gear does not have teeth that completely surround the output shaft. The Hitec HS-81 and the Tower Hobbies TS-75 are examples of servos that cannot be modified. The gears inside these motors are shown here to illustrate this limitation.

The other servos listed in Table 8-5 can be modified into a continuous rotating gearmotor. The internal details of the servo motors are not published, so you will need to take one apart to determine if it can be modified. We'll look at modifying servos after examining some details about their electrical interface.

After you've modified a servo, you will have a small, compact, high-torque gearmotor. These motors are ideal for mini and lightweight sumo robots. There are several robot builders that use them for 3-kilogram sumo robots. They are not as fast as regular gearmotors, but they can provide a lot of torque. For example, the TS-80 servo motor has a massive 343 ounce-inch of torque. That is 21.4 inch-pounds. With a 2-7/8 inch diameter wheel directly connected to the servo, you get 14.9 pounds of pushing force, and this is only one motor! With four of them, you could have a robot that could theoretically push almost 60 pounds!

When using servo motors with 3-kilogram robots, the servo shaft should not be supporting the weight of the robot. This puts too much bending stress on the output shaft of the servo. This is okay

Manufacturer	Servo No.	Dimensions (in.)	Weight (oz)	Transit Time (sec)	Torque (oz-in.)	RPM
Cirrus	CS-21BB	0.86 x 0.43 x 0.77	0.32	0.08	23.0	125
Hitec	HS-81	1.17 x 0.47 x 1.16	0.58	0.09	45.1	111
Tower	TS-63	1.78 x 0.89 x 1.00	1.24	0.40	110.0	25
Futaba	S3004	1.59 x 0.78 x 1.42	1.30	0.19	56.9	53
Cirrus	CS-80 2BBMG	1.69 x 0.78 x 1.60	2.01	0.25	129.9	40
Tower	TS-75	2.33 x 1.13 x 1.96	3.60	0.19	110.0	53
Tower	TS-80	2.59 x 1.18 x 1.26	5.36	0.14	343.0	71

TABLE 8-5 Servo Size, Speed, and Torque Specifications for Some R/C Servos

to do with mini and lightweight sumo robots. With mini sumo robots, you should consider using the Hitec HS-300, Futaba FP-S148, Tower Hobbies TS-53, or Airtronics 94102. If you use larger sized servos, the completed robot will be wider than the 10-centimeter specifications. For the larger sumo robots, you can use any servo motor that suits your needs (as long as it can be modified into a continuous rotation motor).

R/C Control Signals

All R/C systems use an industry-standard, three-wire connection cable to transmit control information from the radio receiver to the various remotely controlled devices, such as servos and ESCs. Although the electrical interface has been standardized, different manufacturers use their own color codes and connectors to attach the cables to the radio receivers and the servos.

FIGURE 8-7 Various types and sizes of R/C servos

CHAPTER 8: It's All About Power: Selecting the Right Motor

Servo Color Codes and Connector Styles

The color codes always follow a similar motif: the ground wire is black or brown; the power line is almost always red; and the signal line is white, yellow, orange, or blue. The order of the wires is the first wire is the signal wire, the center wire is the 5-volt power wire, and the last wire (third wire) is the ground wire. The exception to this wiring scheme is the Airtronics brand connectors, where the first wire is the signal wire, the center wire is the ground wire, and the last wire is the 5-volt power wire. Figure 8-8 shows a schematic of this type of control cable's connector.

Each manufacturer has its own connector style. They have standardized on a 0.100-inch spacing between each of the wires, but the plastic connector housing differs from manufacturer to manufacturer. There will be a tab, notch, or a keyway molded into the housing so that the connectors with fit together in only one way, so that the wiring order cannot be reversed when assembled. This makes connecting different manufacturers' equipment together difficult. Electrically speaking, all of the manufacturers' systems are compatible, so the connectors can be easily cut off and swapped with another style of connector. This means that servos and speed controllers from different manufacturers can work with one another.

Servo Control Signals

Movement commands are encoded using a *pulse position modulation* (PPM) method (many people incorrectly call this pulse width modulation). The pulse signal is based on the movement of a joystick position on a radio transmitter.

When the joystick is in its neutral position, it is centered between the maximum extreme positions of the stick's motion. This position has been encoded as a 1.5-millisecond wide pulse. When the joystick moves from one extreme to another, the encoded pulse width changes from 1.0 to 2.0 milliseconds. The pulse width is proportional with the motion of the joystick. All R/C receivers should output a 1.0- to 2.0-millisecond wide signal, with a neutral position around 1.5 milliseconds. (Airtronic systems signals are from 0.7 to 1.7 milliseconds, with a center position at 1.2 milliseconds.) Some R/C systems will output pulse widths that can be a little wider or shorter in this range, but this is considered the industry standard. The control signal pulses are

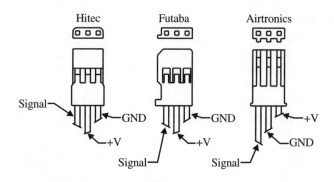

FIGURE 8-8 Geometry and wire positions for Hitec, Futaba, and Airtronics style connectors

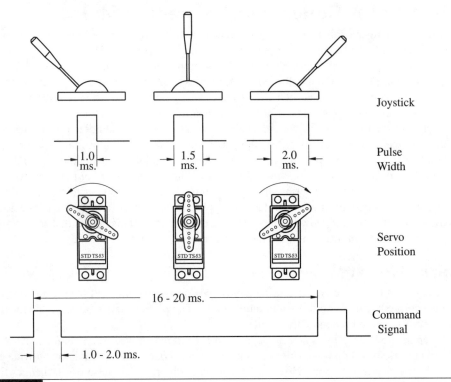

FIGURE 8-9   How the joystick position affects the pulse width signal and servo position

repeated every 15 to 20 milliseconds (50 to 60 Hz). If the servo does not receive a signal within this time frame, it will shut off. Figure 8-9 shows a schematic of how this control signal relates to a joystick position, pulse width, and servo position.

Modifying a Servo

Modifying a servo is a relatively simple task. Inside the servo is a PMDC motor, a set of gears, an electronic motor control circuit, and a position feedback potentiometer (the feedback sensor). The position feedback potentiometer is directly connected to the output shaft of the servo.

To modify the servo, you will need to remove the physical connection between the potentiometer and the output shaft, and you will need to remove a plastic tab from the output shaft. The tab is used to physically prevent the output shaft from rotating past 120 to 180 degrees. This is to protect the potentiometer from becoming damaged if the torque on the output shaft causes the shaft to rotate to its extreme positions. The following sections provide step-by-step instructions for modifying a servo.

Disassembling the Servo

Follow these steps to disassemble the servo:

1. Remove the servo arm, called the *servo horn*, from the servo's output shaft. First, remove the single screw that holds the servo horn in position. Then, with two fingers, pull the horn straight off the shaft. Do not twist off the horn. There are a set of splines that are there to prevent relative rotational motion between the horn and the output shaft.

2. Remove the top of the case. First, remove the four screws from the bottom of the servo. With your thumb on the output shaft's spline and your two forefingers under the front and rear mounting tabs, push down on the spline with your thumb at the same time as you pull up with your fingers. This will cause the top part of the case to come off. You should see a set of four gears on the top of the servo.

3. Carefully lift the top middle gear off the center spindle shaft, and set it down inside the top of the case. Then pull the output shaft/gear from the servo. You might need to pull hard on this gear, since it is connected to the potentiometer. Figure 8-10 illustrates the first three steps.

FIGURE 8-10 R/C servo disassembly

4. You will notice a small plastic tab on the top of the gear. Use a sharp knife to cut off this tab, as shown here. (The tab is colored black in this illustration to make it easy to see.) Make sure you do not get any cuttings caught in the teeth of this gear.

Modifying the Position Feedback Method

The next step is to decide on how you will modify the position feedback method. There are several different methods that can be used to modify the feedback circuits. One method is temporary; the others are permanent.

The temporary method really depends on the type of servo you have. On the bottom of the output shaft on some servos, such as Futaba's FP-S148 servo, there is a small retaining clip that is used to clamp onto the potentiometer's shaft. All that is required to make this servo into a continuous rotating servo is to remove this retaining clip, as shown here. Reinstalling the clip into the output shaft makes the continuous servo a regular R/C position-controlled servo again (so you will want to keep this retaining clip). If the output shaft does not have this retaining clip, you will need to use a permanent modification method.

One permanent method requires you to cut off the shaft of the potentiometer. Use a pair of wire cutters to clip off the end of the shaft flush to the top of the gear support, as illustrated next.

CHAPTER 8: It's All About Power: Selecting the Right Motor

(The shaft is made of brass, so it can be cut with a pair of wire cutters.) After cutting off the end of the potentiometer, place the output shaft back on the servo and rotate it by hand to make sure that it rotates freely. If it does not, you need to cut it down a little shorter.

The other permanent method requires you to the remove the entire potentiometer from the circuit. To do this, you will need to remove the motor and circuit board from the bottom of the servo. Some servos have the motor screwed to the top of the servo case. Make sure you remove any screws before prying the motor and circuit board out of the servo case. Next, remove the potentiometer from the circuit. You can either cut the wires or desolder them. Keep track of the wires, because you will need to replace the three wires from the potentiometer with two 2.2 or 2.7 kΩ resistors. The potentiometer has a total resistance of 5 kΩ, and the two resistors added together will come close to the 5 kΩ value. Figure 8-11 shows how these resistors should be wired together to replace the potentiometer.

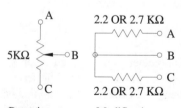

The 5KΩ potentiometer is replaced with two resistors that simulate the potentiometer.
The terminals A-B-C on the modified resistor pair must match the same terminal positions of the potentiometer on the servo circuit board.

FIGURE 8-11 Electrical circuit modifications for R/C servos

Illustrated here is the circuit board before and after this modification. Once this step is completed, place the motor and circuitboard back inside the servo case.

Calibrating the Servo

Before completing the assembly of the servos, they need to be calibrated. This requires a device that can generate a 1.5-millisecond pulse and can repeat that pulse ever 20 milliseconds. Any microcontroller is capable of doing this.

Below is a simple Basic Stamp 2 program that can be used to generate the 1.5-millisecond pulse width for calibration. The servo signal line is connected to pin 0 on the stamp. A 4.8- to 6.0-volt power supply is connected to the power and ground pins on the servo, and the Basic Stamp's ground is connected to the ground of the servo's battery supply.

```
Main:
  Pulsout 0, 750    'Send a 1.5 ms Pulse on a Basic Stamp 2
  Pause 20          'Pause 20 ms
Goto Main
```

To calibrate a servo that had the shaft cut off of the potentiometer, you will need a pair of needle-nose pliers. With the servo connected to the Basic Stamp, run the program. The servo motor should be turning one direction or another. With the pliers, rotate the remaining portion of the potentiometer until the motor stops turning. If the motor is not turning at the start of the calibration, turn the potentiometer until the motor starts turning, and then move it back to its original position. If the motor does not turn, there is a problem with the program, electrical circuit setup, or the servo.

To calibrate a servo that had the potentiometer replaced with a pair of resistors, you will need to modify the Basic Stamp program. First, run the program as it is. If the motor is turning, then change the 750 value until the motor stops turning. The final number that you come up with is the new neutral center position. You need to record this, since it will be different from servo to servo. It is best to mark the servo case with the microsecond number. To convert this number into microseconds, multiply the number by the pulse width period. For a Basic Stamp 2, the pulse width period is 2 microseconds.

Reassembling the Servo

After the servo has been calibrated, reassemble the servo in the reverse order of the disassembly. Make sure there is no foreign debris on the gears. If you need to add some more grease to the gears, use servo grease for this purpose. You can get servo grease at your local hobby store.

Motor Sources

The best places to obtain electric motors are from your local hobby stores or from mail-order hobby stores such as Tower Hobbies (www.towerhobbies.com), Hobby Lobby International (www.hobby-lobby.com), and Hobby People (www.hobbypeople.net). You can also acquire a motor from a robot-parts supplier such as Lynxmotion (www.lynxmotion.com) or Acroname (www.acroname.com). Good electronics parts suppliers for electric motors are Jameco (www.jameco.com), Marlin P. Jones (www.mpja.com), and Electronic Goldmine (www.goldmine-elec.com). Contact information for these companies is included in the appendix.

When you purchase an electric motor, you will want to make sure you get the motor performance specification sheets for the motor. If you have the specification sheets, you won't need to figure out the motor performance parameters yourself.

Final Comments on Motors

The motors are the muscles of your robot. By understanding how the motors work and what happens when you push them to their limits, you will be able to determine the appropriate motors, the types of batteries, and the appropriately sized ESCs for your robot.

When building your sumo robot, the motors are usually the first major component that you need to select. Then the rest of the components start to fall into place. Before you select the motors for your robot, you should read the next chapter, which covers motor controllers. Different motor controllers have different current requirements, as well as different prices, which might affect the motor selection process.

Of course, all of this information is not required to build a sumo robot (or any other type of robot). In fact, many robot builders simply pick a motor and build a robot around it. And in many cases, everything works just fine. If they're not lucky, things break because they inadvertently pushed components past their capabilities. How you choose to build your robot is totally up to you.

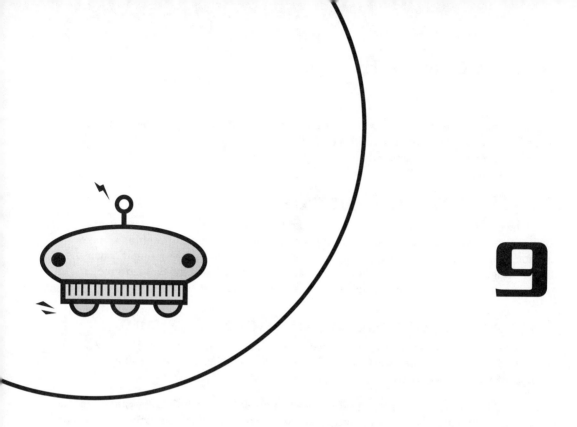

Motor Control Fundamentals

The previous chapters explained how the wheels, gears, transmissions, batteries, and motors relate to one another in making a robot move. The last part of the locomotion system that you need to select is the motor controller. The motor controller is the device that actually controls the speed and direction of the motors.

The simplest form of motor control is to use a switch to turn a motor on or off. By placing a switch between the battery and the motor, you can use the switch to control all of the power going to the motors. Though this type of a switch configuration is not used to control motor performance (such as going forward and reverse), it is used in most robot designs as the main power-supply switch. In other words, a single switch is used to control all of the power going from the batteries to the motors. All robots should have a manual power-supply switch that is used to cut off power to the motors. If there is a short circuit in the motor controller (or some other problem), you will want to be able to quickly cut the power to the motors.

This chapter explains how motor direction and speed are controlled and discusses the various types of motor controllers. It will end with a discussion of a wide variety of low-cost commercially available motor speed controllers that are ideal for sumo robots.

Motor-Direction Control

One of the nice features about PMDC motors is that their rotational direction can be changed by simply reversing the direction of the current flowing through the motors. To control the motor direction, you can use a simple circuit called an *H-bridge*. Figure 9-1 shows a simple schematic of an H-bridge that is commonly used to control the current direction going through a motor. By looking at Figure 9-1 from a distance, you can see why it is called an H-bridge.

Switch Combinations for Motor Actions

The logic table in Figure 9-1 shows what the motor will do based on which switch is closed. This table shows that there are a total of 16 different switch combinations that result in only five different motor actions:

- *Brake* is when the motor rapidly slows down and resists turning. This does not mean that the motor will be mechanically prevented from turning, as when you apply the brakes on your car. The motor can still turn, but turning will require a little more torque.

- *Free wheel* is when the motor freely slow downs and does not resist any applied torque on the motor shaft.

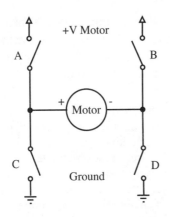

A	B	C	D	Motor Result
0	0	0	0	Free Wheel
0	0	0	1	Free Wheel
0	0	1	0	Free Wheel
0	0	1	1	Brake
0	1	0	0	Free Wheel
0	1	0	1	Short Circuit
0	1	1	0	Reverse
0	1	1	1	Short Circuit
1	0	0	0	Free Wheel
1	0	0	1	Forward
1	0	1	0	Short Circuit
1	0	1	1	Short Circuit
1	1	0	0	Brake
1	1	0	1	Short Circuit
1	1	1	0	Short Circuit
1	1	1	1	Short Circuit

0-Open, 1-Closed

FIGURE 9-1 Simple H-bridge schematic for controlling motor direction

- *Forward* and *reverse* result in the motor shaft rotating clockwise and counterclockwise, respectively.

- *Short circuit* switch positions should be avoided at all costs, since they will directly short the battery to the ground.

Of the 16 different switch combinations, only two will control motor direction—forward and reverse. The rest will result in the motor either not turning (or slowing down) or damaging the motor power circuit.

There are six different combinations of switch positions that can result in a short-circuit condition. Shorting the batteries straight to ground is a dangerous situation; it will destroy the batteries and the motor control circuit, and it might cause the batteries to explode. Therefore, the H-bridge needs to be wired in such a manner that the short-circuit condition cannot happen.

For most robots, the switches that are used to control the direction of the motor are electrically controlled using relays, transistors, or a combination of the two.

Relay Control

Relay-controlled motors are relatively easy and inexpensive to build. Instead of using switches as shown in Figure 9-1, relays are used to control the motor. With a relay, you can use a low-voltage control circuit to control a high-voltage motor. A low voltage is used to turn a relay on or off, and then the relay can be used to pass a higher voltage and/or higher current from the batteries to the motor.

Figure 9-2 shows how to wire a set of four relays and two switches to make a motor control circuit. Notice the H-bridge shape inside the circuit. With this circuit, you can control the forward and reverse directions of the motor, and you can apply motor braking. By using a pair of SPDT switches, the H-bridge can be wired so that a short circuit in the motor control circuit cannot happen, regardless how the switches are oriented.

Relays have three different sets of specifications:

- The electrical circuit description—whether it is an SPDT (single-pole double-throw) or a DPDT (double-pole double-throw) relay
- The coil voltage and current that are required to close the relay
- The contact current rating

Relays have a four-letter designation for the electrical circuit description. The first two letters represent the number of poles, and the second two letters are the number of throws. The number of poles is the number of wires that are separated from each other but are all switched at the same time. A single-pole switch is designated with SP, and a double-pole switch is designated with DP. The number of throws is the number of positions the switch can be switched to, making an independent connection with each pole. A single-throw switch is designated with ST, and double-throw switch is designated with DT.

The relay's coil current and voltage ratings are important to know, since the relay's control (driver) circuit must be able to supply enough voltage and current to open and close the contacts.

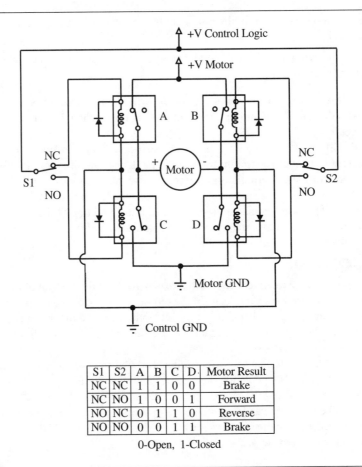

FIGURE 9-2 Simple schematic for a relay-controlled H-bridge

The contact current rating is the amount of current the switch can safely handle when opening and closing. At the instant when the relay first opens or closes, there is a tiny arc that occurs between the contacts. The higher the voltage, the further apart the contacts can be to maintain the arc. The higher the current, the bigger (hotter) the arc. If the current is too high, the arcing will start to pit the surface of the contacts, which makes them less efficient. If the current is too high, the contacts will literally melt together.

Transistor Control

Mechanical switches are generally used to control the motor control relays in tether-controlled robots. For autonomous and most remote-control robots, the relays are controlled by other electrical circuits. This is usually done by using a transistor instead of a mechanical switch to direct current to the relay coils. Typical miniature relay coils require from 20 to 100 milliamps of current to close. Since most microcontrollers cannot supply this much current, a transistor must be used to control the relay.

Figure 9-3 shows two simple schematics for using either a NPN or a PNP bipolar-style transistor to control a relay. You can use just about any type of bipolar transistor, as long as the transistor's current rating is higher than the relay's coil current rating. The resistor that is connected to the transistor's base is there to limit the input current to the transistor.

> **NOTE** *You should always use a transistor between a microcontroller pin and any inductive load, such as relays, motors, and coils.*

With transistors, you can use a simple microprocessor to control a relay-controlled H-bridge. Figure 9-4 shows a schematic of this type of a control system. The H-bridge shown in Figure 9-2 uses a pair of SPDT switches to prevent the circuit from shorting out. The same type of protection can be accomplished using a hex inverter (a Not logical gate). A Not logical gate inverts the input state, so an input of 0 volts will be inverted to a positive 5 volts on the output, and vice versa (Inverting does not mean that a +5 volts become a −5 volts, it just inverts the On-Off states). The logic table in the Figure 9-4 shows how the motor will turn based on the two control signals.

With all inductors (relay coils and motor coils are inductors), a voltage is induced across the inductor that is proportional to the rate of change in the current. When a switch is used to turn off a relay coil, the current does not instantaneously go to zero, but it does go to zero in an extremely short period of time. Since the high rate of change is decreasing, the inductor will induce a

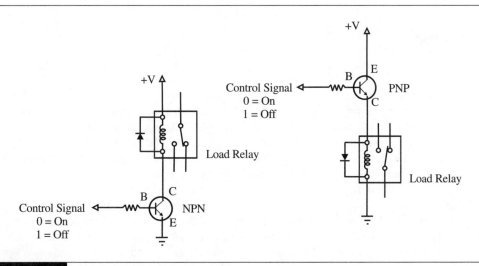

FIGURE 9-3 Bipolar transistor-controlled relay

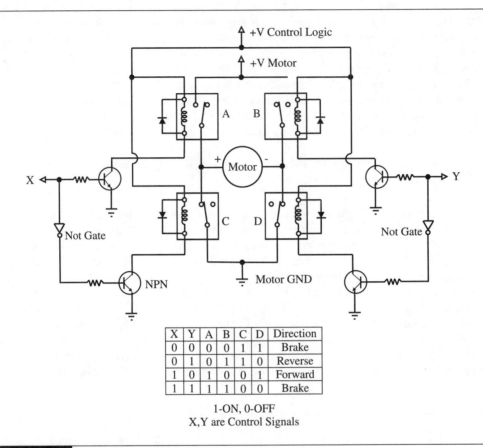

FIGURE 9-4 Transistor-controlled, relay-based motor controller

voltage that can be several hundred volts, flowing in the opposite direction, for an instant. For mechanical switches, the voltage is generally not high enough to arc across an open contact. However, for a transistor, the voltage can be high enough to damage the transistor. (This is more likely to occur if the switch/transistor is between the inductive load and the ground.) Thus, a *flyback diode* is placed across the coil terminals to provide a current path away from the switch/transistor during these shutoff periods. It is common practice to place a diode across all relay coils. (But if the diode is placed in backwards, you will short out the coil!)

When switching high currents, there will always be some arcing across the contacts that will pit the surface, which will eventually wear out the relay. Relays use a small spring to keep the relay open. The time that is required to close a relay is a function of the magnetic field generated in the relay's coil and the stiffness in the contact spring. This puts a limit on how fast the relay can switch on and off. Thus, high-frequency switching is not possible for relays, and low-frequency switching the relay on and off will eventually wear out the contacts and springs.

When this type of switching is required, you should use a solid-state H-bridge, as described in the next section.

Solid-State H-Bridges

An H-bridge using relays is more of a mechanical switching circuit, since relays use mechanical contacts to switch the current paths. Another class of H-bridges is called the *solid-state H-bridge*. A solid-state H-bridge uses transistors instead of relays.

Whether you should use a solid-state H-bridge or a relay H-bridge depends on the application. For constant-speed applications, both systems will work fine. When you need to switch high currents or use high-frequency switching, transistors are preferred over relays.

Figure 9-5 shows two different types of transistor-based H-bridges: one uses MOSFET transistors, and the other uses bipolar transistors. Notice that they look similar to a relay H-bridge (see Figure 9-2).

The biggest differences between these two types of H-bridges are their voltage and current-handling capabilities. Generally speaking, MOSFET transistors have higher current-handling capability than bipolar transistors do. Bipolar transistors can operate at lower voltages than MOSFET transistors. A bipolar transistor's current-handling capability is proportional to the current going into the transistor's base. A MOSFET transistor's current-handling capability is proportional to the voltage going into the transistor's gate. In other words, bipolar transistors are controlled by current, and MOSFET transistors are controlled by voltage.

A bipolar transistor may require more current than what can be supplied directly from a microcontroller. This will require a second transistor to be placed between the microcontroller

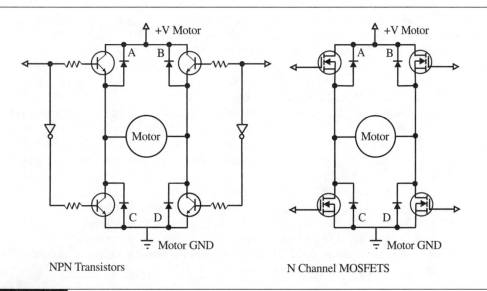

FIGURE 9-5 Bipolar and MOSFET transistor-based H-bridge circuits

and the H-bridge transistor. Most MOSFET transistors require the gate voltage to be 10 volts higher than the source voltage to be fully turned on. Thus, MOSFET H-bridges should be used with motor controllers that have a 12-volt power supply to control the transistors, although the motors can be operated at a lower voltage. There are some H-bridge motor controllers that operate at less than 12 volts, but they have special voltage boosting circuits to control the MOSFETs.

CAUTION *MOSFET transistors are very sensitive to static electricity and short circuits. They short out very easily when misused. On the other hand, bipolar transistors are very tolerant of mistakes. There are two types of people when it comes to custom making MOSFET circuits: those who* have *blown up a MOSFET and those who* will *blow up a MOSFET.*

Whether you should use a bipolar or MOSFET transistor H-bridge depends on the current and voltage requirements for the motor controller. Bipolar H-bridges are well-suited for low-voltage and low-current motors. MOSFET transistor H-bridges should be used for all other

Should You Build Your Own Motor Controller?

Conceptually, a transistor-based H-bridge should be easy to build, but in practice, this is not the case. You need to carefully select all of the components based on their operating parameters, the method of controlling them, and the voltage and current requirements of the motor during an actual match. The voltage and current ratings on the components should be matched and have at least twice the worst-case needs of your robot. All of these parameters have an effect on the final motor controller design.

Learning how to build your own transistor-based motor controller can be time-consuming and frustrating. Before taking on the task of designing your own custom motor controller, you should review all of the low-cost commercially available motor controllers to determine if any of them will fit your needs. Several different types of commercial motor controllers that are ideal for sumo robots are reviewed later in this chapter. If you use one of these, you will have more time to devote to building the rest of your robot.

If you are still interested in building your own motor controller, a great source of information is *Motor Control Electronics Handbook*, by Richard Valentine (McGraw-Hill, 1998), and the Internet-based Open Source Motor Controller (OSMC) project at www.groups.yahoo.com/group/OSMC. Both of these sources provide a lot of information about designing and building motor controllers.

motor controllers that need more than 5 amps. For motor controllers that need between 1 and 5 amps, either type of H-bridge will work.

Integrated-Circuit-Based Solutions

There are several types of integrated-circuit-based motor controllers that work well for low- and medium-powered robots. These circuits include the L293D and the L298N chips from ST Microelectronics. Both of these integrated circuits have two complete internal H-bridge circuits that can be used to independently control two separate motors. The L293D can supply 0.6 amp continuously and 1.2 amps momentarily. The L298N can supply 2 amps continuously and 4 amps momentarily.

Figure 9-6 shows a simple schematic using the L293D. The L293D has the flyback diodes inside the chip, which makes it an easy circuit to build for a low-powered motor controller. The logic table in the figure shows how the motor direction is controlled based on the input signals from a microcontroller.

The Texas Instruments SN754410 quadruple half H-bridge is pin-for-pin compatible with the L293D and has almost twice the current-handling capability. This chip can supply 1.1 amps continuously with 2-amp peak currents. The only difference between this chip and the L293D is that it does not have the internal flyback diodes.

> **CAUTION** *The integrated-circuit chips will overheat very quickly if the motors are drawing more than half the continuous current rating without using any heat sinks attached to the top surface of the chip.*

Shown here is a low-cost L298 board from HVW Technologies (www.hvwtech.com), which can be used to control two motors at the same time.

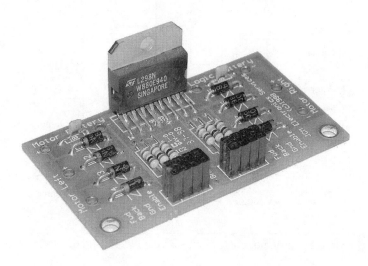

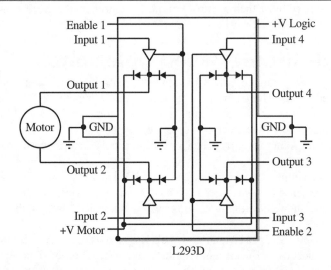

Enable 1	Input 1	Input 2	Motor
H	H	H	Brake
H	H	L	Forward
H	L	H	Reverse
H	L	L	Brake
L	X	X	Free Wheel

X - Doesn't Matter

Enable 2	Input 3	Input 4	Motor
H	H	H	Brake
H	H	L	Forward
H	L	H	Reverse
H	L	L	Brake
L	X	X	Free Wheel

X - Doesn't Matter

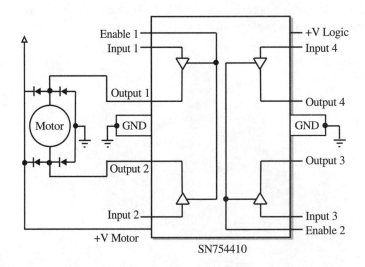

FIGURE 9-6 Simple dual-motor controller using the L293D and the SN754410 dual H-bridge

Variable-Speed Control

The second part of the motor control system is the variable-speed control. For many robots, speed control is an optional feature. Some people want their robots to move only at full speed going forward or in reverse. Some robot builders want to control the speed of their robot, so that, for example, the robot moves at a slower speed when searching for its opponent and then attacks at full speed. Other people want their robots to move with some radius, which requires variable wheel speed, as described at the beginning of Chapter 6.

Chapter 8 showed how motor speed is a function of the applied voltage. So, to make a variable-speed motor, you will need a way to vary the applied voltage to the motor. There are two methods that you can use to accomplish this: a variable resistor or *pulse width modulation* (PWM).

Variable Resistors

Controlling the motor speed using a variable resistor is an old-fashioned but very effective method. A variable resistor is placed in series between the battery and the motor. All of the motor power will go through this resistor. An arm on the resistor moves to change the resistance in the circuit, which results in a voltage drop across the variable resistor and reduces the applied voltage to the motor.

Shown here is a variable-resistor motor speed controller used in older R/C cars and boats. This particular circuit can also reverse the direction of the motor. The resistor arm is moved using a traditional R/C servo motor.

Although a variable resistor in series with the motor power circuit is effective, all of the power that is reduced from the resistive motor controller is converted as heat. These types of motor controllers are not very efficient and will get very hot. This is one of the reasons why an electronic speed controller (ESC) is the preferred method for controlling motor speed.

Pulse Width Modulation

Since electric motors are inductors, cutting off the power doesn't immediately stop the current from flowing through the motor. The induced voltage from the collapsing magnetic field actually helps keep the motor running for a moment. When the voltage is applied as a square wave, the motor will time-average the applied voltage. To the motor, the time-averaged voltage has the same effect as operating at a constant voltage that is equal to the time-averaged voltage. Figure 9-7 shows a PWM signal and the average voltage.

Equation 9-1 shows the definition of the duty cycle of a PWM signal. The *duty cycle* is a percent of the amount of time the applied voltage is on during its frequency cycle.

Equation 1
$$Duty\ Cycle = \frac{t_{ON}}{t_{ON} + t_{OFF}}$$

When the frequency of the square wave is high enough (at least 1,000 Hz for most motors), the duty-cycle percentage will be equal to the average voltage percentage the motor will see, as shown in Equation 9-2.

Equation 2
$$V_{average} = \frac{t_{ON}}{t_{ON} + t_{OFF}} V_{in}$$

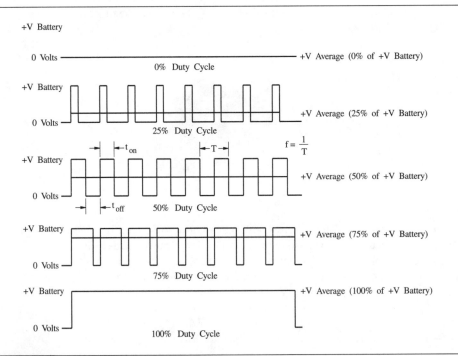

FIGURE 9-7 Typical PWM control signal

The duty cycle will range from 0 (completely off) to 100 percent (completely on). The minimum PWM frequency to "smooth" the average voltage depends on the motor. Some motors require a PWM frequency higher than 20,000 Hz; some require a PWM frequency as low as 100 Hz. Frequencies about 1,000 Hz will work for most motors used in sumo robots.

In order to change the speed of a motor, one of the active transistors in the H-bridge must be oscillated at the PWM frequency at some duty cycle. It does not matter which transistor is oscillated, as long as it is one of the transistors that is used to conduct current through the motor. Figure 9-6, shown earlier, illustrates an integrated circuit version of an H-bridge. To make one of these circuits a variable-speed motor controller, apply the PWM signal to the enable signal line. The two input control lines set the direction of the motor, and the enable line turns on the power to the motor.

One of the drawbacks to using a relay-based H-bridge is that rapidly turning on and off the relay will shorten its life. Also, this type of H-bridge cannot switch fast enough to take full advantage of the voltage averaging being equal to the duty cycle. If a transistor is placed between the H-bridge and the batteries, the relays can be used to set the motor direction, and the transistor can be used to turn on and off the power to the motor. This is analogous to how the integrated circuit versions of the H-bridges work. Figure 9-8 shows a schematic of this type of a circuit using an N-channel MOSFET.

Since the gate voltage on the MOSFET transistor must be 10 volts higher than the source voltage (in this case, the ground) to be completely turned on, a bipolar NPN transistor can be used to supply the correct voltage to the MOSFET. With this circuit, you can build a simple high-current, variable-speed motor controller. An example of a robot that uses this type of a motor controller with cordless drill motors is shown here. This is a champion 3-kilogram robot named LEO, built by Tom Dickens.

NOTE *The new logic-level MOSFET transistors require the gate to be at only 5 volts above the source voltage to fully turn on.*

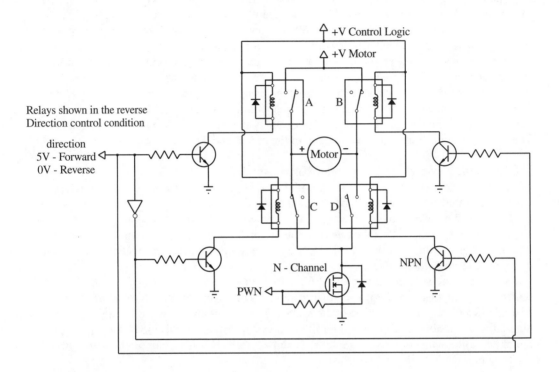

FIGURE 9-8 A variable-speed motor controller using relays for the H-bridge and a MOSFET for the variable-speed control

Commercial Electronic Speed Controllers

Probably the best choice for a motor speed controller is a commercially made ESC, which has several advantages:

- Someone else has spent all of the time and effort to select the right type of components for the motor controller.
- The final performance specifications are documented, which helps you determine which speed controller has the current-handling capability your motor needs.
- It generally costs less than making the same one from scratch.

CHAPTER 9: Motor Control Fundamentals

Commercial ESCs fall into three different categories: H-bridge-only motor controllers, R/C ESCs, and complete variable-speed motor controllers.

Commercial H-Bridges

Several companies sell complete H-bridge circuits that can be used to directly control the speed and direction of your motors. These controller boards use relays, transistors, MOSFETs, or single-chip motor controllers. The motor and power supply directly connect to the motor controller, and a microprocessor is used to set the motor speed and direction. The control signals set the motor direction control pins high or low and send a PWM signal.

Table 9-1 lists several popular commercially available H-bridges. Figure 9-9 shows the L298 H-Bridge from HVW Technologies (www.hvwtech.com), the Maxi Dual H-Bridge from Mondo-tronics (www.robotstore.com), and the ICON H-Bridge from Solutions Cubed (www.solutions-cubed.com). (See the appendix for contact information for other motor controller suppliers.)

All of the H-bridges listed in Table 9-1 are excellent and well-proven motor controllers. The current ratings are the continuous current rating, and the voltage rating is the minimum and maximum for the entire circuit. Although these motor controllers can be used to operate motors at a lower voltage than shown in the table, the minimum voltage is what the control electronics need to operate.

Radio-Control Electronic Speed Controllers

The ESCs that are used for R/C off-road and racing cars are excellent off-the-shelf technology that can be used in sumo robots. Instead of needing to use two to four wires to set the motor direction and one wire for motor speed, an R/C ESC needs only one wire to control both direction and speed. This simplifies some of the wiring and control requirements in the robot.

Name	Voltage	Current (amps)	Circuits	Size (in.)	Type	Manufacturer
ICON H-Bridge	12–46	12	1	1.9 x 2.5	MOSFET	Solutions Cubed
L298	6–46	2	2	1.5 x 2.7	L298	HVW Technologies
Maxi Dual H-Bridge	12–40	10	2	2.1 x 2.3	MOSFET	Mondo-tronics
Mini Dual H-Bridge	5	0.3	2	1.0 x 1.1	Transistor	Mondo-tronics
Magnavision	12–55	3	2	2.0 x 3.5	LMD18200	Acroname

TABLE 9-1 Some Commercial H-Bridges

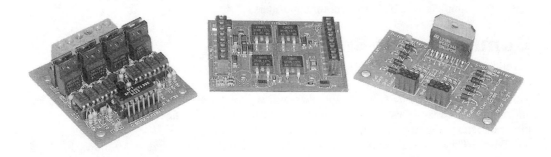

FIGURE 9-9 Commercial H-bridge motor controllers from Mondo-tronics, Solutions Cubed, and HVW Technologies

To control an R/C ESC, you need two wires to connect to it: the ground wire and the signal wire. The signal wire needs to generate a 1- to 2-millisecond (ms) pulse that is repeated every 50 to 60 Hz. Shown here are a pair of Hitec SP-560 R/C ESCs mounted on a 3-kilogram sumo robot.

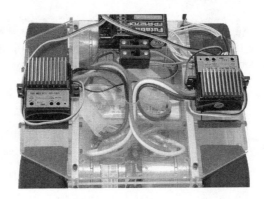

Although there are some complications and special considerations when using R/C car ESCs, they still make great choices for sumo robots. They are small, lightweight, reliable, easy to control with a microcontroller, and reasonably priced. Their main shortcoming is their documentation. They are well-documented for R/C car applications, but very poorly documented for robotics needs. You may need to learn about quirks particular to the ESC you are using, which don't matter in an R/C car but are a major problem with robots. For example, some ESCs shut off after a certain amount of time in reverse or produce a sudden voltage surge to the motor when first turned on.

R/C ESC Speed and Direction Control

As with R/C servos, R/C ESCs use the same 1 to 2 ms pulse width to select speed and direction. A 1.5 ms pulse width stops the motors. A pulse width between 1 and 1.5 ms sets the motor direction to reverse, and a pulse width between 1.5 and 2 ms sets the motor to forward.

The internal microcontroller will proportionally scale the pulse width values between 1.5 to 2 ms to correspond to a duty cycle between about 25 to 100 percent. When the ESC first turns on, the duty cycle will be approximately 25 percent, with a pulse width of approximately 1.5 ms. Then the duty cycle will proportionally increase to 100 percent when the pulse width reaches 2 ms. The reverse direction is handled in a similar manner. The main difference with reverse is that the duty cycle may start with a higher percentage, such as 30 to 50 percent, and it might not reach 100 percent when the pulse width goes down to 1 ms.

At around the 1.5 ms pulse width or less, the ESC uses braking to rapidly slow down and stop the motor. You can think of this zone as a dead band where the motor will not turn. Many ESCs allow you to adjust the width of this braking zone.

There are several points to keep in mind about R/C ESC direction control. One thing to watch out for is the forward-to-reverse time delay. All of these ESCs have a small time delay when the motor changes directions. This can range from a few milliseconds up to a full second. This time delay is intentionally placed in the ESC to protect the ESC, motors, and gears in the R/C car so they do not become damaged when the ESC is trying to slam the motor into forward and reverse all the time. Many of the advanced ESCs allow you to adjust the amount of the time delay. For sumo robots, you will want to use the smallest reverse time delay as possible.

Another thing to watch out for is that some ESCs will shut down when the motors have been running in reverse for a certain amount of time. For example, the Runner Plus Reverse ESC, from LRP Electronic, will shut down after 5 seconds, regardless of the amount of load the motors are placing on the ESC. Many times, this automatic shutdown is not documented in the literature that comes with the ESC. Commanding the ESC to move forward will reset this ESC, so it can move in reverse for another 5 seconds. ESCs that have automatic timed shutdowns should be avoided in sumo robots, since they may be moving in reverse about as much time as they are moving forward.

Also be aware that R/C ESCs do not have equal performance ratings for moving forward and reverse. These ESCs were designed to supply power to cars and trucks that will spend most of their time moving forward. When R/C vehicles move in reverse, speed and power are not required.

R/C Motor and ESC Specifications

The biggest drawback to R/C car motors and ESCs is determining their performance specifications. In the R/C car industry, motors are specified by the number of turns and the number of windings. The number of turns is the number of times a wire is wrapped around one of the motor poles on the armature. The number of windings is the number of wires wrapped around a pole. The windings are single, double, triple, quad, and sometimes quint (five). As a general rule, the greater the number of windings, the faster the motor will turn, and the lower the number of turns, the greater the amount of torque the motor can deliver. Hence, a lower number of turns means the motor will draw more current.

Throughout the R/C car industry, all of the cars, parts, and electronics are based on these turns and windings numbers. Unfortunately, these numbers are qualitative values for comparison, not quantitative numbers. They cannot be directly converted into useful numbers, such as revolutions per minute and current, or motor speed and torque constants.

ESCs are rated by the number of motor turns they can drive and the number of battery cells. The battery cells number can easily be converted into a voltage by multiplying it by 1.2 volts (1.2 volts is a standard voltage for a NiCd battery, which these ESCs are based on using). The number of turns does not tell you how much current the ESC can supply. This number is useful only when you use an R/C motor with an R/C ESC. As long as the R/C motor has more wire turns than the ESC rating, the ESC should be able to supply all of the current the motor will need.

When you read the specifications that are written in the ESC advertisements and packaging, you will see numbers like a current rating of 200 amps. These numbers are misleading. They are the theoretical instantaneous current-handling capability of the MOSFETs inside the ESC. If you draw this much current for a few microseconds, you will let the magic smoke out of the MOSFETs, and they will stop working. In fact, the wires that come with the ESCs cannot tolerate this much current. The useful current rating should be less than one-quarter of the current rating advertised.

Of all of the ESCs that are available, only about one-third of them are reversible. You will want to make sure that the ESC that you purchase is reversible. Also, the reverse current-handling capability for most reversible ESCs is less than the forward direction and usually has half the current-handling capacity.

Table 9-2 lists several popular ESCs that can be used in sumo robots. Notice that the values for the advertised current-handling capability and the number of turns do not correlate among the different manufacturers.

For most sumo robots, the current-handling capability of these ESCs will be sufficient. You will need to be careful in your ESC selection if you are using high-current motors (greater than 20 amps) or voltages greater than 8.4 volts.

R/C ESC Power Supply

All ESCs have a battery eliminator circuit (BEC) that can be used to supply power to other electronics. The BEC gets its power from the batteries that are used to supply power to the motors. The +5 volt wire is in the center of the standard three-wire connector that connects to an R/C receiver. An R/C receiver can use the power from the ESC instead of needing to use a separate battery supply. When using the three-wire connector from the ESC in your own custom electronics, you will need to disconnect this wire if you do not want to use the power in your circuits.

Manufacturer	Model	Number of Cells	Number of Turns	Max. Current
Duratrax	Blast	5–7	27+	140
Futaba	MC230CR	6–7	20+	90
Futaba	MC330CR	6–7	13+	200
HiTec RCD	SP 560	5–7	10+	200
Novak	Reactor	6–7	12+	160
Novak	Rooster	6–7	15+	100
Novak	Super Rooster	6–10	Any	320
Traxxas	XL-1	4–7	17+	100

TABLE 9-2 R/C ESCs That Can Be Used in Sumo Robots

The BEC voltage and current capability vary from ESC to ESC. They can range from 5 to 6 volts and current ratings from 0.5 to 3 amps.

Radio-Control-Compatible Electronic Speed Controllers

Because of the popularity of robot combat events such as BattleBots and Robot Wars, several companies have developed specialized motor controllers that use the standard R/C 1 to 2 ms pulse width control method.

Sozbots (www.sozbots.com) offers a 5-amp dual-motor controller that has fully proportional speed control in both forward and reverse directions. This motor controller can control two independent motors using the standard R/C signal interface. It comes in a 1.5-inch square package and is ideal for mini and lightweight sumo robots.

Vantec's (www.vantec.com) R/C ESCs are the number one choice for many combat robots. Most of the Vantec motor controllers are too large for sumo robots, but several models are small enough. The 2.5-inch square motor controllers work the same way as traditional R/C ESCs, except that their forward and reverse capabilities are identical and fully proportional. The RET411 model can supply a continuous current of 12 amps, the RET512 can supply 18 amps, and the RET713 can supply 33 amps.

Innovation First (www.innovationfirst.com) has developed a true high-powered ESC. Shown here is a Victor 883 motor controller. This motor controller accepts standard 1 to 2 ms pulse widths for fully proportional forward and reverse motor speed control. Additionally, this controller can deliver 60 amps of continuous current! This battle-proven motor controller will be able to supply all the current needed for the most powerful motors.

Advanced Motor Controllers

There are many different types of motor controllers that you can use to control the speed of a motor. Many require a PWM signal or a 1 to 2 ms pulse width signal and a set of wires to select the motor's rotational direction. Most of these motor controllers are not smart motor controllers; they require continuous signal updating from a host microcontroller to continue to control the motor.

Solutions Cubed (www.solutions-cubed.com) has developed a host of motor controllers that can be controlled with simple serial communications. The smallest motor controller is the Motor Mind B and the next smallest is the Motor Mind C, as shown here. Both of these motor controllers are ideal for mini and lightweight sumo robots.

The Motor Mind B is a 1.1-inch square motor controller that can run 5- to 30-volt motors, with current draws up to 2 amps continuous current. A 2,400-baud serial communication will tell the Motor Mind B which direction and speed to turn the motor. What makes this motor controller different from the other motor controllers discussed in the previous sections is that the host microcontroller can read status information back from the motor controller. In addition to status information, the Motor Mind B has a tachometer input that can be used in closed-loop motor controllers to precisely control the speed of a motor.

The Motor Mind C is a 0.7-inch by 2.4-inch long motor controller that can control two separate motors operating up to 24 volts and drawing 4 amps of continuous current. This motor controller is mounted on a standard 40-pin, dual-inline package. It has built-in overcurrent and overtemperature protection circuits, and it can serially communicate with baud rates up to 38.4 kbps. What makes the Motor Mind C unique is that in one package, motor speed and direction control can be accomplished by serial communication, standard 1 to 2 ms R/C style pulses, and 0- to 5-volt analog signals. This motor controller offers many different options for controlling motors.

The heavy-weight motor controller from Solutions Cubed is the ICON H-Bridge motor controller, shown here on its interface board. This MOSFET motor controller can deliver 12 amps continuously up to 40 volts. The motor controller can use serial communication baud rates up to 9,600 or a 0- to 5-volt analog signal to set motor speed and direction. It has programmable overcurrent and temperature setpoints, and it can send serial data to the microcontroller indicating the current temperature of the motor controller and how much current is being drawn through the motor controller.

Final Comments on Motor Speed Controllers

This chapter explained how motor controllers worked and what types are available for sumo robots. So how does all this information help you select the right motor controller? The main requirements are that your motor controller must be able to supply the voltage and current your motor needs.

The motors you want to use will have a maximum current draw requirement, as discussed in Chapter 6. The motor controller must be able to handle more current than the motor will need, or you will burn out the motor controller. Sometimes, you will need to choose a lower current motor to match the current-handling capabilities of the motor controller you have. Also, the operating voltage of the motor you want to use must be within the minimum and maximum voltage rating of the motor controller. Some motor controllers have minimum logic control voltages over 10 volts, but the motor voltage can be much lower. Make sure your batteries can supply the minimum voltage the motor controller logic requires. The L293D and L298 motor control chips are great low-cost solutions for building low-powered motor controllers. A combination relay and MOSFET motor controller for higher-powered applications is fairly simple to take on. However, most experienced competition robot builders will tell you that buying a prebuilt motor controller is the way to go. The higher the motor current requirements, the higher the recommendation becomes. Building your own motor controller from scratch is not recommended for anything that needs more than 3 amps of current. It is difficult to determine all of the correct parts and handle the current direction changes. Custom building a high-current motor controller will even challenge electrical engineers.

Although there are some problems with R/C ESCs, they still make durable, low-cost motor controllers for robots. Also, several new companies, such as Innovation First and Solutions Cubed, make great robotic motor controllers. If you need feedback information from your motor controller, the motor controllers from Solutions Cubed are the way to go.

After reading this and the previous chapters, you should have all the information you need to select the right components and build the drive system in your robot. The next chapter will explain how to use this information to build a remote-control system for your robot. Then the following chapter begins the discussions on how to make a fully autonomous robot.

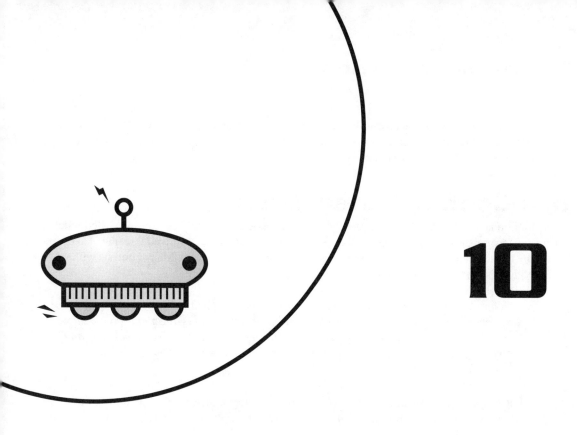

Remotely Controlling Your Robot

There are two different ways to control your robot: You can remotely control it or you can program the robot to run on its own, as an autonomous robot. This chapter will focus on how to remotely control your robot. The following chapters will begin the discussions on building autonomous robots.

If you are building a remote-control robot, you need to consider several aspects of its design. The remote-control system must be robust, reliable, and reasonably immune to external and internal interferences and impacts from other robots. It should have enough range to communicate with your robot, regardless of where it is in the competition area. It should be able to control your robot's direction and speed, and it must be able to shut down your robot at any time without failure. And most important, the remote-control system must be easy operate so that your robot will perform exactly the way you want it to perform.

Most remote-control robots use either radio-control or tether-control systems, although some contests allow other types of control systems, such as sound and infrared systems. This chapter covers radio-control and tether-control systems.

Radio-Control System Basics

The traditional radio-control (R/C) system is primarily used for remotely controlling scale-model cars, boats, airplanes, and helicopters. Because R/C systems are so readily available, they have become the standard type of remote-control system used for sumo robots. Many contests will not allow any other type of remote-control system.

NOTE *Before building a remote-control sumo robot, check with the local sumo contest officials to determine the local rules for remote-control robots. Some organizations do not allow remote-control robots; some organizations have restrictions on radio frequencies; and some organizations have restrictions on tether-controlled or other types of remote controlled robots.*

The R/C system consists of five major subsystems: the transmitter, receiver, batteries, radio frequency crystals, and auxiliary equipment such as electronic speed controllers (ESCs) and servos. Figure 10-1 shows two different style remote-control transmitters.

FIGURE 10-1 Typical pistol-grip and joystick style R/C transmitters

Radio-Control Channels

One of the characteristics an R/C system is rated by is the number of independent control signals the transmitter can send simultaneously to the receiver. These control signals can be used to independently control ESCs. Each one of these control signals is called a *channel*.

Typical R/C systems have two, three, four, six, eight, or nine control channel capabilities. Most of the low-cost R/C systems are the two-channel type. The radio transmitter is used to send command information to the receiver to control both steering and motor speed (or throttle) simultaneously. The next level is the three-channel radio. Most of these radio transmitters use a *pistol-grip* configuration, in which a gun-style finger trigger controls the throttle channel, and a miniature wheel on the side of the transmitter controls the steering channel (see Figure 10-1). The third channel is intended to control a gearshift, air horn, lights, or other on-board accessory. Pistol-grip transmitters are primarily used for R/C cars.

The four-channel radios are commonly used for both ground and aircraft applications. The six-, eight-, and nine-channel radios are predominately used in aircraft applications, but they are gaining popularity in robotic contests such as BattleBots and Robot Wars. These radio transmitters use two sticks, called *joysticks*, to control the first four channels, (see Figure 10-1). Channels 5 through 9 are typically controlled by switches and knobs for extra control capabilities.

Each of the two joysticks controls two channels—one channel with the horizontal motion, and one with the vertical motion. Most of the high-end radio sets also have computerized control interfaces that allow the operator to configure the channel allocation and change mixing settings, and the R/C system can be programmed for custom control sequences.

In robot sumo, you will need as a minimum a two-channel radio. One channel will be used for steering, and one channel will be used for throttle, or both channels can be used for throttle if you are using a differential-steering robot. If your robot has any additional features that need to be controlled, you may need at least one additional channel for every feature your robot will use. (The additional channels depend on how many semi-autonomous features your robot will have.)

Some more complex robots that involve omni-directional wheels or multilegged walking mechanisms may use a microcontroller to assist in the driving control. For example, a six-legged, 12 degree-of-freedom robot that uses two servos to drive each leg will require a total of 12 servos. Controlling 12 servos at one time requires a very complex set of instructions. But a microcontroller can handle all of this work, and all you need to do is tell it the direction and speed via the radio link.

Radio-Control Frequencies

R/C systems use radio waves to transmit the control signals from the transmitter to the receiver. In order to prevent radio interference between the transmitter and non-R/C equipment, the Federal Communications Commission (FCC) has set aside certain frequency bands for remote-control systems. As a result, the R/C radio manufacturers have standardized specific frequencies inside these frequency bands used by hobbyists.

A specific frequency band is divided into a set of intermediate frequencies that are equally spaced with respect to each other. Each of these individual frequencies is given a *channel number*.

> **NOTE** The channel number is not to be confused with the number of channels a radio transmitter has. The channel number represents the frequency the radio is transmitting on. The number of channels represents the number of individual control signals the transmitter can control. Unfortunately, the R/C equipment industry has adopted this somewhat confusing nomenclature.

Frequency bandwidth allocation varies by country; a radio operating on a legal frequency in the United States may not be legal for use in Japan or the United Kingdom, and vice versa. Also, some frequency bandwidths have been allocated specifically for ground use only. Other frequencies are allocated for aircraft use only, and some frequency bandwidths can be used for both air and ground applications. The separation between air and ground use is for safety reasons, so that ground transmitters cannot accidentally interfere with the control of an airplane and cause it to crash.

If your robot is going to use a traditional R/C system, the frequency bands that you are allowed to use by law are 27 MHz, 40 MHz, 50 MHz, and 75 MHz. However, using the 40 MHz frequency is not legal in the United States, but is legal in Japan and England.

27 MHz Radio Frequency Band

Both ground and aircraft vehicles are allowed to use the 27 MHz radio frequency band. This radio band is usually used for small, low-cost R/C toy cars, tanks, boats, and planes. It starts at 26.995 MHz and is divided into six separate channels spaced every 50 kHz. R/C systems that transmit on the 27 MHz band can use either the amplitude modulation (AM) or frequency modulation (FM) transmission method.

These radios are usually sold as two- or three-channel systems and are low-power systems with a limited operating range. Also, this frequency band crosses into the lower channels on the citizens band (CB) radio frequencies, so there is a chance of interference by CB radio operators.

Whenever you are using R/C equipment in areas where other people are using R/C equipment, you must place a flag on the equipment's antenna to let everyone in the area know the frequency on which you are operating. The standard flag colors for channels 1 through 6 are brown, red, orange, yellow, green, and blue, respectively.

72 MHz Radio Frequency Band

The 72 MHz radio band is for aircraft use only. In other words, ground vehicles, including sumo robots, are not allowed to use this frequency band. The 72 MHz frequency band is a much larger band than the 27 MHz band. This band is divided into 50 different channels that start at number 11 and end on channel 60. The frequency band is from 72.010 MHz to 72.990 MHz, and each channel is evenly spaced by 20 kHz. A white streamer and a flag with the written channel number should be attached to the antenna of R/C equipment using this band.

75MHz Radio Frequency Band

The 75 MHz radio band is for ground use only. This band has been divided into 30 different channels that start with a frequency of 75.410 MHz and increment in 20 kHz units to 75.990 MHz. The channel numbers start from 61 and end at 90. The antenna channel number identification is a red streamer with a flag that has the channel number.

50 MHz Radio Frequency Band

The 50 MHz frequency band has been allocated for both ground and air use. This band is divided into ten channels starting from 50.800 MHz and spaced every 20 kHz. The channel numbers are two-digit numbers that range from 00 to 09. The antenna identification markers include a black streamer with a flag marked with the channel number.

Unlike the 27, 72, and 75 MHz radios, the 50 MHz radios require an amateur ham radio license from the FCC to use.

Japanese Radio Frequency Bands

The R/C systems in Japan are limited to the 27 MHz frequency band using narrow bands 1 through 12, and the 40 MHz frequency band using channel numbers 61, 63, 65, 67, 69, 71, 73, and 75. The 12 narrow frequency bands in the 27 MHz are the same channels as the six wide frequency bands used in the United States. These channel numbers are not the same channel numbers that are within the 75 MHz band used in the United States.

United Kingdom Radio Frequency Bands

R/C systems in the United Kingdom are similar to those in the United States and Japan, except that the particular radio frequencies used are different. The remote-control system runs on the 35 MHz band for aircraft use, 40 MHz band for ground use, and 27 MHz band for both ground and air use.

Changing Frequency Channels

Within a frequency band, the individual channel numbers divide the band so that multiple R/C systems can be used simultaneously. For example, there are 30 different radio channel numbers that can be used in the 75 MHz frequency band. The specific channel number frequency is controlled by an oscillator called a *frequency crystal*. The frequency crystals come in a metal can and are sold in pairs: one for the transmitter and one for the receiver. A typical transmitter and receiver radio frequency crystal set is shown here.

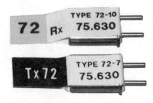

To change the channel number on your radio, you simply replace the frequency crystals. The crystal that is used with the transmitter is marked with a *TX* and the channel number, and the receiver's crystal is marked with an *RX* and the channel number. In order for the radio system to work properly, both of the crystals must have the same radio frequency. Both the transmitter and receiver must use the same channel number, or the system will not work.

> **NOTE** *You cannot change the radio's frequency band by changing the frequency crystals. In other words, you cannot change a 72 MHz band radio into a 75 MHz band radio just by swapping frequency crystals. Although the crystals look identical in size and shape, swapping the crystals between the two radio frequency bands will not work. Switching from one band to another requires retuning the radio, which should be done only by an FCC-licensed technician or special-ordered from the equipment manufacturer.*

When you select an R/C system, make sure you can change the frequency crystals. This is because it is likely that at least one other person at a competition may have their radio operating at the same frequency that your system is using. So you may be required to be able to change the frequencies of your R/C system to avoid frequency conflicts.

Radio Transmission Methods

The actual method in which a transmitter sends the control signals to the receiver is by using radio waves. This information is transmitted either by *amplitude modulation* (AM) or *frequency modulation* (FM).

Most R/C systems use a single radio-carrier frequency to transmit the control information from the transmitter to the receiver. Radios with multiple channels will transmit all the control signals serially, one pulse following another on the same radio signal, and then repeat the signal transmission 50 to 60 times per second.

Amplitude Modulation

An AM radio system will transmit the control signals using a constant radio frequency and vary the amplitude of the transmitted power to encode the control information. The transmitter sends each control channel's position as a time-coded analog pulse width that varies from 1 to 2 milliseconds. The radio frequency is transmitted continuously "on" for the same amount of time as the control signal. At the end of the pulse, the transmitter stops sending a signal for 0.35 millisecond, which is considered an "off" signal. All of the channel signals are sent sequentially with a 0.35 millisecond off period between each on period. At the beginning of each set of transmitted data, there is a special framing pulse that is sent so tell the receiver that this is the start of a sequence of incoming data.

AM radios are inexpensive and easy to implement electrically, but they are highly susceptible to radio interference. Any electrical noise from electric motors, surrounding fluorescent lights, or other radio transmissions can cause unwanted movement in the robot, since the electrical noise can be added to the original AM signal. Because AM receivers interpret the intensity of the incoming radio signal as command signals, they do not have a way to distinguish electrical noise from the actual transmitted signals. This results in the receiver sending false command signals to the robot's control systems. For safety purposes, AM radios should not be used on robots that are strong enough to cause damage or injure people.

Frequency Modulation

Modulating the frequency of the transmitted signal is a more reliable method for transmitting control signals. Unlike with the AM method, with FM, the amplitude of the signal is held constant. The transmitted signal is modulated about the carrier frequency to encode the data, and the FM receivers are tuned to lock onto the specific carrier frequency. By being able to lock onto a specific frequency, the system can ignore all other signals. Power intensity does not affect the signal frequency. This does not mean that FM systems are immune to radio interference, because all radio systems are subject to radio interference. However, FM radio signals are far less susceptible to radio interference than AM radio signals are.

Pulse-Code Modulation

To further improve the reliability of FM radios, an advanced signal transmission method known as *pulse code modulation* (PCM) can be used. The normal time-coded control pulses that traditional R/C systems use are converted into a digital signal. After the data is digitized, an error *checksum* is also created. Both of these signals are then transmitted using an FM transmission method. The microprocessor-controlled receiver receives the data and compares it with the received checksum value. If the two blocks of data do not agree with each other, the data set is ignored. Then the receiver uses the previously correct control signal. If the receiver continues to receive "bad" data, it will enter failsafe mode because it has lost reliable communication with the transmitter. In failsafe mode, the receiver will start outputting a preprogrammed set of safe operating commands. For robots, this means shutting down the motors.

PCM radios do not guarantee 100 percent reliable transmissions, but they do significantly decrease the chance of interpreting bad transmission signals as good signals. There will always be some radio interference in all signal transmissions. A PCM radio will ignore signals that don't pass the checksum test and eventually end up in failsafe mode. It is better to have a robot shut down than run off in an uncontrolled manner.

Most PCM radios are known as *computer radios*, and they can be customized. Since the command signals are being digitized and encoded before being transmitted, it is easy for the internal computer to perform custom mixing and scaling operations on the data before being transmitted. This can help simplify advanced control algorithms, such as conducting differential-steering control with only one stick on the transmitter (see the "Servo Mixing for Steering Controls" section later in this chapter for details on this approach.)

R/C Control Signals

The actual control signals that are transmitted from the transmitter to the receiver are a series of sequential pulses that range between 1.0 to 2.0 milliseconds. Depending on the R/C system, the transmitter can be transmitting from two to nine different signals. Each signal is transmitted immediately after the previous signal is completed. If you hook up a two-channel oscilloscope to the first two channels of a receiver, you will be able to see two pulses: channel 2 transmitted right after channel 1, no signal for about 16 milliseconds, then the two pulses repeated again.

If you hook up the oscilloscope to channels 1 and 3, you will notice a dead band of approximately 1.5 milliseconds between the two measured signals. This dead band is actually the channel 2 signal that is not being measured. The channel signals are sequentially transmitted right after one another. There is no dead band between sequential signals, regardless of the pulse width.

This is basically all that the transmitter does: Convert joystick, switch, and knob positions into pulse width signals, add a synchronization signal, and transmit them to a receiver. The receiver converts the radio signal back into the series of 1.0 to 2.0 millisecond pulses, and then distributes the individual pulses to individual channel ports. Because the control signal is just a 1.0 to 2.0 millisecond pulse, it can be easily read in by a microcontroller, or created by a microcontroller, which makes traditional R/C systems ideal for robotic control.

Radio Interference

Originally, R/C equipment was designed to operate outdoors to control airplanes several thousand feet away, not indoors to control high-powered robots. Remotely controlling any vehicle indoors is a challenge due to all of the radio, reflection, and electrical noise interference that does not normally exist outdoors with the same level of intensity. Because of this, a robot that is a few feet away from you could suddenly go out of control. In order to safely remotely control your robot, you need to understand the various sources for radio interference and how to minimize its effects on your robot.

Reducing Electrical Noise from Motors

Most radio interference comes from the electric motors. A high-frequency electrical noise is created by the arcing when the electrical current flow is being switched in the motor's armature windings as the commutator segments rotate past each of the motor's brushes. The radio receiver can pick up this noise, which could result in interpreting the noise as a command signal, or the noise can be strong enough to jam the unit and prevent it from receiving any command signals from the transmitter.

Using Capacitors

To minimize the amount of electrical noise escaping from the motor, short-circuit the noise before it leaves the motor's case. To do this, place small capacitors between the motor's terminals and motor case to filter the noise from the brushes. Connect the capacitors as close to the actual brushes as possible, ideally inside the motor case itself.

For this purpose, it is best to use ceramic capacitors that have a 0.01 to 0.1 µF range. Do not use polarized capacitors, because they will be destroyed by the motor and may damage the motor. Do not add capacitors to motors that already have capacitors built inside them.

When using one capacitor, solder it between the two motor terminals. In a two-capacitor setup, solder one end of the capacitor to a motor terminal and the other end of the capacitor to the motor's case. You might also use a third capacitor to further reduce the electrical noise. The third capacitor's leads should be soldered across both of the motor's terminals. Shown here is an electric motor with three small capacitors soldered to it.

CAUTION *Some R/C car experts will suggest that you use Schottky diodes to help reduce the electrical noise. Whatever you do, do not attach diodes across the terminals on the motor. Diodes are to be used on motors that will spin in only one direction. Reversing the direction of a motor with diodes in place can destroy the diodes and damage the motor and the ESC.*

Twisting Motor Wires

Another source of electrical noise is from the motor's wires. Any electrical noise from the motor can migrate into the electrical power lines. The wires will act like a broadcast antenna and transmit the electrical noise to the receiver's antenna. To reduce this problem, twist the motor power wires together (yes, leave the insulation on the wires!).

Also, as the high current travels through the motor wire, a circular magnetic field is generated around the wire. A small amount of current can be induced in a second wire that is parallel to the high current wire if it is within this magnetic field. (This is a result of the physics between electricity and magnetism, which is why electric motors and power transformers work in the first place.) Twisting the motor power leads helps cancel out the magnetic fields, so they don't interfere with other electrical signals. In addition, do not run any electrical control lines near and parallel to high-current wires.

Isolating the Motors

For electronic circuits to work correctly, there must be a common ground between all of the components. But electrical noise can be transmitted through the ground wires. Since electric motors are the biggest source of electrical noise in a robot, the ideal way to minimize the amount of noise going from the motors to the radio receiver or other control electronics is to electrically isolate the motors from the rest of the control circuits.

To isolate the electrical connection between motor and the rest of the electrical circuits, place optical isolators between them. Many ESCs use built-in optical isolators. You can also use mechanical relays to isolate circuits. Optical isolators work in a similar manner to mechanical relays, except that a light source, rather than a mechanical lever, is used to switch the relay.

For low-powered robots, electrical isolation is usually not needed. But for high-powered robots, you should select an ESC that has a built-in optical isolation circuit.

Reducing Interference from Other Radios

Another common source of radio interference is when two radios are transmitting on the same frequency. The receivers on both robots won't know which is the correct transmission, and they will fail to respond to any command, accept the commands from the transmitter that has the higher output power level, or oscillate between these two choices. This can be a very dangerous situation because your robot can suddenly start or stop moving in the middle of a tournament. At this point, the robot is no longer controllable. This is also why you display the frequency number flags on your transmitter's antenna to let everyone else know which frequency you are currently using.

In large tournaments, frequency conflicts are always a major concern. The best way to handle this situation is to impound all transmitters throughout the tournament. When it is your time to compete, the organizers will give your transmitter back to you with an appropriate set of radio crystals for your transmitter and receiver. This is why the receiver on your robot must allow for easy replacement of the crystals. After the match is over, you turn your transmitter over to the officials again, regardless of whether you win or lose. At the end of the tournament, the officials return the transmitters to their owners. This may seem harsh, but it is the only way to make sure that there is no chance for a frequency conflict. This is the method that has been successfully used in Japan, where hundreds of remote-control robots show up at tournaments.

At any remote-control tournament, you should have at least two different sets of frequency crystals for every sumo ring that is being run to avoid any frequency conflicts during the competition. Most tournaments have only one sumo ring operating at a time. Some of the larger contests will have two or three rings running at the same time. Any contest that will be running more than three rings at the same time must clearly spell out the rules for avoiding frequency conflict.

R/C Antenna Setup

One source of poor radio communications is an improper antenna setup. Antennas transmit and receive data from the transmitter to the radio receiver on your robot. Without the antenna, you will not be able to communicate with your robot.

All electromagnetic waves, including light and radio waves, travel with a speed of 300 million meters per second, regardless of the frequency on which they are operating. The wavelength of a particular radio frequency (λ) is the speed of the electromagnetic wave (v) divided by the frequency (f) of the wavelength. Equation 10-1 shows this relationship.

Equation 1

$$\lambda = \frac{v}{f}$$

Antenna Length

The ideal antenna configuration would be a vertical wire that has a length equal to one wavelength of the radio wave used for communication. For a 75 MHz transmitter, this works out to be a 157-inch antenna, which is not very practical for a robot. Thus, most 72 MHz and 75 MHz radios come with a one-quarter-wave antenna instead of a full-wave antenna. This cuts the length of the antenna by a factor of four, or to a length in the range of 37 to 42 inches. For many robots, this is still too long.

The way to correct this problem is to use a *base-loaded* antenna. A base-loaded antenna uses a shorter antenna length with a tuned resonance circuit attached at the base of the antenna. Base-loaded antennas are tuned for specific frequency bands. A 72 or 75 MHz base-loaded antenna can be as short as 6.5 inches. The base of the base-loaded antenna should be at least one inch away from any metal part surface, and the wire from the antenna to the radio receiver should be as short as possible and not run near any motor power lines. W.S. Deans manufactures a base-loaded antenna that you can find at most hobby stores.

If you are going to use the antenna that comes with the receiver, it is better to pile antenna wire loose inside the robot with only a portion of the antenna protruding on top of the robot than it is to cut the antenna wire short. Cutting the antenna will affect how well the receiver will receive the transmission. A coiled-up antenna with a proper length will work better than an antenna with an improper length. The antenna wire should rise vertically above the robot. To add stability to the antenna, run the antenna wire inside a small-diameter plastic tube. Do not use a metal or an electrically conductive tube. While the radio reception will be far from ideal, a sumo match is typically less than 10 feet away from the transmitter, so a poorly set up antenna will probably still work.

Antenna Orientation

All antennas—both transmitting and receiving antennas—should be oriented vertically because of how antennas transmit their energy. The energy pattern of antennas have a donut shape, where the antenna passes through the center of the donut. The strongest part of the transmitted signal travels radially away from the antenna, and the weakest part of the transmitted signal radiates in the direction of the antenna.

The strongest reception occurs when the receiving antenna is parallel with the transmitting antenna, and the weakest reception occurs when the transmitting antenna is pointing at the receiving antenna. If the transmitter's and the receiver's antennas were placed in space, where there are no reflections, no signal would be created if both antennas were pointed at each other.

On Earth, especially in a room with a metal floor, the signals bounce around, and reception can be accomplished with almost any orientation. However, these reflections are far weaker than a direct signal.

One of the biggest mistakes novice radio operators make is pointing the antenna straight at their robot, thinking that this will improve the reception; in fact, it is the exact opposite. The antenna on your robot and your hand-held transmitter should always point straight up for optimum signal transmission and reception. From a safety point of view, holding the antenna horizontal to the ground could result in poking someone in the eye, so antennas should be held vertically for this reason alone.

R/C Configurations for Sumo Robots

The radio-control robot is the easiest form of a sumo robot to build from a controls standpoint. This is because most, if not all, of the components can be purchased and then plugged together. All the control equipment you need is a transmitter, receiver, battery, and two ESCs. Figure 10-2 shows a simple schematic of an R/C robot using differential steering.

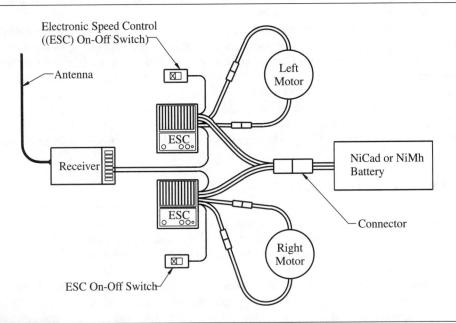

FIGURE 10-2 A general-purpose schematic of an R/C style sumo robot using ESCs for motor speed control

CHAPTER 10: Remotely Controlling Your Robot

If the ESCs do not have a *battery eliminator circuit* (BEC), you will need a separate battery to supply power to the receiver. Both ESCs should be identical. If you are using a separate battery for the receiver and the ESCs have BECs, you must remove the positive voltage pin from the connector that plugs into the receiver.

Figure 10-3 shows a schematic of how this type of a remote control works in a robot. This is a classic differential-steering control method. (Bulldozers, tractors, and tanks are controlled in a similar manner, except they are not remotely controlled.) Each stick on the remote controls a motor. When a stick moves forward, the corresponding wheel moves froward. Pull the stick backwards, and the wheel reverses direction. This is a very popular and intuitive method of controlling robots.

Illustrated here is the actual wiring setup (shown in Figure 10-2) used to convert the Terminator robot from Lynxmotion (www.lynxmotion.com) into an R/C sumo robot. The speed controllers are Hitec SP-560 (www.hitecrcd.com), which are also available from Lynxmotion.

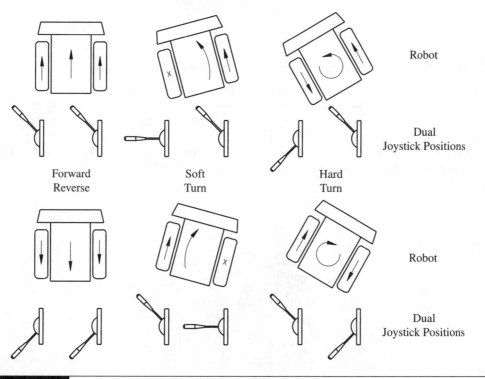

FIGURE 10-3 Differential steering using two independent joysticks for motion control

The electric motors are 6-volt, high-torque gearmotors from Lynxmotion. The transmitter is a four-channel Futaba T4VF transmitter, and the model number for the Futaba receiver is FP-R127DF. As you can see, the required wiring is a relatively simple task.

Mini Sumo Robot R/C Configuration

The basic R/C configuration for mini sumo robots is the same as for full-sized robots, except when modified servos are used for the motors. Using modified servos makes this configuration very simple to fabricate, as shown in Figure 10-4. All that you need is an R/C transmitter, a receiver, a battery, and two modified R/C servos.

Shown here are all of these components mounted on the Marvin Slyder base kit, developed by Sine Robotics (www.sinerobotics.com).

CHAPTER 10: Remotely Controlling Your Robot

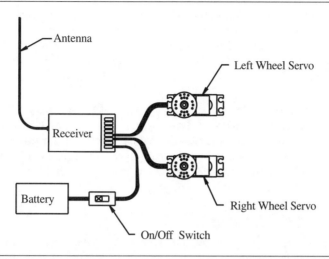

FIGURE 10-4 An R/C mini sumo robot schematic using modified servos for motors

Servo Mixing for Steering Controls

The steering method using two sticks on a transmitter is a very intuitive method of remotely controlling a robot, but it has its drawbacks. One of them is that it requires you to use both hands to control the robot. If your robot does not have any special features—such as deploying arms or a lifting lever to flip your opponent on its back—a two-stick transmitter will work just fine. But in cases where you want to control the robot with only one stick so you have a hand free to do other things, you will need a different control method. Also, if you are using a pistol-grip style transmitter, you have a trigger for throttle and a steering wheel for steering. This type of transmitter is not easy to use for differential steering without some additional control algorithms.

What is needed for smooth driving control is the ability to convert the two independent controls (two sticks or a throttle trigger and steering wheel) into a pair of drive controls that are linked together. This is done by mixing the two input signals together and then splitting the results into two drive commands. A single stick is just like a throttle trigger and a steering wheel. Moving the stick forwards and backwards controls the speed and has the same motion requirements of a throttle trigger on a pistol-grip transmitter. Moving the stick left and right is for steering. The direction you move the stick is the direction you want the robot to turn. This has the same function as the steering wheel on the pistol-grip transmitter.

Mixing these two signals together is generically called *servo mixing*. This term comes from R/C airplane flyers. These airplanes use servos to control the flaps, elevators, ailerons, and the rudder. In certain flying maneuvers, moving the rudder requires a certain movement in the ailerons. These special movements have resulted in the development of custom circuits that are used to mix the desired input move commands together and output a pair of servo commands for the ailerons and rudders. These circuits are commonly called *V-tail mixers* or *elevon mixers*.

For a robot, the mixing is needed so that when the throttle stick is moved forward, both wheels start to move forward, and when the throttle stick is moved backward, both wheels move in that direction. As you learned in Chapter 6, to make a differential steering robot turn, the outside wheel must turn faster than the inside wheel. So, to turn, the mixer must add some amount to the throttle command on one wheel and subtract the same amount from the throttle command on the second wheel.

Some advanced R/C transmitters have built-in servo-mixing capabilities. Top-of-the-line programmable transmitters can be programmed with custom mixing commands. Many transmitters have built-in V-tail mixing algorithms that work well for differential steering methods. If your transmitter does not have these built-in capabilities, one solution is to place a servo-mixing circuit between the R/C receiver and the ESCs. Another solution is to build your own mixing circuit.

Adding a V-tail Mixer

V-tail mixers have two inputs that plug into the two output channels to be mixed from the R/C receiver, and two output channels that have the mixed signal results. Servos or ESCs connect directly to the two output channels.

Prepackaged mixing circuits are a nice plug-and-play solution to obtaining differential steering with a pistol-grip style transmitter or using a single joystick. Illustrated here is a V-tail mixer made by Graupner (www.graupner.com).

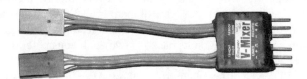

When assembling the mixing circuit into the robot-control system, you need to make sure that the steering channel plugs into rudder channel on the V-tail mixer, and the throttle is connected to the aileron channel on the mixer. The left motor control connects to the left servo channel on the mixer, and the right motor control connects to the right servo channel on the mixer. Since left and right motors face in opposite directions, the wiring for one of the motors will be reversed with respect to the other motor. Figure 10-5 shows how the mixer is placed in the remote-control circuit for the robot.

If your V-tail mixer does not have obvious markings like this, you will need to experiment with the wiring setup. When you move the joystick/throttle forward, both sets of wheels should move forward together. When moving the steering knob/stick to an extreme position, the motors should go in opposite directions. Moving the joystick/throttle all the way forward and the steering all the way to the left/clockwise should increase the width of the signal pulse. An easy way to test this is to attach a servo to the receiver. The servo will rotate clockwise as the width of the command pulse increases. If moving one of the sticks on the transmitter results in a

CHAPTER 10: Remotely Controlling Your Robot

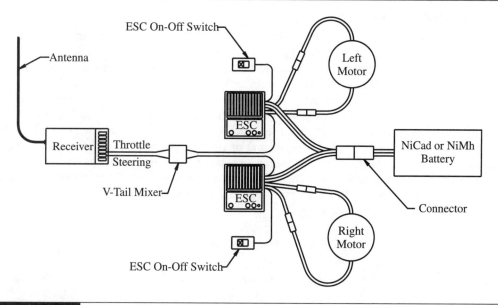

FIGURE 10-5 Fitting the V-tail mixer between the receiver and ESCs

counter-clockwise movement when you're expecting a clockwise movement, flip the servo-reversing switch for that channel.

When setting up the V-tail mixer, be prepared to switch the wiring around until it is set up correctly. This includes the wires between the mixer and the R/C receiver and the ESCs, and the power wires connecting the motors. When the transmitter's controls are in their neutral positions, the motors should not be turning. You may need to adjust the trim dials on the transmitter to zero the movements in the motors.

The biggest drawback to using a V-tail mixer is that it is not customizable. Using the trim dials to zero the motor motion works, but a slight movement of the stick will result in a movement in the motors. Adjusting a dead band on the mixer is not possible. (A *dead band* is a small area in which moving the joystick yields no change in the output results of the robot.)

Some people prefer to use an exponential control on the steering and throttle. *Exponential control* means that as the stick is moved from the neutral position to a maximum position, the beginning movements yield very little motion in the robot. As the stick approaches the maximum position, the robot will see the greatest rate of change in performance. With this type of control, the robot is less sensitive at slow speeds and more sensitive at high speeds. Other people may want to be able to use an inverse exponential control in their robots. Standard V-tail mixing circuits do not allow exponential control. For customizable servo mixing, you can build your own mixing circuit.

Software Mixing

To build a mixing circuit, you need a microprocessor and software to program the microcontroller. (The next chapter begins the discussions of microcontrollers.) You can use any microcontroller. Figure 10-6 shows a schematic drawing of how to wire a Basic Stamp 2 microprocessor from Parallax Inc. (www.parallaxinc.com) to the receiver and to two R/C ESCs.

Equations 10-2 and 10-3 show the governing relationships in how mixing is mathematically performed. Each of the variables are pulse widths with a range from 1.0 to 2.0 milliseconds. The *Throttle* value is some speed you want for robot movement. This is controlled by the throttle stick position. The *Steering* value is an amount of steering you want. It is an additive and a subtractive process. For example, if you want your robot to turn toward the right, the left motor must spin faster than the right motor. The *Steering* value is added to the *Throttle* value to increase the left motor speed, and the *Steering* value is subtracted from the *Throttle* value to decrease or even reverse the right motor speed.

Equation 2   *Left Motor = Throttle + Steering - 1.5ms*

Equation 3   *Right Motor = Throttle - Steering + 1.5ms*

The following is a simple working program that reads in the two signals from the receiver, compares the values with a dead band and checks if they are outside the 1.0 to 2.0 millisecond boundary for a normal signal, performs the mixing, and then outputs the results to the servos.

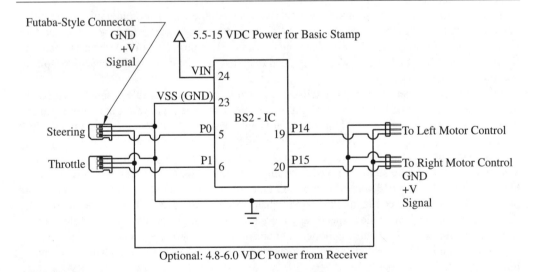

FIGURE 10-6   A mixing circuit using the Basic Stamp microcontroller

CHAPTER 10: Remotely Controlling Your Robot

This program can be modified to add features such as transforming the linear motion input signals into nonlinear output signals, so that it could be more responsive at low or high speeds.

```
'Program Mixer.bs2
'This program reads in two R/C servo commands then mixes
'them together to output two control signals for differential
'steering control with a single joystick or pistol-grip style
'R/C transmitter. Pulsin and Pulsout commands are based on
'integer multiples of 2us time periods, so a 1.5 ms pulse width
'will have a 750 Pulsin or Pulsout command.
'This program has a built-in dead band algorithm of +/-10% of
'the full motion. Any measured pulses within this threshold
'will be treated as a neutral position or a 1.5 ms position.
'Also, the program will treat any signals that are outside the
'1.0 ms - 2.0 ms as invalid and will shut down the motors by
'setting them to 1.5 ms neutral position for the ESC.
'
'Steering signal line is connected to Pin 0 on the Stamp.
'Throttle signal line is connected to Pin 1 on the stamp.
'Left_Motor output command signals is connected to Pin 14.
'Right_Motor output command signals in is connected Pin 15.
'
'Remember to tie all grounds together.

'{$STAMP BS2}            'Stamp 2 Directive


Steering            Var     Word    'Steering position from Transmitter
Throttle            Var     Word    'Throttle position from Transmitter
Left_Motor          Var     Word    'Output command for the Left Motor
Right_Motor         Var     Word    'Output command for the Right Motor


Loop:

    pulsin 0,1,Steering         'Read in Steering pulse width
    pulsin 1,1,Throttle         'Read in Throttle pulse width

    if Steering < 725 or Steering > 775 then skip1
        Steering = 750
```

```
skip1:
   if Throttle < 725 or Throttle > 775 then skip2
      Throttle = 750
skip2:
   if Steering < 1000 and Steering > 500 then skip3
      Throttle = 750
      Steering = 750
skip3:
   if Throttle < 1000 and Throttle > 500 then Mix
      Throttle = 750
      Steering = 750

Mix:

   Left_Motor = Throttle + Steering - 750
   Right_Motor = Throttle - Steering + 750


   pulsout 14, Left_Motor      'Output result on pin 14
   pulsout 15, Right_Motor     'Output result on Pin 15
goto Loop
```

Tether-Control Systems

So far, this chapter has focused on R/C systems for remote-control robots. A tether is another remote-control method for robots. Some large robot sumo tournaments, such as the Northwest Robot Sumo Tournament and the Western Canadian Games, allow tether-controlled robots to compete against radio-controlled robots. Before building a tether-controlled robot, make sure that the rules of the local contests you would like to enter allow this style.

A tether-controlled system uses a long, small-diameter cable that runs between the control box and the robot. The cable is used to transmit control signals from the control box to the robot. The long cable is called a *tether*.

One of the big advantages of a tether-controlled robot is that it is not susceptible to any of the interference issues that are present with R/C systems. The drawback to this method is that it requires the operator to move around the dohyo during a match so that the tether does not get tangled up with the other robot or the robot's cable. You'll need to control the tether carefully. The one thing you do not want to do with the tether is pull back on it enough to cause your robot to tip over.

General Specifications for Tether-Controlled Robots

For a tether-controlled robot, all of the power that the control box uses must come from the robot. Any computational logic must come from the robot, not from the tether control box. The control box contains only the switches and knobs that are used to direct motion control on the robot, and no power sources.

The weight of the tether and control box is not included in the total weight of the robot. The one exception to this is if you want the control box to contain any power sources or computational logic. In this case, the weight of the tether and control box will be included as part of the weight of the robot.

The robot must contain a thin, stiff wire pole for the tether to run up on, so that the tether does not slide on the ground. Other than this, a tether-controlled robot is the same as a radio-controlled robot. Figure 10-7 shows an effective diamond-wedge-style tether sumo robot design. This robot, named Spade, was built by Bill Harrison of Sine Robotics.

Simple Tether-Control Circuits

The minimum control signals that are needed to control a robot are forward, reverse, turn left, and turn right. Figure 10-8 shows a simple schematic drawing of a relay-controlled robot that is

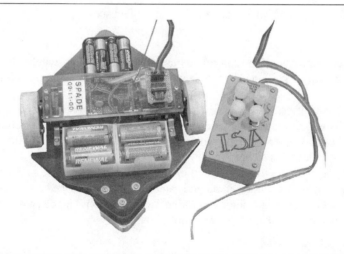

FIGURE 10-7 A tether-controlled sumo robot design (built by Bill Harrison of Sine Robotics)

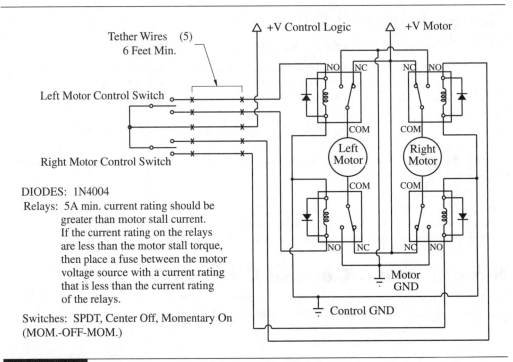

FIGURE 10-8   Schematic drawing of a tether-controlled sumo robot

used with a tether-control box. The control box uses two large toggle switches to simulate the same feel as with a joystick robot. The toggle switches must be normally open center-off style SPDT switches.

This type of control system offers only maximum forward and reverse speed control in the robot. Therefore, the gear reductions in the motors need to be high enough to allow full control of the robot. If the robot is too fast, you will not be able to accurately control the robot, which would make it difficult to win a contest.

An alternative approach is to use joysticks or variable resistors, and use a motor speed control board that accepts an analog voltage input and can output variable-speed control. The Motor Mind C and the ICON H-Bridge motor controllers from Solutions Cubed (www.solutions-cubed.com) are examples of these types of control boards.

Failsafe Control

With all robots, safety is a primary concern. You need to make sure that the robot has failsafe control in case you lose radio communication with your robot. Generally, in failsafe mode, all of the motors will shut down.

Many competitions require the robot to pass a safety inspection prior to competing. One of the safety tests is demonstrating that the robot will stop moving if the transmitter is shut off. If the robot continues to move after being shut off, it is considered an unsafe robot.

Most R/C servos and ESCs will shut down when they stop receiving valid command pulses. ESCs that use a servo to move a variable resistor for the speed controller will not pass a safety test, since the servo will remain in its last position when the transmitter is shut off. If the variable resistor was in an on position at the time the servo shut down, the motors will continue to run because nothing physically shut them off.

If you use a custom microcontroller in an R/C robot, it must be programmed to shut down the motors if it loses contact with the transmitter. A servo-controlled variable resistor speed controller combined with a microcontroller can pass the safety test if the microcontroller is directly controlling the servo and commands the servo to move to an off position when the robot loses communication with the transmitter.

Advanced PCM radio systems have custom-programmable receivers. If the receivers receive bad data or lose contact with the transmitter, they will start outputting preprogrammed failsafe commands. These commands can be used to tell the ESCs to stop the motors, or they can move a servo-controlled, resistor-type speed controller to an off position without the need of an intermediate microcontroller.

Final Comments on Remote-Control Robots

At this point, you should have all of the information required to put together a remote-control robot. One of the best ways to develop a good autonomous sumo robot is to first build a remote-control robot. The remote-control robot could be the core frame used to test all of the mechanical aspects of the robot, or it can be used as a practice robot against the autonomous robot. By using a remote-control robot against an autonomous robot, you can develop new programming strategies that can be used to beat the human brain that is controlling the remote-control robot.

One other advantage a remote-control robot has is that it can be used in environments that are outside the sumo ring. You can use a remote-control robot in other types of contests, or just run it around the house.

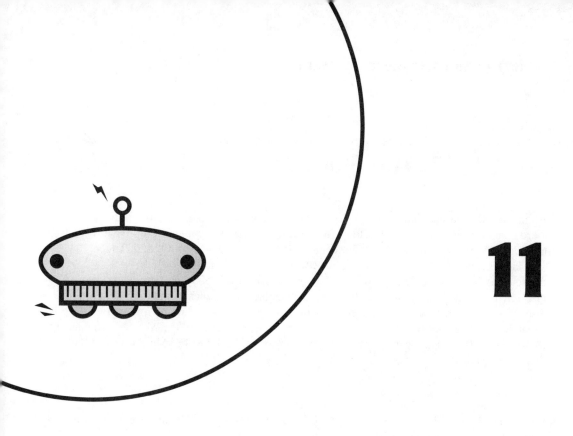

Introduction to Microcontrollers

All robots have some level of intelligence. Webster's dictionary defines intelligence as "the ability to learn or understand from experience, the ability to acquire and retain knowledge; mental ability." Based on this definition, robotic intelligence is a loosely used term. Robots have brains, and they react differently to different input stimuli, but how they respond to different input stimuli is usually preprogrammed. It can be argued that humans are preprogrammed to respond in a certain way to different input conditions. However, humans can independently change how they respond to a situation, whereas robots generally must rely on a new program from a human.

For the most part, sumo robots themselves do not learn as they compete during a contest. They respond to their environment either by external remote control or by some form of internal logic. Internal logic can take on many different forms, such as simple switches, vacuum tubes, neuro-networks, programmable microcontrollers, or any combinations of these. Though vacuum tubes are not very practical, Grey Walter did show that vacuum tubes can be used to make

BEAM Robots

Along with the artificial intellligence (AI) research that is being done at various universities, there is another interesting technology for creating robots that can learn: the BEAM (Biology, Electronics, Aesthetics, and Mechanics) technology, invented by Mark Tilden.

The heart of a BEAM robot is a neuro-network. Mark Tilden has shown that a few discrete components can be used to create biologically inspired robots that can learn from their environment. These robots can respond to changing environments, and they can even respond to being damaged without a single program. This is a very new and impressive technology. There are several BEAM-based sumo robots competing at the Western Canadian Robot Games (www.robotgames.com). For more information about BEAM robots, visit the Solarbotics website (www.solarbotics.com).

autonomous robots in 1950. In fact, the first mobile autonomous robot, the tortoise named Elsie, was built using vacuum tubes for all of the internal logic.

Most sumo robots use either remote-control systems or microcontrollers to control their robots. Chapter 10 discussed remote-control systems. This chapter focuses on microcontrollers. Microcontrollers are necessary for autonomous sumo robots, but they are also very helpful in assisting with the driving control for remote-control sumo robots.

What Is a Microcontroller?

Many people call a microcontroller a microcomputer. Although they have many characteristics in common—both are programmable, can accept inputs, can control other devices, have memory, and can make calculations—they are not the same thing.

A *computer* and a *microcomputer* are devices that accept inputs directly from humans and will output their results for humans to use. Much of their electronics are devoted to converting electrical signals into signals that humans can understand and vice versa.

A *microcontroller* is a slave device that accepts inputs from one electrical device, conducts some operation on the signals, and then outputs the results to control other electronic devices. Not much of a difference between a microcontroller and a microcomputer, right? Actually there is a big difference: A computer will have dozens of embedded microcontrollers inside it to control specific internal systems, but a microcontroller will not have a computer inside it.

A microcontroller is often used to interpret inputs from a human and convert them into electrical signals for the rest of the system to use. Microcontrollers are used in just about every electronic consumer product available today. They are inside our computers, printers, scanners, toasters, coffee machines, microwave ovens, refrigerators, alarm clocks, radios, television sets, video cassette recorders, battery rechargers, telephones, cell phones, electronic locks, video games, electronic toys, and automobiles.

Basic Microcontroller Features

The basic features of a microcontroller are the number of input/output (I/O) pins, the amount of electrically erasable programmable read-only memory (EEPROM), the amount of random access memory (RAM), a central processing unit (CPU), and clock speed.

EEPROM and RAM

The EEPROM is where the programs and permanent data are stored. The RAM is where all of the temporary data that the microcontroller uses is stored. The amount of RAM a microcontroller has puts a limit on the number of variables you can use in your programs. The amount of EEPROM sets the limit on how large a program you can use.

CPU

The CPU is the internal core of the microcontroller. This is used to accept the input data, execute the programs, and output the results. Generally, the CPU will add data, move and compare data, execute loops, read and store data, read and modify internal status registers, and increment counters.

I/O Pins

All microcontrollers have a certain number in I/O pins. They can have as few as six pins to more than 60 pins. Figure 11-1 shows several different sizes of microcontrollers from Microchip (www.microchip.com). Depending on the microcontroller, some I/O pins are input only or output only. Most microcontrollers have bidirectional I/O pins. The software tells the microcontroller to make a certain pin an input pin or an output pin. Most I/O pins are digital I/O pins. They will output or input either a 0-volt or a 5-volt signal. Some I/O pins will input a 0-5 volt analog signal, and some output pins will output a continuous PWM signal.

Clock Speed

Clock speed is the speed of an external crystal oscillator (a clock) that is used to control how fast the microcontroller executes internal instructions. All microcontrollers use an external oscillator, and some microcontrollers have the option of using a slower built-in clock. Typical clock speeds are 4 megahertz (MHz), 8 MHz, 20 MHz, and 40 MHz; some microcontrollers run up to 75 MHz. Microcontrollers can operate with a wide range of speeds, from zero up to their maximum rating.

Clock speed is a general indicator on how fast a microcontroller is, but it shouldn't be used as an absolute method to determine which microcontroller is faster. Different microcontrollers do different things internally with the clock speed. For example, the Microchip microcontrollers will

FIGURE 11-1 8-, 18-, 28-, and 40-pin microcontrollers from Microchip

divide the external clock speed by four to use for the internal clock speed. Thus, the internal instructions are running at one-quarter the external clock speed.

Different microcontrollers will handle similar instructions differently. The number of internal clock cycles required to execute an instruction can be different from microcontroller to microcontroller. And for a given microcontroller, the time required to execute a given instruction will be different for different programming language compilers. Because of these factors, it is difficult to say which microcontroller is the fastest.

NOTE *A compiler is used to convert a program into the native language the microcontroller understands, which is roughly equivalent to assembly language (discussed in the "Programming Environments" section later in this chapter).*

Unlike computers, fast microcontrollers don't mean robots will work any better. Regular computers spend most of their time waiting for you to tell them to do something. Microcontrollers are the same way. They monitor input pin states and will do nothing until they register a change in one of the input pins. Then they will act on the change.

Special Microcontroller Features

All microcontrollers have special features that make them ideal for specific applications. The special features include the type of memory, internal timers, hardware PWM generation, external and internal interrupt capability, analog-to-digital (A/D) converters, hardware serial communications, internal oscillators, and operating voltage range. Different microcontrollers have different types of special features. They are created to fill very specific needs, and these needs can be adapted for your robot.

Memory

The different types of memory include flash and one-time programmable (OTP) memory. Flash memory allows you to electrically reprogram the memory thousands of times. The other type of erasable memory microcontroller has a clear window on top of the chip, and placing the exposed window under an ultraviolet (UV) light for several minutes will erase the chip. OTP memory, as its name implies, cannot be reprogrammed once it has been programmed.

TIP *When using UV erasable chips, put a piece of white or black tape over the window, so memory does not become corrupted from the ambient light.*

Hardware Communications

Many microcontrollers will have a built-in Universal Asynchronous Receive/Transmit (UART) serial communication. Other microcontrollers will have inter-integrated circuit (I^2C) or Serial Peripheral Interface (SPI) serial communication systems built in. And some microcontrollers will have built-in parallel communication hardware. If the microcontroller doesn't have hardware communication features, the communication procedures must be handled in software.

Voltage & Other Features

Some microcontrollers will operate at voltages down to 2.7 volts, which will allow for some low-voltage applications. Other microcontrollers will have a built-in clock, so an external clock will not be needed. An internal A/D converter will also save on external component needs. The resolution of the A/D converter in rated by the number of bits in the digital result; the greater the number of bits, the finer the resolution the A/D converter will be.

Programming Environments

There are two types of programming environments for microcontrollers: low-level and high-level. Low-level programming is using the specific assembly language for the microcontroller. Assembly language is microcontroller-specific and is generally not fully transferrable from one microcontroller to another, especially from one manufacturer to another. The advantage of programming in assembly language is that you have absolute control over how the microcontroller will perform. You will know all of the timing sequences, and you can make very fast and efficient programs. The drawback to assembly language programming is that it is not an intuitive language, and it usually takes longer to write a bug-free program.

The high-level languages include Basic, C, C++, Java, and various other languages and variants on these languages. The advantages to high-level languages are that they are fairly universal and you can generally understand what the program will do by reading the source code. Basic, C, C++, and Java are fairly standard languages, so when you learn one of them, you usually can quickly learn how to use a variant.

For example, the Basic language is a standard language used on most personal computers. It has a specific syntax structure (grammar) and a set of commands that was standardized more than 30 years ago. The Basic Stamp modules from Parallax use a modified version of this language called PBasic. If you are familiar with traditional Basic, you will be able to use PBasic, since these languages are very similar. All you need to understand is the subtle differences between the two languages, so the learning curve is relatively short.

Types of Microcontrollers

There are many different types of microcontrollers that are available for robotic applications. They come in different shapes, sizes, and capabilities. Figure 11-2 shows a wide variety of microcontrollers.

Here, you will learn about some different types of microcontrollers that are well-suited for robot sumo applications. The descriptions will help you to understand the differences between the types of microcontrollers and to choose the microcontroller that will be best for you. At the company's websites, you will find specific details about the capabilities, software, programming instructions, and sample programs.

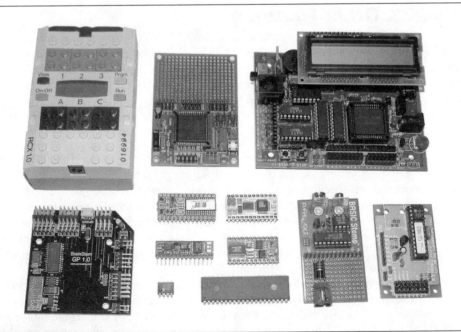

FIGURE 11-2 Various types and sizes of microcontrollers

> **TIP** *There are many more types of microcontrollers than those discussed in this chapter (the topic could fill an entire book). For more information about other types of microcontrollers, search the Internet or read through a good electronics magazine such as* Nuts and Volts.

Parallax Basic Stamp Modules

The Basic Stamp is one of the oldest and widely used stand-alone microcontroller modules on the market. The first Basic Stamp (Basic Stamp 1 Rev. D) module was released in 1992, and since then, Parallax has been developing and releasing new Basic Stamp modules. They are so popular that they have become the standard for comparison with other microcontrollers.

Parallax has a strong educational philosophy; teaching electronics, programming, and how to use microcontrollers is a major part of the company's mission. The Parallax website (www.parallaxinc.com) is packed with free information about how to use microcontrollers and electronics to solve everyday challenges. Parallax has chosen to keep its product line simple yet powerful, so that its microcontrollers are easy to use.

A Basic Stamp module has a microcontroller, clock oscillator, voltage regulator, program EEPROM, and various other support components, all located on a single 0.6 x 1.2-inch, 24-pin module. Figure 11-3 shows three different popular modules. The preprogrammed microcontroller contains a PBasic language interpreter. In essence, the PBasic language interpreter chip (the microcontroller) is just like a miniature operating system. It reads in the program from the EEPROM, one instruction at a time, and then the operating system converts the PBasic program instruction into a set of actions the microcontroller will execute. Once the instruction has been executed, the next instruction is read in, and the process is repeated.

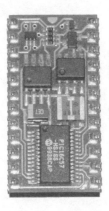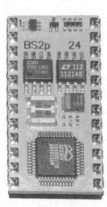

FIGURE 11-3 Basic Stamp 1, Basic Stamp 2, and Basic Stamp 2p modules

Available Basic Stamp Modules

Table 11-1 lists several different Basic Stamp (BS) modules, with a summary of their capabilities. There are several subtle differences between each of the modules. The BS-1 and BS-2 use a microcontroller from Microchip. The other modules use Scenix microcontrollers from Ubicom (www.ubicom.com). The program speeds listed are approximate average speeds. Each program instruction takes a different amount of time to execute. The numbers listed here are pretty good estimates for how fast the programs are running. The "+2" specified with the number of I/O pins means that, in addition to the regular number of I/O pins, there are two dedicated serial communication pins. The EEPROM is listed in bytes, and "8 x 2K" means that you can have eight different programs up to 2 kilobytes (KB) each.

In Table 11-1, you can see that the BS2p-24 and BS2p-40 modules are about three times faster than the BS-2 module, but the processor speed is the same. As discussed earlier, clock speed doesn't tell you everything about how fast the microcontroller executes program instructions. When the Scenix SX48AC is operating in turbo mode, its internal clock does not divide the external clock cycles by four, as do the Microchip microcontrollers. This enables the Scenix microcontrollers to execute lines of code faster than the Microchip microcontroller does (but not four times faster, since there are internal programming differences between the microcontroller brands).

In Table 11-1, also notice that the BS2p-24 and BS2p-40 modules support more program instructions than the other modules do. Added to the BS2p modules are a new set of commands that can be used to directly control liquid crystal display (LCD) modules. They also have the I^2C and Dallas one-wire serial communication protocols, developed by Phillips Semiconductor and Dallas Semiconductor, respectively. The new serial communication protocols are in addition to the RS-232 serial communication method used in all Basic Stamp modules. The BS2p

	BS-1	BS-2	BS-2e	BS-2sx	BS2p-24	BS2p-40
Microcontroller	16C56	16C57	SX28AC	SX28AC	SX48AC	SX48AC
Processor speed	4 MHz	20 MHz	20 MHz	50 MHz	20 MHz	20 MHz
Program speed (approximate)	2,000	4,000	4,000	10,000	12,000	12,000
Number of I/O pins	8	16+2	16+2	16+2	16+2	32+2
EEPROM	256	2K	8 x 2K	8 x 2K	8 x 2K	8 x 2K
RAM	16	32	96	96	166	166
Commands	32	36	39	39	55	55

TABLE 11-1 Basic Stamp Module Features

modules have an added set of polling commands to simulate interrupt capabilities other microcontrollers have.

Basic Stamp Programming

The following is an example of a Basic Stamp program created in the Basic Stamp editor environment (version 1.32), as shown in Figure 11-4. This is a simple test program that can be used to measure the pulse width of a radio-control signal from a remote-control transmitter. This program is useful for developing and calibrating radio-control sumo robots.

```
'Program Name: RCTEST
'This program is used to measure the pulse width of a R/C control signal.
'The signal line from the R/C receiver is connected to pin 0 on the Basic
'Stamp. The R/C control signal ground is connected to the Stamp ground.
'This program will measure the pulse width on Pin 0 and output the results
'to the Debug Terminal. This program can be used with different Basic
'Stamps. All that is needed to be done is change the Period variable to
'correspond to the pulse width period for your stamp. See Below.
ChA      var    word       'pulse width in terms of measured period count
time     var    word       'This is the pulse width in microseconds
Period   var    word       'Stamp pulse width period variable


Period = 75                'For BS2p: Period = 75
                           'For BS2sx: Period = 80
                           'For BS2e: Period = 200
                           'For BS2: Period = 200
Main:
   Pulsin 0,1,ChA          'Read in R/C signal
   time = (ChA/100*Period)+(ChA//100*Period/100)   'Convert to microseconds
   Debug CLS, "Period Counts = ", DEC ChA, CR      'Display period count
   Debug "Pulse Width in Microseconds = ", DEC time 'Display in microseconds
goto Main
```

This program outputs its results to the Debug terminal, as shown in Figure 11-5.

Parallax has produced several different types of development boards for the Basic Stamp modules, which simplify both programming and circuit development. This type of a package combines a microcontroller and a motor controller into one package that is well-suited for robot motion control. Shown here is Parallax's Board of Education and Solution Cubes Motor Mind C Stamp Carrier Board. The Basic Stamp can be programmed directly on the Motor Mind C board,

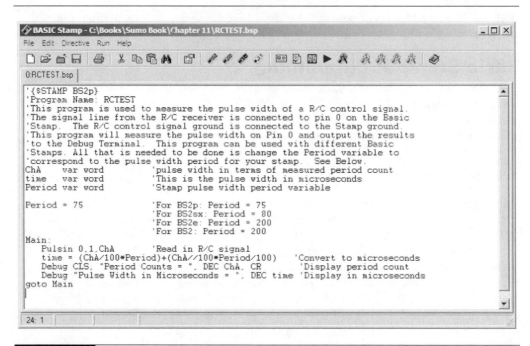

FIGURE 11-4 The Basic Stamp editor version 1.32

and there are output pins for the stamp to control other devices, in addition to two motors using the Motor Mind C module.

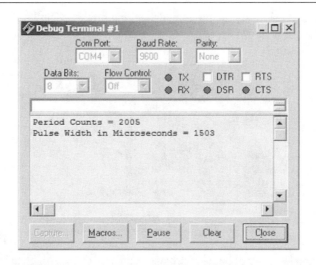

FIGURE 11-5   The Debug terminal for software testing

Several other companies have developed top-quality Basic Stamp-based development boards, such as Lynxmotion's Next Step development board. Also, some companies have included a socket for using a Basic Stamp in their products.

The Parallax Javelin Stamp Module

Parallax has recently released a new module that is programmed in Java, instead of their traditional PBasic, called the Javelin Stamp. The Javelin Stamp is pin-for-pin compatible with the regular Basic Stamps. The Javelin Stamp uses a subset of Sun Microsystem's Java programming language, a popular programming environment for the Internet. Java is a structured, object-oriented programming language, which is similar to the C and C++ programming languages.

Javelin Stamp Features

The Javelin Stamp has 32 KB of RAM and 32 KB of EEPROM. It is equipped with its own integrated development environment (IDE). The IDE can also be used as an in-circuit debugging system, so you can monitor how the Javelin Stamp is performing while it is in the actual final circuit, which can greatly simplify the program debugging process. The Javelin Stamp mounted in its development board is shown here.

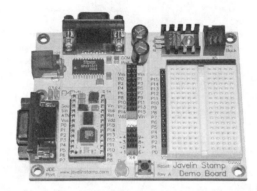

The Javelin Stamp has a set of built-in virtual peripherals (VPs), which are time-dependent applications that don't require continuous monitoring and updating in the main program. The Javelin Stamp has two types of VPs: background and foreground.

The background VPs include a UART, PWM, delta/sigma A/D converter, 1-bit digital-to-analog converter (DAC), and a 32-bit timer. Up to six different background VPs can run simultaneously. This can really simplify programming sumo robots. For example, two different PWM signals can be used to control two motor speeds. The Javelin outputs the PWM commands, and the motors will start running until told differently. Meanwhile, the Javelin is monitoring its sensors to see where the opponent is and working on an attack strategy.

The foreground VPs include pulse counting, pulse-width measuring, pulse generation, remote-control timing, and SPI master control. The pulse-measurement and pulse-generation VPs are similar to the Basic Stamp's Pulsin and Pulsout commands. These are not regular Java commands, so Parallax created a set of special foreground VPs to give the Javelin the same functionality as the Basic Stamp modules.

Javelin Stamp Programming

The following Javelin program works in the same way as the Basic Stamp Pulsout command works. In this example, a 1.5 millisecond (ms) pulse is being sent to pin 12 on the Javelin. Then a 20 ms delay is executed before the loop repeats.

```
//Pulsout.java
//Sample program to send a 1.5 ms pulse to pin 12
//to drive a servo. The pulsout period is 8.68 us.
//1.5 ms / 8.68 us = 172.81 time periods. CPU.Delay
```

CHAPTER 11: Introduction to Microcontrollers

```
//pauses program in 100 us periods.
//20 ms / 100 us = 200 delay periods

import stamp.core.*;


public class pulsout {
   public static void main() {
   CPU.writePin(CPU.pin12,false);
   while (true) {
   CPU.pulseOut(173, CPU.pin12);
   CPU.delay(200);
   }
   }
}
```

For more information about the Javelin Stamp, visit the web site at www.javelinstamp.com.

Basic Micro Basic Atom Modules

In 1996, Basic Micro, Inc., started to manufacture programmers and software compilers for Microchip's PIC microcontrollers. And in 2001, Basic Micro introduced a new set of microcontroller modules called the Basic Atom. There are three different versions of the Basic Atom modules: Atom 24, Atom 28, and Atom 40. The numbers represent the number of pins on the module. Shown here is the Basic Atom 28.

The Basic Atom 24 is pin-for-pin compatible with the Basic Stamp 2 module. All three modules have 8 KB of programming space, 368 bytes of RAM, and 256 bytes of EEPROM. The microcontroller inside the Basic Atom 24 and 28 pin modules is the Microchip 16F876. The Microchip 16F877 microcontroller is used in the Basic Atom 40. The Basic Atom 24 has 16 I/O pins, the Atom 28 has 20 I/O pins, and the Basic Atom 40 has 32 I/O pins. These modules run at 20 MHz, which enables them to execute 33,000 instructions per second.

The Basic Atom modules are programmed using a Basic programming language that is about 90 percent compatible with the Basic Stamp modules. Many Basic Stamp programs can run on the Basic Atom modules without any modifications. What makes the Basic Atoms different from the Basic Stamps are the Basic Atom's special hardware functions and software capabilities.

The Basic Atoms have two hardware PWM modules that can be used to control the speed of motors. Since this is a hardware PWM signal, once it is turned on, the microcontroller can do other things without needing to worry about constantly updating the PWM signal. The Basic Atom has four hardware A/D converters, so you don't need an external resistor and capacitor or an external A/D converter to measure analog voltage. There are two built-in capture-and-compare modules that can be used to count pulses in the background. You can use this capability to monitor encoders on a wheel for closed loop velocity control. The Basic Atoms have three internal hardware timers and true interrupt capability.

The special software capabilities include true IF-Then-Else-ElseIF-EndIF conditional statements and 32-bit math, including 32-bit multiplication. The Basic Atom modules also support floating-point number mathematics and follow the traditional mathematical order of operations (parentheses first, then multiply and divide, and then add and subtract from left to right).

For more information about the Basic Atom products, visit the Basic Micro website at www.basicmicro.com.

The Basic Micro MBasic Compiler

Basic Micro also offers a Basic language compiler. The first Basic compiler Basic Micro produced was called Pic n' Basic, which was a Basic Compiler for the Microchip PIC microcontrollers. Over the years, the Basic Compiler has evolved into the current compiler called MBasic.

Why would you want to use a Basic compiler instead of using either a Basic Stamp or a Basic Atom module? One reason is cost. The microcontroller inside the Basic Stamp and Basic Atom modules costs about one-tenth as much as the entire module. These modules contain a lot of support electronics and software that you may or may not need. Using the internal microcontroller with a dedicated program may be the most cost-effective and space-effective solution.

Another reason to use a Basic compiler is the functionality of the microcontroller. Microchip manufactures more than 100 different types of microcontrollers. Each has its own unique set of capabilities. Why use a 40-pin microcontroller in an application that requires only 4 I/O pins?

Instead of having one microcontroller do everything in your robot, you can use several smaller slave microcontrollers, where each slave microcontroller is selected based on its unique capabilities. For example, you can use a pair of Microchip 8-pin 12C671 microcontrollers as dedicated motor-speed controllers to control the PWM speed and directions of a pair of motors, and serially communicate with the main processor. A dedicated microcontroller for operating infrared sensors can serially communicate with the master controller, and the master controller can be used to monitor and send controls signals to each of the slave processors. This approach can simplify the logic timing, as well as save you a lot of money, because you don't need to use multiple Basic Stamp or Basic Atom modules in the same robot.

The MBasic compiler uses the same programming structure as the Basic Atom modules. The Basic Micro development boards for the Basic Atoms and for the 28-pin and 40-pin microcontrollers are shown here.

Many Basic Stamp programs can be compiled into stand-alone programs for PIC microcontrollers. All of the MBasic commands work with Microchip's PIC 16F876 and 16F877 microcontrollers. With other microcontrollers, the MBasic commands that are available depend on the microcontroller's features. For example, the 16F84 microcontroller does not have any A/D I/O pins, so none of the compiler's A/D commands will work. However, most of the compiler's capabilities can be used with all of the microcontrollers.

Directly programming microcontrollers is one of many options that can be used to build a robot, as discussed in more detail in the "Single-Chip Solutions" section later in this chapter. For more information about the MBasic compiler, visit www.basicmicro.com.

The Acroname BrainStem

Acroname is a small company that was created in 1994 to make robotics easier for people to learn and use. One of the company's unique products is a small microcontroller called the BrainStem.

The BrainStem uses a 40 MHz PIC 18C252 as its core microcontroller. It has four dedicated R/C servo ports that control position and speed. It also has a dedicated port for controlling the Sharp GP2D02 infrared range sensor. The BrainStem has five general purpose digital I/O pins and five A/D inputs. With its 16 KB of memory and the ability to run four different programs simultaneously, the BrainStem is an advanced robotics motion controller.

The BrainStem physical package, shown here, was designed so that stacking multiple BrainStem modules directly on top of one another will not interfere with any of the I/O pins. With the I^{2}C port, up to 126 different BrainStems (or other I^{2}C devices) can be combined into one large robot brain.

The BrainStem uses a programming language called TEA (Tiny Embedded Application), which is a subset of the industry standard ANSI C programming language used on most computer systems. BrainStem's TEA virtual machine programming environment runs on Windows, WinCE, Linux, MacOS, and PalmOS platforms. This allows you to use your favorite computer system to program the BrainStem, instead of needing to use a Windows operating system, as most other microcontrollers require.

For more information about the BrainStem, visit the Acroname website at www.acroname.com.

The LEGO RCX Brick

One of the most popular microcontrollers for beginning robot sumo builders is the LEGO Mindstorms Invention System RCX brick. The RCX brick began as the programmable LEGO brick that was originally developed by Fred Martin, Randy Sargent, and Brian Silverman from the Massachusetts Institute of Technology (MIT) Media Laboratory. The programmable brick was developed to support the MIT autonomous robot class competition that uses LEGOs to build robots to solve a new challenge each year. Since then, the programmable brick has evolved into

CHAPTER 11: Introduction to Microcontrollers

the LEGO RCX brick, and LEGO has developed an entire robotics product line. The LEGO sumo robots compete in mini sumo, lightweight sumo, and 3-kilogram sumo. Some clubs have LEGO-only sumo contests.

The RCX brick has three input and three output ports, an LCD for displaying the program number and status, and four buttons (for turning on/off the RCX, running and selecting programs, and viewing the status of the RCX brick). The standard LEGO sensors include touch sensors, angular-rotation sensors, light-reflective sensors, and a temperature sensor. The outputs are used to drive high-efficiency DC motors. Here is a photograph of the LEGO RCX brick with two touch sensors, a light sensor, and two motors.

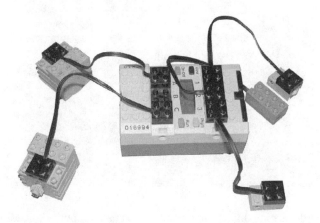

Inside the LEGO RCX is a 16 MHz Hitachi HD6433292 microcontroller with 32 KB of RAM. The actual microcontroller has 43 I/O pins, eight A/D converters, eight input-only pins, and a dedicated serial communication port. The internal microcontroller has far more I/O capability than the regular RCX brick.

The standard programming environment is graphically based. Program functions are shown on a computer screen as different types of LEGO bricks that are assembled like a LEGO robot. In practice, programming the RCX brick is just like building a logic flowchart for the robot. It is a very simple programming environment to learn.

Although the graphical based programming language is very effective in making the RCX brick a good robot controller, it does not provide access to all of the capabilities of the internal microcontroller or advanced logic and math routines. Several third-party programming languages have been developed for the LEGO RCX brick. Some of the popular languages include *Not Quite C (NQC)* by Dave Baum and *LEGOs* by Markus Noga. There have been dozens of books written about advanced programming techniques and methods for expanding the I/O capability of the LEGO RCX brick. A search on the Internet using the key words "LEGO RCX" will yield hundreds of websites explaining all of the different things you can do with the LEGO Mindstorms RCX microcontroller.

Handy Board

The Handy Board, another product developed by Fred Martin of MIT, was developed to be a general-purpose robotic microcontroller. The Handy Board, shown here, uses a 2 MHz Motorola 68HC11 as the core microcontroller, with 32 KB of RAM.

The Handy Board has nine digital inputs and seven analog inputs. An additional I/O pair is used for a 38 kHz infrared reflective sensor. The Handy Board has four built-in motor controllers using the L293D motor-controller circuits and a buzzer for sound effects. It also has a built-in 16-by-2 character LCD for displaying data and robot status.

The Handy Board is programmed using the Interactive C programming language, a subset of the traditional C programming language. The Interactive C language can support 32-bit integer and floating-point numbers and advanced math functions. The Interactive C programming language is a full multitasking environment that can be used to monitor several different sensors and control multiple motors at the same time.

The Handy Board is an old design, but it is still a very popular and effective robot controller. For more information about this microcontroller, visit Gleason Research at www.gleasonresearch.com.

The BotBoard Plus

The BotBoard was developed by Marvin Green, Karl Lunt, Tom Dickens, and Keven Ross to be a small, but powerful, robot microcontroller. It uses the 8 MHz Motorola 68HC811 microcontroller, with 2 KB of programming space.

As shown here, the BotBoard Plus is smaller than the Handy Board because it does not have all of the built-in support hardware features the Handy Board has. It does have a small prototyping area that can be used to add special hardware for robot controls, such as infrared sensors and a L293D motor controller chip.

Three programming languages can be used with the BotBoard: assembly, C, and SBasic. SBasic, developed by Karl Lunt, is a simple but powerful programming environment using the Basic programming language. You can download it for free from Karl's website at www.seanet.com/~karllunt/.

The BotBoard and the SBasic programming language are one of the popular combinations used in many sumo robots throughout the United States. For more information about the BotBoard Plus, visit www.kevinro.com.

Single-Chip Solutions

All of the microcontrollers mentioned in this chapter have a single microcontroller at their core. When building robots, you can use a prepackaged microcontroller board/module, or you can build your robot using a single-chip solution, the core microcontroller.

To get most microcontrollers to work, all you need is the microcontroller, a resistor, and a clock resonator (or a clock crystal and two capacitors). Then all of the other components that you will add will be based on what type of sensors, motor controllers, and other types of I/O you want to use. Figure 11-6 shows a simple microcontroller schematic.

For most robotic applications, just about every single microcontroller made will work, so the selection is more of a personal taste. First, consider the programming environment. Does the microcontroller have a programming language that you are familiar with? If not, do you want to learn a new language so that you can program it? If the programming language that you would like to use is not available directly from the microchip manufacturer, in many cases, it will be available from third parties. For example, Microchip does not have a Basic compiler for its

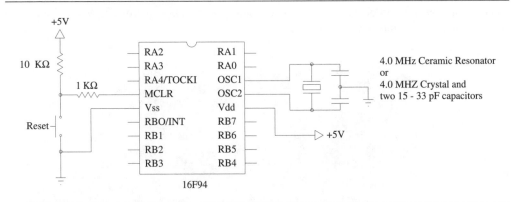

FIGURE 11-6 A simple 16F84 circuit

microcontrollers, but you can get a good Basic compiler from Basic Micro. On the other hand, Atmel does have a Basic compiler for its AVR chips.

The other factor that you should consider is the cost of the programmers and the programming software. Many times, a search on the Internet can yield free compilers for different types of microcontrollers, and third-party companies may sell programmers cheaper than the microcontroller manufacturer. With a little homework, you can select the brand of microcontroller that suits your needs.

The next step is to select the actual microcontroller from the many different types available. First, determine how many I/O pins the microcontroller must have to control your robot. Then determine if there are any special features you want, such as hardware PWM. Then look at the list of available microcontrollers that have these features. Usually, you will have several to choose from. Some people will pick the smallest microcontroller; others will choose the bigger one that has extra capacity. Again, your choice depends on how you want to design your robot.

The company with the largest selection of microcontrollers is Microchip (www.microchip.com), with more than 100 different types. Atmel (www.atmel.com) has an excellent set of microcontrollers that are gaining popularity in the robotics community. Ubicom (www.ubicom.com) sells the extremely fast 75 MHz Senix chips. Cypress Microsystems (www.cypressmicro.com) sells a line of microcontrollers that have a very nice programming environment called Program System on a Chip (PSoC). Motorola (www.motorola.com) offers a wide variety on microcontrollers. A search on the Internet will yield dozens of other microchip manufacturers.

Closing Comments on Microcontrollers

The fastest way to start a microcontroller holy war is to log on to one of the robotic club's e-mail servers and ask "What is the best microcontroller?" You will get about as many different answers as there are different microcontrollers out there. Then you will get to watch the arguments about which is better than another.

For the most part, all microcontrollers will work for sumo robots. Some microcontrollers may require software to execute a function such as serial communication. Some might require external hardware such as a resistor and capacitor to measure an analog voltage.

Robotics is a combination of electronics, mechanics, and software. All of these are combined together in different ways to make the robot. The only answer to "What is the best microcontroller?" is the microcontroller that you know how to use.

When looking for a microcontroller, consider what you will use it for. If your program is only 500 bytes long, who cares if another microcontroller has 32 KB of memory? Also, there is no point in spending more money for a faster microcontroller if it is going to spend most of its time doing nothing. There are applications where speed is critical, such as real-time vision processing, which pushes even the top-of-the-line microcontrollers to their limits.

Sumo robots do not require super computers to compete and win. Quite often, the robot with the simplest microcontroller wins a tournament. In 2002, a LEGO RCX sumo robot made it to the semifinals at the International Robot Sumo Tournament in San Francisco. Though physically outclassed by all of its competition, with some good programming and some luck, it was able to compete well against some of the world's best sumo robots.

If you are new to microcontrollers, I recommend that you start with a prepackaged microcontroller module such as the Basic Stamps. Basic Stamps are used in many champion sumo robots and other combat robots, such as BattleBot class robots. Parallax's smallest microcontroller, the Basic Stamp 1, is still currently being used in several champion sumo robots in both mini and 3-kilogram class robots. Yes, there are other microcontrollers that are faster, have more I/O, have more memory, have advanced mathematics routines, and can multitask. But all of these features do not necessarily make a better sumo robot.

Parallax has designed their software, hardware, documentation, and support products to teach people how to get started with microcontrollers. Once you are comfortable with programming and interfacing with sensors and actuators, you might want to consider using a more advanced microcontroller, such as the Basic Atom, the BotBoard, or even an Atmel MegaAVR chip to control your robot.

Chapter 9 showed what the microcontroller needs to do to control the motors. Here, you learned about the types of microcontrollers. The next chapter covers sensors and will give you a good idea as to what a microcontroller needs to do in order to analyze sensor data. With all of this information, you will be able to select the right type of microcontroller, along with all of the support electronics. For most robots, this is a not a difficult decision, since most microcontrollers will work. Then you can implement your ideas in the software and hardware you've selected. What makes building sumo robots fun is using your creativity, and programming a microcontroller is usually one of the most creative parts, since it helps create your robot's personality.

Sensors: Letting Your Robot See the World

In order for an autonomous robot to be able to effectively compete in sumo, it needs the ability to sense its environment so that it will know where to move and where not to move. Without this ability, your robot will run off the sumo ring on its own. Sensors are not limited to autonomous robots. They can be used in remotely controlled robots to help simplify some of the control actions of the robot.

Sumo robots sensors generally fall into three categories: dohyo edge-detection sensors, opponent-detection sensors, and internal-system-monitoring sensors. Your robot may have only one of these types of sensors, or it might have several sensors in each category working together. This chapter covers dohyo edge-detection and opponent-detection sensors. Internal-system-monitoring sensors monitor the internal functioning of your robot, such as battery voltage, motor temperature, motor current draw, or wheel slippage. Internal system monitoring is not done with most sumo robots because the matches last such a short time.

However, some robot builders still like to know what is going on inside their robot. A search of the Internet using key words such as "Voltage Indicators", "Temperature Sensors", "Current Sensors", and "Wheel Encoders" will provide many different sources of information about these types of sensors, and links to other similar websites.

Edge-Detection Sensors

The most important type of sensor for your robot is the edge-detection sensor. These sensors are used to alert your robot that it has come to the edge of the dohyo, so that it can take evasive actions. Otherwise, it will run out of the dohyo, without any help from its opponent. As a minimum, a sumo robot should have at least one edge detector. This way, the robot will be able to stay on the dohyo by itself.

Edge detectors can be mechanical or optical. Mechanical detectors actually feel for the edge of the sumo ring. Optical detectors look for the edge. This is why the border edge of the dohyo is painted white.

Mechanical Edge-Detection Sensors

A mechanical edge detector is a simple system. It has three components: a mechanical switch, an actuator, and a resistor. Some mechanical switches have a built-in actuator. When wiring a mechanical switch to a microcontroller, you need to consider the true state of a microcontroller input pin. Here, we will look at some simple designs for mechanical edge-detection sensors, and then discuss how pull-up and pull-down resistors can be used to ensure proper voltage on input pins.

A Simple Lever-Action Switch

Figure 12-1 shows how a lever-action switch can be used to detect the edge of the sumo ring. A typical SPDT lever switch is used in this configuration. The roller helps the end of the lever slide across the surface of the sumo ring. The switch should be placed at the bottom of the robot, so that the plunger is fully depressed when the robot is sitting flat on the surface of the dohyo. When the front of the robot goes off the edge of the dohyo, the lever will spring open. The corresponding electrical circuits are shown under the robot.

Most robots will use two switches: one placed at each of the front corners of the robot. This tells the microprocessor which side of the robot went across the edge of the dohyo.

A Knitting Needle Sensor for Long-Range Detection

In the example shown in Figure 12-1, part of the robot must go over the edge of the dohyo in order for the switch to detect the edge. This places your robot at risk of being pushed off the dohyo if the other robot hits your robot from behind. Having a long-range edge-detection sensor will help prevent this from happening.

CHAPTER 12: Sensors: Letting Your Robot See the World

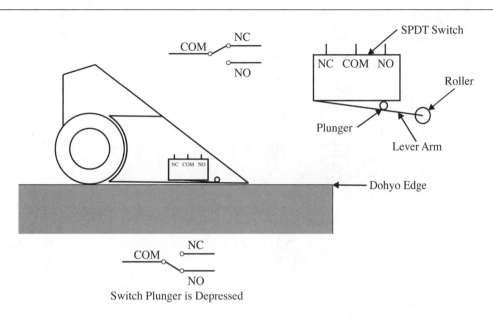

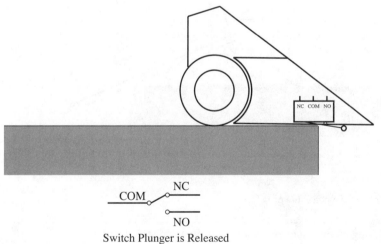

FIGURE 12-1 Schematic of a robot edge-detection switch

A simple sensor idea that Bill Harrison came up with uses an aluminum knitting needle and a lever switch to detect the edge of the dohyo. Figure 12-2 shows a schematic for implementing this type of a sensor.

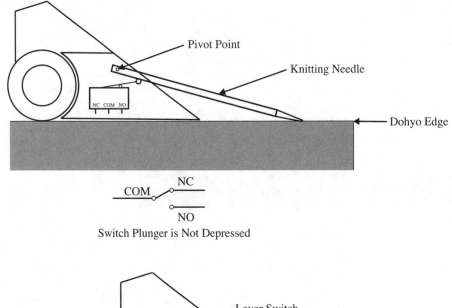

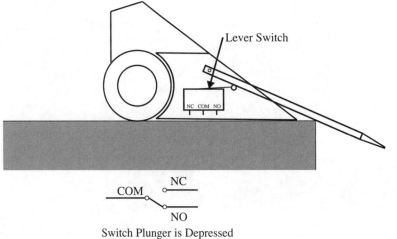

FIGURE 12-2 Using a knitting needle for a long-range edge detector

A small hole is drilled through the knitting needle, near its rear. Then a smaller diameter pin, or a screw with two nuts on both sides of the needle, is used as a small pivot point for the needle to rotate on. A switch is placed below the needle, so that when the tip of the needle is touching the surface of the dohyo, the lever on the switch is not pressed. Then, when the robot moves toward the edge of the dohyo, the tip of the needle starts to slide off the edge of the ring, and the

CHAPTER 12: Sensors: Letting Your Robot See the World

needle rotates downward, pressing on the switch. When the switch is closed, the robot can turn around to continue.

The key considerations with this type of a sensor are that the tip of the needle doesn't touch the ground outside the dohyo and the needle is angled toward the ground (as shown in Figure 12-2). Because the knitting needle is located in front of the robot, it will need to start the contest pointing upwards, so that it will fit inside the size constraints for the sumo weight class. The easiest way to knock the knitting needle forward is to have your robot make a short, sudden movement backward. This will cause the needle to fall forward; then your robot can start moving forward. The electrical setup for this approach is the same as the circuit shown in Figure 12-1.

Proper Voltage for Input Pins

When using microcontrollers, you should never leave an input pin floating. In other words, you should not leave the voltage of an input pin in an unknown state. If nothing is connected to an input pin, that doesn't mean that the voltage at the input pin is zero. In fact, it actually jumps around a couple volts all by itself. There is always a small amount of current (a few micro-amps) that leaks through an input pin.

Since microcontrollers interpret a logic one (high state) as approximately 2 volts (the actual voltage depends on the microcontroller), it is easy for the software to misinterpret a floating input pin state. To avoid this problem, you can use either a pull-up or pull-down resistor. A pull-up resistor is used to hold the input pin in a high state (about 5 volts) when the switch is open. A pull-down resistor is used to hold the input pin in a low state (about 0 volts) when the switch is open. Closing the switch inverts the input pin state.

Figure 12-3 shows how a 10 kilo-ohm (kΩ) resistor is used as a pull-up or pull-down resistor. The 10 kΩ resistor is used to minimize the voltage drop across the resistor so that the input pin will see approximately 0 or 5 volts.

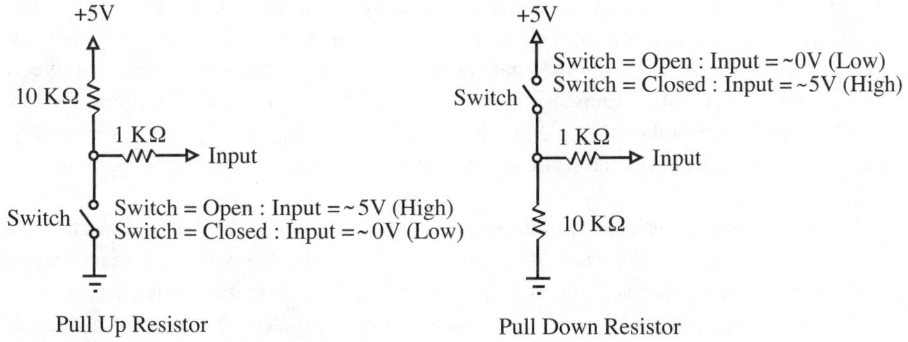

FIGURE 12-3 Pull-up and pull-down resistors are used to ensure proper voltage states to microcontroller input pins.

The 1 kΩ resistor is optional but highly recommended when building test circuits. This resistor is an insurance policy against programming errors. For example, suppose that we are using the pull-up resistor in Figure 12-3 and we accidentally program the input pin as an output pin (a very common occurrence), and the switch closes. Without the 1 kΩ resistor, we will have a direct short to ground, which will usually destroy our microcontroller. The 1 kΩ resistor is there to limit the current from the microcontroller to 5 ma (5 volts / 1 kΩ = 0.005 amp). A few pennies is pretty cheap insurance for a few 1 kΩ resistors to protect a $50 microcontroller.

Optical Edge-Detection Sensors

Optical edge detectors "see" the edge of the dohyo. Since the edge of the dohyo is painted white, all that an optical sensor needs to do is look for the change of color on the surface of sumo ring. This is done by shining a small light onto the surface of the dohyo and measuring the intensity of the reflected light.

> **NOTE** *All colors will reflect light with different amounts of intensity, because different colors will absorb different amounts of the light spectrum. This is why we see different colors. Black and white represent the opposite ends of the reflectance spectrum. White reflects the most light, and black reflects the least amount of light.*

The general operation of this type of sensor is to shine a light directly down onto the surface of the dohyo. The light source is typically a light-emitting diode (LED). Right next to the LED is a sensor that can measure the intensity of the reflected light. The three common types of sensors used in this application are photoresistors, photodiodes, and phototransistors. Here, we will look at designs for photoresistor and phototransistor detectors.

A Photoresistor-Based Edge Detector

A photoresistor is a variable resistor. Its resistance changes based on the intensity of the light that is shining on it. These sensors are commonly called cadmium sulfide (CdS) cells or photocells.

The internal resistance of a photocell is inversely proportional to the light intensity. When these cells are in total darkness, the internal resistance is very high—somewhere between 100 kΩ to 10 mega-ohm (MΩ) (depending on the particular cell). When the photocell is in complete sunlight, its resistance drops down to around 20 kΩ. Figure 12-4 shows how to use a photoresistor to measure reflected light intensity using a voltage-divider circuit.

> **NOTE** *The 470 Ω resistor is used to limit the current going to the LED. Using a smaller resistor will increase the brightness of the LED, thus increasing the reflection sensitivity range. If the microcontroller is supplying the current, the absolute minimum value should be 180 Ω. A smaller value would result in the microcontroller supplying more than 20 ma of current, which can damage the microcontroller.*

CHAPTER 12: Sensors: Letting Your Robot See the World

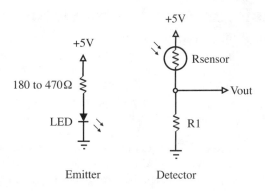

FIGURE 12-4 Using a photocell and an LED to detect the edge of a sumo ring

Equation 12-1 shows what the output voltage (V_{out}) is as a function of the input voltage (V_{in}), the photoresistor (R_{SENSOR}), and the other resistor (R_1). To use this circuit, you will need a microcontroller that has an analog-to-digital (A/D) converter, or use an external A/D converter, such as the ADC0831, a serial A/D converter from National Semiconductor (www.national.com). Using an A/D converter is discussed in the "Sharp Infrared Range Sensors" section later in this chapter.

Equation 1

$$V_{out} = \frac{R_1}{R_1 + R_{SENSOR}} V_{in}$$

The R_1 resistor value is arbitrary. If R_1 is small, the output voltage will not be as sensitive to changing light conditions as it would be if R_1 were large. You will need to experiment with determining the appropriate value for R_1. It is best to start with a variable resistor for R_1, say around 50 to 100 kΩ, and measure the output voltage as you pass the sensor over a black-and-white color transition. Adjust R_1 up or down to get the greatest range in the change in voltages from the sensor. Then measure the value for R_1, and substitute the variable resistor with a fixed resistor that is near the measured value. When the sensor is over the black surface, you will want the voltage to be less than 1 volt, and when the sensor if over the white surface, your will want the output voltage to be greater than 3 volts.

You will want to use a visible-light LED with this type of sensor, since photocells are not very sensitive to infrared light. Also, you will want to shield the sensor from receiving any ambient light. Too much ambient lighting on your sensor can cause your sensors to not to detect the sumo edge correctly or to give false readings from outside sources, such as a camera flash.

The main drawback to using a photoresistor is that it has a relative slow response time when compared to the fast phototransistors. For slower sumo robots, the photoresistors will work fine. Fast robots might go off the edge of the sumo ring before the sensor can detect it.

A Phototransistor-Based Edge Detector

Instead of using a photoresistor, you can use a phototransistor in a similar circuit to detect the edge of the sumo ring. The amount of current that flows through a phototransistor is a function of the amount of light it receives. With a photoresistor, the internal resistance will get smaller and smaller as the light intensity increases. With a phototransistor, the current will increase until the transistor is saturated. At this point, it is like a closed switch, and it cannot conduct any more current by further increasing the light intensity. Figure 12-5 shows a schematic of a phototransistor sensor.

When the transistor is over the black area of the sumo ring, the transistor should not be conducting any current. In this case, the 10 kΩ resistor is operating as a pull-up resistor, so the output voltage is approximately 5 volts. When the sensor is over the white area of the sumo ring, the phototransistor should be fully saturated, so the output voltage drops to zero. However, this is an ideal situation that may be difficult to accomplish.

The 1 kΩ variable resistor is used to adjust the intensity of the LED. By increasing or decreasing the intensity of the LED, the output voltage will increase or decrease over both the black and white areas. You can use an A/D converter to read in the raw voltage, or you could set the output voltage to be at least 3 volts over the black area and 0 volts over the white area, and directly connect to a microcontroller. Most microcontrollers will interpret voltages above 1.5 volts as a logic high signal, and voltages below 1.5 volts as a logic low signal. So, if the sensor is outputting at least 3 volts over the black area and less than 1 volt over the white area, the microcontroller can use this change as the indicator that the sensor has passed over a black-to-white color transition.

The advantage of a phototransistor circuit is that it can detect the color change very quickly—in microseconds. This is in contrast to the photoresistor method, which can take a few milliseconds to detect the change. Also, the results for both methods will change based on the distance between the sensor and the surface of the dohyo. This can be a good thing or a bad thing, depending on how much sensitivity you want. For a very sensitive and narrow response

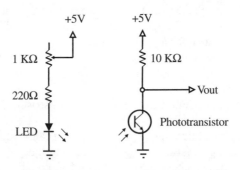

FIGURE 12-5   A simple phototransistor setup for edge detection

band, place the sensors close to the dohyo; increasing the distance decreases the sensitivity. It is usually best if you can set your sensors between 1/16 to 1/8 inch from the surface of the dohyo (this spacing is also dependent on the type of sensor you are using).

Infrared Edge Detectors

With phototransistors, unlike photoresistors, you can use either infrared sensors or visible-light sensors. Infrared sensors have the advantage of being less susceptible to ambient light interference than visible-light sensors. Their disadvantage is that you cannot visually see infrared light, thus it is difficult to determine if the circuit is on or off. In order to see infrared light, you need some sort of a detector. Digital cameras are sensitive to infrared light, so if you see a bluish color light from an LED on the camera and you can't see any light coming from the LED, you will know the infrared LED is working. Another method for detecting infrared light is to use an infrared sensor card. These cards have a special coating that changes color when infrared light hits it. You can obtain infrared sensor cards at Radio Shack (part number 276-1099).

Remember that you can only use infrared LEDs with infrared phototransistors, and visible-light LEDs with visible light phototransistors. But mixing the two types of LEDs will result in poor performance from the sensors.

When your buy infrared LEDs and sensors, make sure they are both tuned to the same infrared wavelength. Most infrared LEDs and sensors operate at either 880 nano-meter (nm) or 940 nm. Although they will work together if they are on different wavelengths, the peak sensitivity is when the LED and the phototransistor are tuned to the same wavelength.

Instead of using a separate infrared LED and phototransistor, you can use a prepackaged phototransistor and LED that are molded together in a single package. These packages are small and very convenient to use in a sumo robot. They come in a wide range of sizes and configurations. Figure 12-6 shows a photoresistor, a phototransistor, an LED, and an infrared phototransistor-LED package. The phototransistor-LED package shown here is made by QT Optoelectronics *www.qtopto.com* (part number QRD 1114) and can be used in the circuit shown in Figure 12-5. All you need are the resistors to complete the circuit.

Lynxmotion (www.lynxmotion.com) sells a complete infrared reflection sensor (part number SLD-01), as shown here. This package uses an Optek OPB745 infrared reflective sensor and a differential comparator to output either a 0 or a 5-volt signal. This sensor has a small, red LED that is used to indicate if the sensor is detecting a white or black surface, and it has a variable resistor to adjust the sensitivity of the sensor. Acroname (www.acroname.com) sells a similar sensor package.

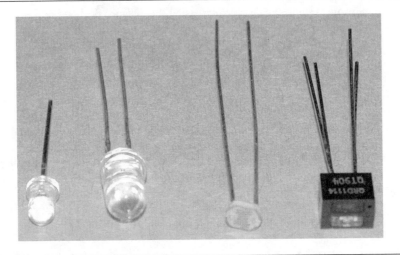

FIGURE 12-6 Reflective edge sensors using a phototransistor, an LED, photoresistor, and a prepackaged phototransistor-LED sensor. The phototransistor looks like a miniature, clear-color LED.

Opponent-Detection Sensors

A robot can randomly run back and forth across the sumo ring. As long as it stays in the dohyo, it will eventually make contact with its opponent or vice versa. However, since the goal of robot sumo is to find your opponent and push it out of the sumo ring, it would be better if your robot could see its opponent. Ideally, you will want to know where the other robot is before your robot comes into contact with it. By knowing where the other robot is, you can position your robot to gain the competitive pushing advantage.

There are many different methods that you can use to find and locate your opponent. As with the edge detectors, there are optical and mechanical opponent detectors.

Mechanical Opponent-Detection Sensors

Although optical methods for detecting opponents are the most popular, mechanical methods are very reliable and effective. Mechanical opponent detection is more like feeling for your opponents rather than seeing your opponent, so some initial contact is required.

One mechanical detection method is to place lever switches behind a hinged front scoop on your robot. With this configuration, your robot can move slowly across the dohyo, looking for its opponent. When the front scoop moves because it hit another robot (thus activating a switch under the front scoop), your robot can then apply maximum power to the motors to push its opponent out of the ring. Many robots use switches under the front scoop as a backup method to detect an opponent if the optical sensors fail to detect the other robot.

The knitting needle-style sensor, similar to the one described earlier for edge detection, offers one of the best ways to feel for your opponent. Instead of using only one switch to detect for the edge of the dohyo, you can add a second switch to detect for opponents when the knitting needle hits an opponent. To do this, the hole in the knitting needle that is used for the pivot must be larger than the pivoting pin/screw. The needle needs to have some lateral slope when moving on the pivot. Shown here is a two-switch configuration used in a robot built by Bill Harrison. The vertical switch is for detecting the edge of the sumo ring, and the horizontal switch is for detecting an opponent.

In this configuration, the horizontal switch can only detect an opponent that strikes one side of the knitting needle. When a robot hits one side of the needle, the end of the needle will rotate into the switch, thus closing it. When a robot hits the other side of the needle, the end of the needle will rotate away from the robot. This isn't a bad thing. Your robot can spin around in one direction, and when the needle strikes the opponent, it will activate the horizontal switch. Then your robot will stop spinning and rush forward to push its opponent out of the sumo ring. The knitting needle not only detects where the opponent is, it helps capture the opponent.

An alternative version of this type of configuration is to use two arms that rotate sideways, instead of one arm rotating forward. With mirror-image switch configurations, your robot can spin clockwise or counterclockwise and will be able to detect its opponent with both arms. And with two arms, it will be easier to corral the opponent off the sumo ring.

Infrared Reflective Sensors

One of the most popular methods for finding an opponent is to detect reflected light from the opponent. This process requires a light source and a sensor facing in the same direction. The light is transmitted in front of the robot, and the detector is looking for any reflected light. Since light will reflect off almost any surface, a sensor can be used to detect the reflected light.

NOTE *Reflective sensors won't work if the surface can absorb the light or the light can pass through the surface. All materials will absorb certain light wavelengths, which is why all materials have a certain color. Other materials are transparent to all light (for example, a clear glass window), a percentage of the light (semitransparent materials), or transparent to specific wavelengths of the light spectrum.*

Because ambient light usually contains all of the visible light wavelengths, including the infrared wavelengths, it is difficult for a sensor to distinguish between the surrounding ambient light and the reflected light the sensor is seeking. To solve this problem, the transmitted light is modulated at some frequency that is not naturally found in the ambient light conditions. Then the sensor is hooked up to a band-pass filter that is tuned to ignore all of the light except for the specific modulated light frequency.

Virtually any light color can be used in a reflecting light sensor configuration, as long as the sensor can detect the particular wavelength of light. Modulated infrared light is very popular because of the remote-control industry. Most remote-control TVs, VCRs, DVD players, radios, and personal data assistants use modulated infrared sensors to control their operation. All of these devices have a sensor that is designed to receive a specific modulated infrared light signal. All handheld remote-control units transmit a modulated infrared light signal. All of these sensors work well with direct modulated light and reflected modulated light from the remote-control unit. You probably have noticed that you can change the channel on your TV set by pointing the remote-control unit at the wall. The reflected light off the wall can control your TV set.

Inside these devices is a simple infrared detector package. The package contains all of the band-pass filters, amplifiers, and control logic to act as a stand-alone infrared detector. Figure 12-7 shows several different versions of the same type of a sensor made by Sharp, LiteOn, and Panasonic. Depending on the model number, these detectors are tuned so that their peak sensitivity (also known as the *center frequency*) is at 36.7 kHz, 38 kHz, 40 kHz, or 56.9 kHz. Because these sensors are very common, they can be found at most electronic parts stores and are easy to implement. They are very popular for robotic proximity/object detection systems.

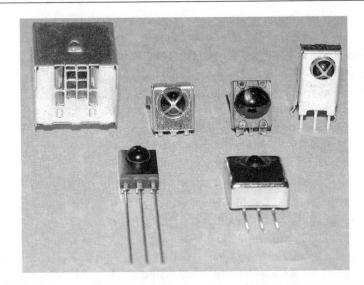

FIGURE 12-7 Various modulated infrared sensor packages from Sharp, LiteOn, and Panasonic

CHAPTER 12: Sensors: Letting Your Robot See the World

An Infrared Circuit

An infrared object detector consists of an infrared LED, a modulated frequency generator (an oscillator), and an infrared receiver module. Figure 12-8 shows a schematic for a simple infrared object detector using a few commonly found components. This circuit uses a single 74HC04 CMOS hex inverter to generate the 40 kHz modulated signal, and the switches for turning on and off the modulated infrared LEDs. The potentiometer R1 is used to adjust the modulated frequency. Resistors R4 and R5 can be decreased in value to increase the range of the detector or vice versa.

CAUTION *Remember to make sure that the infrared LEDs and infrared receiver module that you select are both sensitive to the same infrared wavelength. Also remember not to exceed the maximum current limits of the LEDs.*

The Sharp detectors that come inside a metal container must have the case grounded to the rest of the circuit. The Sharp GP1U52 and GP1U58 series infrared receiver modules are the most commonly used sensors. The Panasonic 4602 series infrared receiver modules are becoming more popular, since they are less sensitive to visible light than the Sharp detectors are, and they are less than half the size of the Sharp detectors.

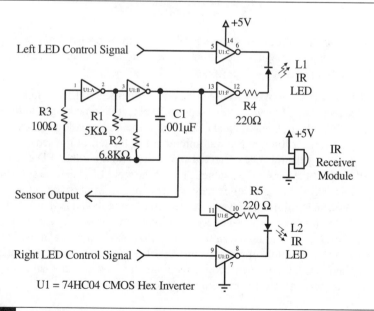

FIGURE 12-8 Schematic of an infrared object detection system

Once the circuit is built, you should put a small tube around the infrared LEDs, if you are not using narrow-focus (+/- 10 degree) infrared LEDs. This will prevent any infrared light from going straight from the LED to the detector. Although most of the infrared light is projected in front of the LED, a small fraction of the light goes sideways, which can cause false readings from the sensor. The ideal tubes are the aluminum or brass tubes you can obtain at your local hobby store. Choose a tube that is slightly larger in diameter than the diameter of the LEDs you are using.

Shown here is this detector circuit used in a mini sumo robot. Notice the brass covers over the LEDs. This circuit uses the Sharp IS1U60 38 kHz infrared receiver from HVW Technologies (www.hvwtech.com). A small amount of aluminum foil tape is covering the back and sides of this sensor to further help reduce the sensor from detecting stray infrared signals for other sources. The Sharp IS1U60 detector requires a polarized 47 µF capacitor to be placed between the positive and negative terminals from the sensor, and a 47 Ω resistor between the 5-volt external battery supply and the Vcc on the sensor.

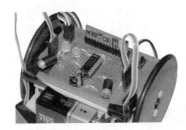

The output voltage from these sensors is 5 volts when the sensor is not detecting anything. As soon as it detects a proper signal, the output voltage will drop to zero until the sensor stops receiving a proper modulated signal.

Using the Infrared Circuit   To use this circuit, aim the two LEDs slightly apart so that you will have a wider field of view for your sensor. To detect if something is to the left of the detector, turn on the left LED, and then check to see if the output voltage from the receiver module dropped to zero. If it did, there was something to the left of the sensor. Then repeat this for the right LED. If the left LED detected something and the right LED did not, there is something to the left of the sensor. The same is true if the right LED detects something. If both LEDs detect something at the same time, there is something in front of the detector.

Depending on which infrared receiver module you use, the modulated infrared LED must be on between 400 to 600 µs to allow the receiver module to stabilize. A shorter LED on period is more likely to result in a false signal. In the real world, there is a lot of "noise" in all signals. Thus, it is better to take a sample of readings instead of relying on a single measurement. One method of sampling is to take five consecutive readings. If you get more than three hits, there is a greater probability that an object is in front of the sensor.

```
'IRPD Demo
'This is a Basic Stamp 2 demonstration for using the infrared
```

CHAPTER 12: Sensors: Letting Your Robot See the World

```
'proximity detector. The infrared receiver module's output sensor
'line is connected to the Basic Stamp's I/O pin number 2.
'When the detector does not see a signal, the input result is always
'high. When the detector detects the reflected signal, the input result
'is zero. This program samples each side of the detector 5 times. A
'result of 4 or 5 means there is nothing there. A result of 0, 1, or 2
'means that there is something on that side of the detector.

LeftLED     con    0       'Stamp I/O pin number where the left infrared
                           'control line is connected to
RightLED    con    1       'Stamp I/O pin number where the right infrared
                           'control line is connected to
Delay       con    750     'Use 300 for 600 ms delay on BS2 and BS2e
                           'use 750 for 600 ms delay on BS2sx
                           'use 800 for 600 ms delay on BS2p
LeftCount   var    word    'counter for the left side detection
RightCount  var    word    'counter for the right side detection
I           var    word    'loop counter variable

Main:      'main loop
    low LeftLED             'Set left LED output to zero
    low RightLED            'Set right LED output to zero
    gosub ReadLeft          'Check left side of the detector
    gosub ReadRight         'Check right side of the detector
    Debug cls, Dec ? LeftCount    'output left side results
    Debug Dec ? RightCount  'output right side resuts
    Pause 250               'Pause to allow time to read the results
goto main

ReadLeft:     'Check left side to the detector
    LeftCount = 0           'Set left side counter to zero
    for i = 1 to 5          'Do this 5 times
       pulsout LeftLED,Delay  'Send a 600 ms pulse out to the left
                              'infrared LED
    LeftCount = LeftCount + in2  'Increment counter with detector
                                 'result
    next
return

ReadRight:    'Check Right side to the detector
    RightCount = 0
    for i = 1 to 5
       pulsout RightLED,Delay
```

```
      RightCount = RightCount + in2
   next
return
```

Adjusting the Circuit Sensitivity   The peak sensitivity of an infrared receiver module is at its rated frequency. For example, a 40 kHz infrared receiver module has its peak sensitivity at 40 kHz. All of the sensors are still functional when receiving a signal that is plus or minus 5 kHz from the center frequency. The further away the actual modulated frequency is from the center frequency, the less sensitive the sensor becomes. Thus, a higher intensity light source will be required to trigger the sensor. With this knowledge, you can adjust the sensitivity of this circuit by shifting the modulated infrared LED frequency away from the center frequency.

The reason it may be important to be able to adjust the sensor's sensitivity is that the detector circuit will detect white objects that are much farther away than black objects. Also, the ambient lighting at an actual competition is usually different from the ambient lighting where the robot was built and tested. Thus, the sensors usually respond differently in the different ambient lighting conditions. Having the ability to adjust the sensitivity of the detectors improves the overall performance of the robot.

Infrared Sensor Variations

Lynxmotion sells two different versions of infrared sensors, which are shown here. The older version, the IRPD-1 (IRPD stands for Infrared Proximity Detector), is a small kit that uses the same circuit as shown in Figure 12-8. This circuit uses the Sharp GP1U581Y infrared reflective sensor that has a center frequency of 38 kHz. Lynxmotion's new sensor, the IRPD-2, is a preassembled circuit board using surface mount components, so its physical size is half that of the original sensor. It has two red LEDs to indicate which infrared LED is currently sensing an object. This board uses the 74HC00 Quad 2-Input NAND gate instead of the traditional 74HC04 hex inverter.

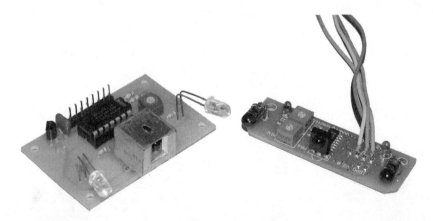

Functionally, both of Lynxmotion's infrared sensors work the same way. The difference is that in the original IRPD-1, the infrared LED modulation frequency is the only adjustment that

CHAPTER 12: Sensors: Letting Your Robot See the World

can be made with the variable resistor. The new sensor has individual control for both LED intensity (brightness) and the frequency sensitivity of the infrared receiver module.

Using a 74HC04 is not the only way to generate a 40 kHz modulation signal for the infrared LEDs. You can use any source that can generate this signal. Many people use the old fashioned 555 timer chip from National Semiconductor, or they program a 16F84 microcontroller from Microchip (www.microchip.com).

TIP *The Parallax website (www.parallaxinc.com) has some documentation on how to wire and program an infrared proximity detector using the 555 timer chip. Go to the Parallax documentation page and download the robotics manual version 1.4 (version 1.5 or newer uses the Freqout function to generate the 40 kHz signal instead of the 555 timer solution).*

One of the advantages of custom programming a dedicated microcontroller for this type of work is that the microcontroller can generate the 40 kHz signal, alternate which LEDs are active, read in the results, conduct statistical analysis on the measurements, and output the results to a main microcontroller. This can save the main microcontroller from doing all of the work, so it can do other things.

A search on the Internet using the keywords "infrared proximity detector" will yield hundreds of websites. You'll find various techniques for object detection using the infrared receiver modules as described here, including places that have programs and schematics for programming a custom microcontroller-controlled infrared detector.

Sharp Infrared Proximity Detectors

The infrared sensors discussed in the previous section are used to detect whether there is a reflected infrared signal hitting the receiver module. These types of circuits are very effective and are used in many different robots. However, the receiver modules used in these circuits were not designed for these applications.

The Sharp GP2D05, GP2D15, and GP2Y0D02YK sensors, shown here, were designed for reflective object detection applications. All of these sensors can be obtained at Acroname (www.acroname.com). These sensors are similar to the infrared proximity sensors described in the previous section. They use a modulated infrared LED, and the LED and the receiver sensor are side by side. They use a band-pass filter to filter out unwanted signals, and the output signal is binary (either on or off). The main difference in these sensors is that they use a small lens to focus the reflective signal on a sensor array.

These sensors have an adjustable maximum range for detecting objects. (The maximum range adjustment is controlled by a small screw at the back of the sensor.) Having a maximum detection range prevents the robot from focusing on objects that are outside the sumo ring. Table 12-1 shows the range of these three sensors, as well as their factory preset maximum range.

 The maximum input signal voltage for the GP2D05 is 3 volts. Exceeding this voltage will damage the sensor.

If two sensors are mounted side by side, and only one sensor is detecting a return signal, how do you determine which sensor's infrared LED is hitting a target, since the reflected light could be hitting the other sensor? Having the ability to turn on and off the sensor is desirable in these situations, since you can alternate which sensor is on at any one time. The GP2D05 can be turned on and off by a microcontroller; the other two sensors are always on.

Wiring and Testing the Sharp Infrared Sensors

Figure 12-9 shows how to wire a GP2D05 to a Basic Stamp 2. The two 1 kΩ resistors are used as a voltage divider (see Figure 12-4), so that the input voltage to the sensor does not exceed 3 volts.

Equation 12-1 shows that the output voltage from the voltage divider will be 2.5 volts, since the voltage from the Basic Stamp pin is 5 volts. This 2.5-volt signal is less than the 3-volt maximum input voltage for the sensor. To use this sensor, the input signal voltage (V_{in}) must be set high (2.5 volts) for a minimum of 1 millisecond, then dropped low (to 0 volts) for 56 milliseconds. At the end of the 56 milliseconds, the output voltage from the sensor (V_{out}) is checked. If there was an object that reflected the infrared light, the V_{out} will be 0 volts. If no object was present, the output voltage will be approximately 5 volts. The following program is used to test the GP2D05 sensor.

Sensor	Range	Preset	Notes
GP2D05	10 to 80 cm 3.94 to 31.50 in.	24 cm 9.45 in.	Externally turned on and off
GP2D15	10 to 80 cm 3.94 to 31.50 in.	24 cm 9.45 in.	Always on
GP2Y0D02YK	20 to 150 cm 7.87 to 59.06 in.	80 cm 31.50 in.	Always on

TABLE 12-1 General Specifications for the Sharp Binary Distance-Measuring Sensors

CHAPTER 12: Sensors: Letting Your Robot See the World

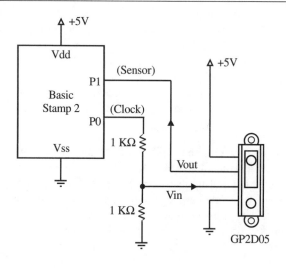

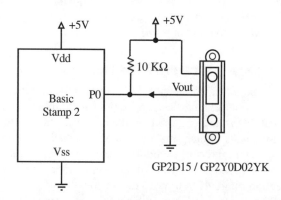

FIGURE 12-9 Schematic drawings for hooking up the GP2D05, GP2D15, and GP2Y0D02YK sensors

```
ready    con   0    'Vin (clock) pin on sensor
sensor   con   1    'Signal pin from sensor
Main:
  high ready      'Set clock pin high for 2 ms to tell sensor to
  pause 2         'start taking a measurement
  low ready       'Set clock low and wait at least 56 ms for
```

```
   pause 56            'measurement
   if in1 = 0 then Detected   'If the sensor detected something the
   debug CLS                  'output signal voltage will be zero
   goto Main           'take another measurement
Detected:              'Detected an object loop
   debug "Detected", CR   'Output to debug screen the result
   pause 250              'pause 1/4 second to read result
goto main              'take another measurement
```

Figure 12-9 also shows how to wire the GP2D15 and the GP2Y0D02YK range sensors. Since these sensors are always on, all you need to do is periodically check the signal coming from the sensor. A 10 to 12 kΩ pull-up resistor must be used on the output signal line from the sensor. The sensor's signal is updated every 38 milliseconds, so there is no point is checking the output from these sensors more frequently than this. When there is no object within its detection range, the output signal is at 0 volts; when the sensor detects an object, the output signal jumps up to 5 volts.

> **TIP** *The black plastic case in which these sensors are mounted is actually electrically conductive. If your sensors start giving you some strange random results, grounding the case will usually clean up the results.*

Using the Sharp Infrared Sensors

These Sharp sensors are small, fast, and reliable for detecting the presence of an opponent near your robot. You can use the shorter range sensors on mini sumo robots, since there is no point checking for an opponent that is farther away than the diameter of the mini sumo ring.

Note that these sensors have a much smaller field of view than the infrared proximity sensors built using the TV remote-control receiver modules, described earlier in this chapter. This means that you will need to use more sensors or you will need to do more scanning. Also, these sensors have a minimum detection range. If you try to detect something that is inside the minimum detection range, you will get some random results that do not correlate to the object's position. This can mislead the controls inside your robot. Thus, you should place these sensors on your robot in such a way that an object cannot get inside the minimum detection range (see the "Sensor Placement" section later in this chapter).

Also, these sensors require very tiny connector plugs for connecting to the outside world. These connectors are made by the Japan Solderless Terminal Manufacturing Company. The GP2D05 uses the four-wire connector (part number S4B-ZR), and the other two sensors use the same three-wire connector (part number S3B-PH). Fortunately, you can obtain all of these connectors at Acroname as well. Their part numbers are R9-JSTCON and R47-JSTCON-2, respectively.

CHAPTER 12: Sensors: Letting Your Robot See the World

Infrared Proximity Sensors as Long-Range Edge Detectors

The infrared proximity sensors actually make pretty good long-range edge detectors. These sensors are sensitive enough to notice a difference between a black and white object. If you point the sensor (both LED and receiver module) down to about 45 degrees from the surface, these detectors will detect the black-to-white color transition. You might need to increase the sensitivity or the intensity of the infrared LEDs to improve the reliability of these sensors. You will be amazed at how well this works.

One thing to keep in mind when using this approach is that you need a way to distinguish between the white line and an opponent. You don't want your robot to run away from an opponent thinking it is a white line. You will probably need to have a second sensor that looks forward, so when the forward detector detects nothing, the edge detector is used for detecting the edge. If the forward sensor detects an opponent, edge sensor results could be ignored.

Sharp Infrared Range Sensors

The Sharp infrared proximity sensors discussed in the previous section will only tell you if there is an object within their detection range. They do not provide any information about how far the object is from your robot. With Sharp's range sensors, you can get the information you need to determine how far your opponent is from your robot. This can be very handy in competition. If your opponent is at the other side of the dohyo, your robot could execute a flanking maneuver to get behind the other robot. If the sensor detects that the opponent is only a few inches away from your robot, you will want your robot to quickly spin around to face its opponent.

Sharp's set of range sensors—part numbers GP2D02, GP2D12, GP2D120, and GP2Y0A02YK—will output range information. These sensors don't actually calculate distance; instead, they output different voltages that can be converted into a distance inside the robot's microcontroller. Their physical appearance is identical to the Sharp infrared proximity sensors shown in the previous section (the GP2D120 and GP2D12 look alike, but the GP2D120 is designed for short-range detection applications). Table 12-2 shows the range specifications for these sensors.

The input signal voltage to the GP2D02 must not exceed 3 volts, or the unit will be damaged.

How the Range Sensors Work

The range sensors use an optical triangulation method to detect the range of an object. Figure 12-10 illustrates how this works. A modulated infrared light emitted from the sensor strikes the object and bounces back toward the sensor. The returning infrared light goes through a lens that focuses

Sensor	Range	Notes
GP2D02	10 to 80 cm 3.94 to 31.50 in.	Externally turned on and off, and the output is an 8-bit serial number
GP2D12	10 to 80 cm 3.94 to 31.50 in.	Always on, and the output is an analog voltage
GP2D120	4 to 30 cm 1.57 to 11.81 in.	Always on, and the output is an analog voltage
GP2Y0D02YK	20 to 150 cm 7.87 to 59.06 in.	Always on, and the output is an analog voltage

TABLE 12-2 Range Specifications for the Sharp Analog Distance-Measuring Sensors

the light onto a position sensitive device (PSD). The light beam that is emitted and reflected to and from the sensor forms a triangle.

By knowing the distance between the LED and the sensor, and the angle of the LED with respect to the PSD, you can determine the distance of the object, because a triangle with two equal sides is created. The sensor outputs different voltages for the reflected distances.

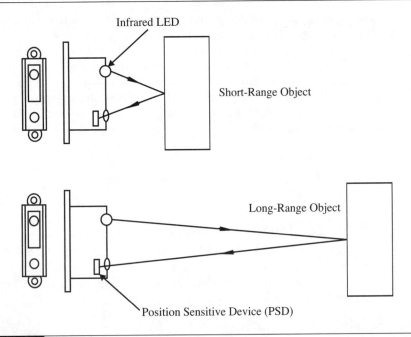

FIGURE 12-10 Sharp range sensors use optical triangulation to detect the range of an object.

Figure 12-11 shows how the output voltage from the GP2D12 changes with the reflected distance. The sensor will output a voltage when the object is between 0 to 4 inches (10 centimeters) from the sensor. These voltages are invalid, and the sensor should be placed on the robot in such a way to avoid being able to detect anything in this range. The output voltages becomes constant (approximately 0.4 volt) when the distance of the object exceeds 31 inches. Between 4 and 31 inches, the output voltage drops off as the distance increases.

Calculating Distance

Converting the voltage into a distance can be challenging, since the output voltage and distances are not linearly proportional. One method that you can use to convert voltages into distances is through a lookup table. Another method is to plot the distance and voltages in a spreadsheet program, such as Microsoft Excel, and create a trendline equation using the Add Trendline option in Excel. For example, Equation 12-2 calculates distance (in inches) as a function of the output voltage (V_o). This equation was generated from published GP2D12 data sheets. This equation is not easy to use. A lookup table is probably the easiest method to implement.

Equation 2

$$Distance = \frac{11.366}{V_0^{1.191}}$$

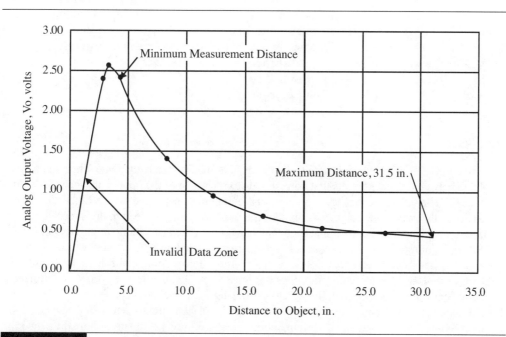

FIGURE 12-11 Voltage and distance values for the GP2D12 sensor

The GP2D02 sensor internally converts the voltage into an 8-bit digital signal that can be read into a microcontroller like serial data. This sensor wires up to the Basic Stamp, just like the GP2D05 (see Figure 12-9). Remember to use a voltage divider to reduce the clock voltage going into the GP2D02 to less than 3.0 volts. The following program uses the Basic Stamp's Shiftin function to read the data from the sensor and output the results to a Debug window on a computer screen.

```
'GP2D02 Demo
'This program will read in the 8 bit digital data from the Sharp GP2D02
'Sensor. The input voltage signal to the sensor is connected to the Stamp
'as the clock line and the sensors output data line is connected to the
'Stamp's sensor line. The output results will be approximately 220 for
'objects at 4 inches, and about 60 for objects that are out at the
'maximum range.
clock       con   0         'Sensor clock line on I/O pin 0
sensor      con   1         'Sensor data line on I/O pin 1
Distance var   byte         'Variable for storing the distance result

Main: 'Main program
    high clock              'Set clock high to wake up the sensor
    pause 2                 'Keep signal high for a minimum of 1 ms
    low clock               'Set low to tell the sensor to take a measurement
loop:                       'It will take approximately 70 ms to take a
    if in1 = 0 then loop    'measurement and convert it into digital format
    shiftin sensor, clock, 2, [Distance]  'Read in data
    high clock              'Set clock high again
    debug DEC ? Distance    'Output result
    pause 250               'Pause 1/ 4 second to make it easier to see
                            'the result
goto Main                   'Start the process over again
```

The GP2D12, GP2D120, and GP2Y0A02YK sensors are electrically the same. They will output a voltage between 0 and 2.5 volts that corresponds to the range of the opponent. The GP2D120 and the GP2Y0A02YK sensors have a very similar output plot shape (see Figure 12-11). The main difference is the maximum range distance. For the most part, all you need to do is rescale the horizontal axis so that the maximum range matches up with the last data point on the plot.

Since the output from these sensors is a voltage, you need an A/D converter to read the measured values. Some microcontrollers, such as Basic Micro's Atom controllers, have built-in A/D converters. If the microcontroller does not have an A/D converter built in, you will need a separate converter. For example, the ADC0831 (a single-channel A/D converter from National Semiconductor) is an 8-pin integrated circuit that can be placed directly between the sensor and the microcontroller without any additional components. Figure 12-12 shows how the sensor and the ADC0831 are wired together with a Basic Stamp.

CHAPTER 12: Sensors: Letting Your Robot See the World

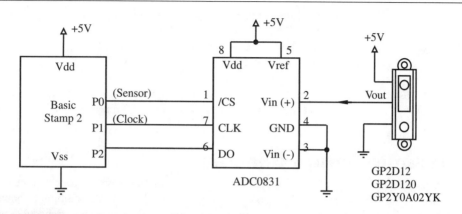

FIGURE 12-12 The ADC0831 A/D converter can be used to monitor the output voltages from the GP2D12, GP2D120, and GP2Y0A02YK sensors.

The following program shows how easy it is to use the serial ADC0831 converter with a Basic Stamp 2.

```
'GP2D12 Demo using the ADC0831 Analog to Digital converter
'The output voltage pin from the GP2D12 is connected
'to pin 2 {Vin(+)} on the ADC. The output from the GP2D12 will be
'between 0 and 2.5 volts. Thus the value collected will be between
'0 and 127. Values at maximum range are approximately 20. At distances
'past that, the values will drop to 1 or 2. Minimum distance values
'should be approximately 120-130. These values can be converted to
'a voltage by multiplying the DIST result by 5 and then dividing by 255.

CS         con  0         'ADC CS pin is connected to I/O pin 0 on the Stamp
CLK        con  1         'ADC CLK pin is connected to I/O pin 1 on the Stamp
DO         con  2         'ADC Data pin is connected to I/O pin 2 on the Stamp
Dist       var  byte      'Measured result (value between 0 and 255)

high       CS             'Make sure the ADC is not being used
low        CLK            'Set clock line low

Main:
    low CS                'Start the conversion process
    shiftin DO,CLK,2,[dist\9]   'Read in 9 bits. The first bit is a
                                'dummy bit. The first clock cycle is
                                'used to initialize and convert the
```

```
                          'voltage into a digital form in the
                          'ADC0831.
     high CS              'Stop ADC
     low CLK              'Reset clock line
     debug dec ? dist     'output result to Debug screen
     pause 250            'pause for 1/4 second to read the
                          'result. Omitted in the actual program
 goto Main                'Take another measurement
```

Ultrasonic Range Sensors

Long-range detection using optical methods is not the only way to see where your opponent is. Ultrasonic methods also work well. Instead of sending out a modulated infrared light to bounce off the opponent, a high-frequency sound wave is transmitted, and an ultrasonic transducer is used to detect the sound. Most ultrasonic transducers will transmit the sound wave at 40 kHz, which is outside the range of hearing for humans.

If you send out a pulse of sound, it will travel outward. Any object in the way will reflect a portion of the sound back to the sender. By measuring the time it takes for the sound to travel out and back, you can calculate the distance to the object. Since the speed of sound at sea level is approximately 13,560 inches (34,442 centimeters) per second, the distance can be calculated by multiplying the time by the speed of sound and dividing by two. The reason you divide the result by two is that the measured time includes the time of the sound wave going out and returning. The time will be equal going both ways.

The time required for a sound wave to travel 1 inch is 0.000074 second; the time to travel 1 foot is 0.00088 second; and the time to travel 1 mile is 4.67 seconds. Since most microcontrollers can measure time periods on the order of microseconds, they are fast enough to measure the time difference between sending and receiving a reflected pulse.

Originally, the ultrasonic range sensors that were used for robots were hacked out of Polaroid cameras that had autofocus features. In recent years, several companies have started manufacturing ultrasonic range sensors that can be used for noncamera applications, such as robotics.

Ultrasonic Transducer Systems

Shown next is an ultrasonic transducer system made by Milford Instruments, Ltd. (www.milinst.com) in the United Kingdom. This sensor package has a serial port that can be connected to a microcontroller or a computer for controlling the position of the transducer element and transmitting distance measurements to the microcontroller. This system has a detection range between 0.5 to 8.5 feet (0.15 to 2.6 meters) with a 10-millimeter resolution, and it has the option of outputting the results as a 0- to 5-volt analog distance that corresponds to a 0 to 8.5-feet (0 to 2.6-meter) range. The system comes with an R/C servo, for panning the ultrasonic sensor so that it can scan a wide area, and a software package for graphically seeing the results in a computer. This system uses a single ultrasonic transducer that transmits and receives the same ultrasonic signal.

CHAPTER 12: Sensors: Letting Your Robot See the World

This type of a configuration can be placed on top of a robot and used to scan the surrounding area. Acroname sells several different versions of this type of a system using the different types of Polaroid ultrasonic transducers. The Milford Instruments system was obtained at Mondo-tronics (www.robotstore.com).

The ultrasonic transducer that is taking the robotics community by storm is the SRF04 Ultrasonic Range Finder, as shown here. This small (0.8 x 1.7 x 0.6 inch) transducer is made by Devantech, also known as Robot Electronics (www.robot-electronics.co.uk) in the United Kingdom. This sensor operates at 5 volts and has a maximum current draw of 50 ma. The detection range is from 1.2 to 118 inches (3 to 300 centimeters), which is plenty of distance to see across a sumo ring. This detector is so sensitive that it can detect a 1-inch diameter broom handle at 6 feet.

Using Ultrasonic Sensors

Figure 12-13 shows a schematic for wiring an ultrasonic sensor to a Basic Stamp. To use this sensor, the Basic Stamp must output a 10 μs pulse to the pulse trigger input line on the transducer. The transducer will then transmit an eight-cycle ultrasonic chirp. The transducer will wait 100 μs before starting the timing sequence. This time delay is to make sure there is no cross communication between the transmitting and receiving elements of the transducer. At the end of the 100 μs period, the echo output will go high, and the Basic Stamp will measure the time the echo output signal is in the high state. The echo will go low after 36 milliseconds if no return signal is detected.

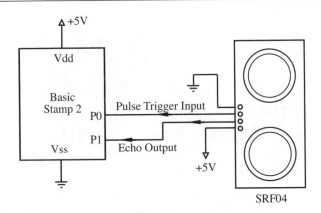

FIGURE 12-13 Schematic for wiring a SRF04 Ultrasonic Range Finder to a Basic Stamp

The following program is used to test the SRF04 sensor. This is a relatively simple program and can be converted for use on most microcontrollers. One of the key points of this program is converting the measured time into distance. The Pulsin command counts the number of time periods that elapse during a measurement sequence. To create a conversion factor, multiply the pulse width period by the speed of sound and divide by two (remember that the measured time is a round-trip time). This number will be less than one. Then multiply this number by the number of periods that was measured with the Basic Stamp. Since the Basic Stamps use integer math (whole numbers), you can invert the conversion factor and round it up or down to the nearest whole number. Then divide the number of measured periods by the new conversion factor. The program listing shown below has these already worked out for the Basic Stamp 2 family of microcontrollers.

```
'Devantech SRF04 test program for the Basic Stamp.
'This program will output a 10 microsecond (min) pulse to the sensor, and
'then measure the return time, in pulsin periods. The time is the round
'trip time, so the actual distance is half the round trip time.
'The conversion factor is used to convert the number of measured
'pulsin periods into distance. The speed of sound is ~13560 in/sec.
'Below lists the conversion factors for different types of Basic
'Stamps.

Init      con   0       'Pulse Trigger line on I/O Pin 0
Echo      con   1       'Echo Signal line on I/O Pin 1
Trip      var   word    'Number of measured time periods for round
```

CHAPTER 12: Sensors: Letting Your Robot See the World

```
Dist      var    word    'trip
                         'Measured distance between the object and
                         'SRF04

Convfac   con    197     'Use 74 for BS2 and BS2e Microcontrollers
                         'Use 184 for BS2sx Microcontrollers
                         'Use 197 for BS2p-24 and BS2p-40

Main:
   pulsout Init, 15       'Initialization pulse of 10 microsecond minimum
   pulsin Echo, 1, Trip   'Measure number of time periods for round
                          'trip
   Dist = Trip/Convfac    'Convert round trip periods into distance
   pause 10               '10 ms minimum delay between measurements
   debug DEC ? Dist       'Display the distance
   pause 250              'Pause for 1/4 second to see the distance,
                          ' remove when not using the Debug
goto Main 'Take another measurement
```

The SRF04 Ultrasonic Range Finder is very accurate and fast. The worst-case cycle time is 47 ms for this sensor. These sensors have a much larger field of view than the GP2D*xx* class of optical sensors. The field of view for the SRF04 is approximately 30 degrees; the GP2D*xx* class sensors have a field of view of a couple inches wide.

Sensor Placement

So where do you place the sensors on your robot? The answer depends on what type of sensor you are using and what you are using the sensor for.

Edge detectors need to be at the bottom and toward the front of your robot. You want to know as soon as possible that your robot has come to the edge of the dohyo, so it can back away and minimize the amount of time the robot is vulnerable to an attack from the rear. If you are using long-range edge-detection sensors, such as the knitting needle approach, the position is not as critical.

The most common place for opponent-detection sensors is the front of the robot. This location will tell the internal microcontroller whether there is an opponent in front of your robot. This configuration requires your robot to turn around to scan the environment. As soon as your robot detects its opponent, it will charge forward.

Placing sensors on the sides of your robot can also be handy. They can tell the internal controls that your robot passed another robot, so you can initiate a quick U-turn for the attack. They might also alert your robot that the enemy is coming in for an attack from the side, so you can initiate a defensive 90-degree turn to face the opponent.

Place your sensors low to the ground. There is no point in having a sensor that looks over the top of your opponent.

Protecting Your Robot's Rear

The one place where most robot builders do not place sensors is at the rear of the robot. This is the most vulnerable spot on the robot, and it is often left unprotected.

Imagine that your robot is heading toward the edge of the sumo ring, and your opponent is following right behind. When your robot gets to the edge of the dohyo, it will stop, and the other robot will crash right into it. Because the other robot has the momentum advantage (it is moving and your robot is not), it will easily push your robot off the edge of the ring, and it doesn't need to push very far. In most robot sumo contests, when you see this happen (a very common situation), everyone will say, "Game over!" because you are going to lose.

The simplest thing you can do to avoid this situation is to place an object detector facing behind in your robot. If it detects something there, your robot can make a 180-degree spin move to gain the advantage.

Sensor Range Issues

Many of the sensors that you will use in your robot will have a minimum detection range. For example, most of the Sharp sensors have a 4-inch (10-centimeter) minimum range. In robot sumo, it is expected that the other robot will come in contact with your robot. If your sensor is at the very front of your robot, and the other robot is closer than the minimum detection range, the sensor may not see it. To solve this problem, recess the sensor more toward the center, or the rear, of the robot, so the minimum detection distance is physically protected by the body of your robot.

You also need to consider the sensor's maximum range. You do not want your sensors to detect objects outside the sumo ring. It could be very embarrassing to watch your robot focus on the guy across the room with the white T-shirt and go after him, instead of the other robot in sumo ring. This might sound funny, but it does happen. All infrared sensors will detect a white object that is farther away than a black object. Because of this, you may need to reduce the sensitivity or the LED intensity to shorten the range, so your robot is not fooled by white objects outside the dohyo.

There is no point in having sensors on a sumo robot that can detect objects that are farther away than the diameter of the sumo ring. For example, the SRF04 can detect objects out to 118 inches. A mini sumo ring is only 30 inches in diameter, and the 3-kilogram sumo ring is 60 inches in diameter. You can still use these sensors on your robot, but you should program an algorithm to ignore all signals that have a range greater than the diameter of the sumo ring.

Using Multiple Sensors

Most of the time, your opponent will not be directly in front of your robot. So how do you determine where it is so your robot can turn towards it? The answer is to use more than one sensor. Also, different sensors have different uses. Some sensors are good for short-range applications; others have good long-range capabilities. Some sensors have a very narrow field of view; other sensors have a very wide field of view. By combining the strengths and weaknesses of different sensors, you can create a strong sensor array that is difficult to evade.

Sensor Patterns for Opponent Location Detection

Figure 12-14 shows three sensor configurations that can be used to determine the location of your opponent relative to the front of your robot. In this figure, a 30-degree beam pattern is being projected forward of each sensor.

For opponent-position detection, placing two sensors on both sides of the robot facing forward works well. To use your sensors, first detect for an opponent with the left sensor, and then detect with the right sensor. If the left sensor returned a hit and the right sensor did not, there is an opponent more toward the left side of the robot, and your robot should start turning left. The opposite is true if the right sensor detects an opponent. If both sensors detect something, the opponent is in the center of the two beam patterns, and your robot should move forward.

Although this approach works, it does have a limitation of the overall width of the detection range. Your robot can easily pass by its opponent, because the sensors have a narrow field of view. This is a very common occurrence. To increase the width of the sensing area, you can point the sensors outward. This increases the width of the detection range, but decreases the center detection range. This configuration is also one of the more commonly used sensor layouts.

The third approach, the crossed-beam pattern, increases the width of the overall detection range while maximizing the center detection areas. The main drawback to this approach is that it decreases the forward projection range of the sensors.

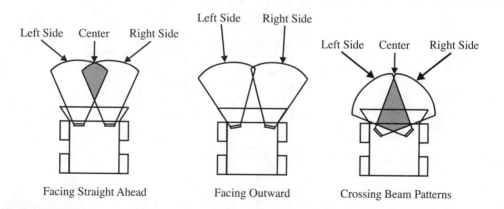

FIGURE 12-14 Sensor placement to determine the location of the opponent

Different Types of Sensors

A good sumo robot will have several different types of sensors on it. At the very least, you will have an edge detector and at least one object detector. Figure 12-15 shows a conceptual sensor layout for a robot.

The robot in Figure 12-15 uses the following sensors:

- A pair of range sensors in the front to detect the distance of the opponent. The reason for the range sensors could be that the robot normally moves at a slow speed to find its opponent. Once the robot has a "lock" on its opponent, it can accelerate in for the victory.
- The rear sensor is a binary (digital) proximity sensor. The binary sensor will tell the robot to spin around now, to prevent it from getting pushed out of the ring.
- The two side binary (digital) sensors can be long-range, narrow-field-of-view sensors that tell the robot that there is something to its side, and when it is time to negotiate a 90-degree turn.

You have probably noticed that this robot has almost 360 degrees of sensor protection, with only five sensors. As a general rule, the more sensors your robot has, the better it can see its environment.

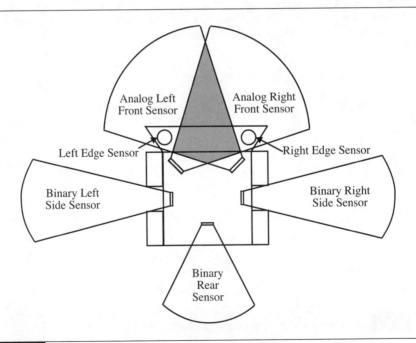

FIGURE 12-15 Sensor layout to give a robot almost 360-degree visibility

Cross-Communication Interference

When using multiple sensors, you need to be able to distinguish between which detector is detecting which emitter signal. When a detector detects a signal, it doesn't care where the signal came from, it just reports that it detected a signal. If the sensors are left continuously on, and when one of the detectors detects a signal, it difficult to tell which emitter sent the signal. This is known as *cross-communication interference*. You will need to either alternate which emitters are on at a time or orient the sensors so that it is not possible for a reflected signal to hit the wrong sensor.

Various approaches to positioning multiple sensors have been used successfully in many sumo robots. There really isn't one method that is any better than the other. It comes down to how you write your programs to handle multiple sensors, how you handle cross-communication interference between sensors, and which sensors you use.

Discovering Your Opponent's Location

The sensors presented in this chapter will either tell you that there is something there or the range of some object. They do not tell you the location of the object. Sometimes, it is desirable to know exactly where the opponent is relative to your robot. There are two ways to do this: Use a range sensor with a panning method or use multiple narrow-field-of-view sensors.

A Panning Sensor

You can place a range sensor on a servo that pans back and forth. When it detects an object, it will continue to pan back and forth, and the microcontroller will start recording the object range with the servo position. When the distance reading is at its minimum and the distance reading increases when the servo rotates to both sides of the object, the servo position at the minimum detection range is pointing right at the object. Then your robot will know exactly where and how far the object is. This is similar to how radar works.

A variant to this approach is to use a single narrow-field-of-view optical sensor in front of the robot, and have the entire robot's body pan back and forth looking for the opponent. This is an alternative to making the sensor do all of the work of moving back and forth.

Multiple Narrow-Field-of-View Sensors

Figure 12-16 illustrates how to place five sensors close together to make an almost solid detection field of view. With this type of a layout, any object inside the detection field will be detected by one of the sensors. If the sensors are range sensors, you will be able to determine your opponent's location and range.

Figure 12-16 also illustrates the approximate field of view for the Sharp GP2D*xx* class of sensors. You can see that these have a wider field of view when the sensors are horizontal to the ground, and a narrower field of view when they are vertical to the ground (the diagram is not drawn to scale to make it easier to see the field of view). This shows how you can adjust the orientation of the sensor to get a wider or narrower field of view.

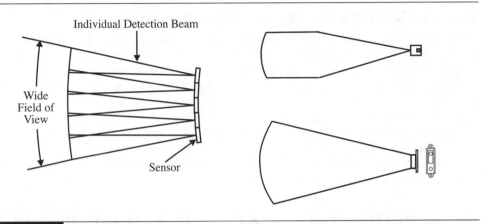

FIGURE 12-16 Using multiple Sharp detectors to determine the opponent's position

Sensor Accuracy Issues

The accuracy of any sensor will be affected by its environment. The real world is not a friendly place for sensors. It is constantly introducing new types of noise or interference that make your sensors work differently from place to place.

Drawbacks to Reflective Sensors

Most of the sensors discussed in this chapter are reflective-style sensors. Although these sensors work quite well, they are not perfect. In order for reflective sensors to work, the reflected signal needs to bounce back to the detector element on the sensor. This doesn't always happen.

In the ideal world, the angle of the reflected beam from a surface will be equal to the angle of the original beam that struck the surface. So if your sensor is pointed straight at a wall, the signal that bounces off the wall will come right back at the sensor. But if the wall is angled away from the sensor, the sensor will not receive a returned signal, even though there was an object that reflected the signal. If the detector element never receives a reflected signal, it will assume that there is nothing there.

Figure 12-17 illustrates how signals reflect off flat surfaces. The left side of the figure illustrates how a wide-angle beam can reflect off a surface and not return to the sensor. In this example, if the beam-spreading angle were much larger, a portion of the reflected beam would have returned back to the sensor. But for the most part, all of the beam reflected away from the sensor. This is why stealth fighter aircraft have flat angular surfaces—so that they can control the direction of reflected signals.

On the right side of Figure 12-17 is a real surface that reflects a signal. Most of the energy will follow the rule that angle out equals angle in, but there is a percentage of the energy that will

CHAPTER 12: Sensors: Letting Your Robot See the World

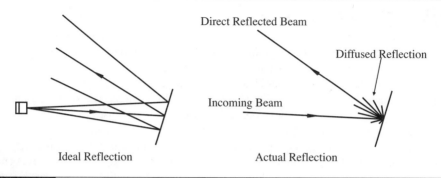

FIGURE 12-17   How signals reflect off flat surfaces

reflect in all other directions. This is primarily due to the roughness of the surface the signal is striking. So a small fraction of the signal will return to the sensor. This small fraction is called the *diffuse* part of the reflection.

If the detector element inside your sensor is sensitive enough, it will detect the diffuse part of the reflection. Increasing the intensity of the transmitted signal or light will increase the chances of the detector detecting the diffuse part of the reflection. The drawback to increasing the sensitivity is that it increases the range of the sensor. Robots that have a lot of parts hanging loose, round edges, and many different surfaces are much easier to detect than robots with single flat surfaces.

Because these sensors could fail to detect a robot that is directly in front of them, it is a good idea to have backup mechanical sensors, so you will be able to detect if another robot has actually struck your robot.

It's a Noisy World Out There

Many robot builders will come to a contest thinking they have the perfect robot, only to find that their sensors don't work the same way as they did where the robot was built. This is because we live in a noisy world.

Lighting and sound conditions at home are always different from the conditions at a competition site. The electrical noise from a high-powered light source or a generator can cause electrical interference in your sensors, so that they can become confused. Also, other people may be using the same sensors that you are using, so your detectors might be detecting the other robot's emitters. The other robot could be your robot's opponent or a robot on the other side of the room.

Infrared sensors work great indoors, but they don't work very well in direct sunlight. Even though the sensors are looking for a modulated signal that does not occur in nature, the intensity of the sunlight saturates the sensor, so that it cannot distinguish anything.

One way to minimize the impacts of interference is to calibrate your sensors in their immediate environment.

Calibrating Sensors

It is to your advantage to use sensors and programs that you can calibrate at the place where you will be competing. The calibration could include adjusting a potentiometer to change a sensor sensitivity level, adjusting a sensor location, or adjusting a program variable.

All of the programs presented in this chapter are single stand-alone programs that test a sensor. These programs can also be used to calibrate the sensors. When your write your programs, you should use a variable for the calibration constants that your program uses to convert the sensor data, such as the distance, into real working numbers. Then, when you are calibrating your robot, all you will need to do is change a couple variables at the beginning of the program, instead of searching through the entire program, making sure that you have made all of the correct changes.

Ideally, you will want to use sensors that you can mechanically calibrate. Changing a program at a contest can be very risky. If you made the wrong change, you will usually find out about it during a contest, and that is the worst time to discover that you forgot to convert inches into centimeters. As the designers of a recent NASA Mars explorer spacecraft found out, a simple programming conversion factor mistake can lead to disastrous results.

Strategies That Don't Rely on Sensors

This chapter has explored the types and advantages of various types of sensors. However, using sensors doesn't mean that you will win a game or even see your opponent. Some robots do not rely on sensors.

For example, one robot design strategy is to not use any sensors. This strategy relies on a very fast and strong robot that can scream across the dohyo in less than a second and slam into its opponent, hoping that the impact will knock the other robot out of the dohyo. This is an all-or-nothing approach. If it fails to knock the other robot out of the ring in the first hit or misses the other robot altogether, it will run out of the ring on its own.

There are also some sumo robots that rely on stealth technology to avoid being seen. Shown here is a stealth mini sumo robot, named Black Widow, built by Kristina Miles. This robot uses only an infrared edge-detection sensor. During a match, it randomly moves across the sumo ring. All of its opponents go right by it, because their sensors do not see it. Eventually,

CHAPTER 12: Sensors: Letting Your Robot See the World   269

the two robots run into each other, and then it becomes a pushing contest. Black Widow's high-traction wheels result in an 80 percent victory rate, which is pretty good for a blind robot.

Remember that using or not using sensors does not guarantee a victory. There are many other factors, including luck, involved with winning a game.

Closing Remarks on Sensors

At this point, you should have enough information about sensors to be able to build highly competitive sumo robots. Sensors are the eyes of your robot, and with good vision, you will be able to better respond to the environment.

This chapter presented basic, proven sensors that are used at some level or another in a majority of the sumo robots that compete in contests. All of the sensor data presented here can be used in all robots, not just sumo robots or autonomous robots. Many remote-control robots can use this information to make semi-autonomous features that can help with driving the robot. For example, a BattleBot class robot could use some of these sensors for an automatic weapon system.

There are many other types of sensors that you can use in your robot. A search on the Internet using the keywords "Robot Sensors" will yield thousands of websites. Some of the sensors you find will be similar to the ones presented here; others will be totally different. Your choice of sensors for your robot really comes down to the design of the robot, in both appearance and functionality.

Once you've selected a sensor, you should test it before putting it in your robot. You need to understand what the outputs from the sensor mean based on the input conditions. If you place the sensor straight into the robot without first testing it, and the robot does not perform as expected, you won't know whether the sensor or the program the robot is using to interpret the signals is the problem.

Write a simple little test program, like the ones shown in this chapter. Place the sensor on a sumo ring, and then simulate the environment in which the sensor is going to detect. Constantly watch the output signals so you can understand what the numbers mean and how fast the numbers respond to the environment. Some sensors are faster than other sensors. Once you have a good understanding how well the sensor is going to work for your robot, place it in your robot, and then test your robot with the sensor.

This completes the discussion of sumo robot components. The next chapter will show how all of the information presented in this book up to this point is put together to build actual working sumo robots.

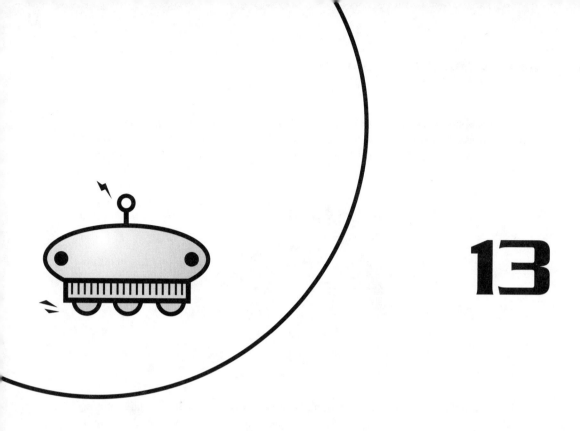

Building the Robot

At this point, you have all the information you need to build your own competitive sumo robot. In fact, all of the information presented up to this point can be used to build any type of robot. This chapter will take you through the steps required to build two complete, working autonomous or remote-control sumo robots: a 3-kilogram robot and a mini sumo robot.

Full-Sized Sumo Robots

Many people who are first getting started in building 3-kilogram sumo robots become frustrated at the difficulty in finding the right parts. Because you have this book, finding parts shouldn't be a problem. Many different sources for sumo robot parts have been referenced throughout the book, and the appendix lists all these resources.

One of the best ways to get started in building a 3-kilogram sumo robot is to obtain one of the base sumo kits that Lynxmotion (www.lynxmotion.com) sells. These kits include an easy-to-assemble plastic frame that uses aluminum bars to join all of the edges together with #4-40 screws, as shown here. Because all the parts are screwed together, assembly, disassembly, and maintenance work is quick and easy. Jim Frye of Lynxmotion did a great job at designing these kits.

The Lynxmotion kits also include a set of high-torque gearmotors, wheels, and mounting hubs. Mounting the motor inside a robot, mounting wheels to the motor, and building a frame have been taken care of for you. All you need to do is figure out what type of motor controllers, sensors, and brains to put into the robot.

Here, we will go through all the steps to build a sumo robot using one of the Lynxmotion kits—the Viper—and take a look at another interesting kit—the Terminator.

Viper Sumo Robot

We will build a 3-kilogram robot using the two-wheel, wedge-shaped Viper sumo kit from Lynxmotion. The elliptical holes in the sides of this robot are sized for holding a standard 7.2-volt or 8.4-volt NiCd or NiMh battery pack used in R/C cars. This basic kit can be assembled in less than an hour with a Phillips screwdriver and a 3/32 Allen wrench. Also included in the kit are a pair of capacitors that are soldered across the motor terminals to reduce electrical noise from the motors, as described in Chapter 10. Figure 13-1 shows four general assembly phases of building this robot.

Viper Performance Specifications

The specifications for the 6-volt motors that came with this kit are that the no-load shaft speed is 186 rpm, with a stall torque of 75 ounce-inches. The 2-7/8 inch diameter wheels will result in a top speed of 1,700 inches per minute, or 28 inches per second. The combined stall torque for both of these motors is 150 ounce-inches. This equates to a maximum pushing force (assuming the wheels do not slip on the sumo ring surface) of 104.3 ounces, or 6.5 pounds, which is the maximum weight of a 3-kilogram sumo.

FIGURE 13-1 General assembly steps in building the Lynxmotion Viper kit

To get better performance, we will use an 8.4-volt battery pack in this robot, so that the maximum output speed will be 39.2 inches per second, and the maximum theoretical pushing force will increase to 9.1 pounds. These specifications are good enough to push any 3-kilogram sumo robot out of the dohyo.

The Motor Controller

The 6-volt stall current for these motors is 2.05 amps, thus at 8.4 volts, the stall current will increase to about 2.9 amps. To be on the safe side, the motor controller must be able to supply at least 3 amps of continuous current for each motor. There are a wide variety of motor controllers that can deliver this much current. For this robot, we will use two R/C ESCs, because we can control them with simple 1- to 2-millisecond pulse-width commands.

Lynxmotion sells two types of motor controllers that will work just fine for this application: the SozBot dual-speed controller (www.sozbots.com) and the Hitec SP-560 ESC (www.hitecrcd.com). The Sozbot controller requires active cooling if the motors are drawing more than 3 amps of continuous current, but these motors will be drawing a maximum of 2.9 amps of current. An advantage of the Sozbot controller is that it can control two different motors at the same time. The Hitec SP-560 controller has the advantage of a higher current-handling capability.

For this robot, we will use the Hitec SP-560 ESCs. This decision was arbitrary, since either controller will work. The illustration here shows how the ESCs are mounted in the back of the Viper kit.

Electrical Wiring

Figure 13-2 shows the main power circuit wiring diagram. To simplify the wiring requirements, a six-pole terminal block is fastened to the bottom of the circuit-mounting base (the plate that is above the motors). The motors, ESCs, and battery are wired to the terminal block as shown in the figure. Unless a motor or an ESC burns out, these components will be semipermanent installations in the robot.

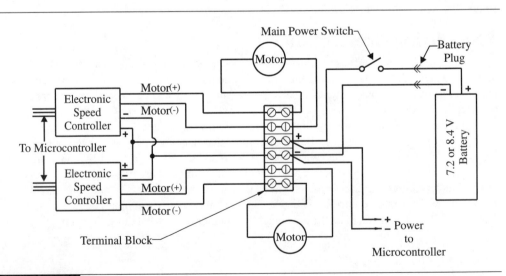

FIGURE 13-2 Main power wiring diagram

You can purchase the terminal block and the crimp wire terminals at any electronics store such as Radio Shack (www.radioshack.com). The crimp terminals can also be found at most automotive parts stores. Make sure you get the #6 ring-style insulated terminals, or the outside diameter of the terminal will not fit in the terminal block.

Shown here is the power circuit wiring. This robot is upside-down with the bottom plate removed. Here, the wiring goes around both sides of the motor.

With the terminal block, you can cut the wires to length so that the wiring follows an easy-to-maintain path from component to component. Being able to see where all the wires are going makes diagnosing problems easier. Before cutting the wires to length, lay out all of the wires to determine the best path for them. Consider how accessible the wires will be if you need to get to them to fix a problem. You can rearrange your wire layout until you come up with the most maintainable scheme, and then cut the wires to length.

The main power switch is located next to the terminal block. The switch protrudes through the circuit-mounting base. This way, when the robot is right side up, the switch is on top of the robot, and the operator can easily turn the robot on or off.

The last step in the wiring process is to cut off the external on/off switches from the Hitec SP-560 ESCs (other ESCs may not have external on/off switches). Leave about 1 inch of the two-wire cable remaining from the ESC. Strip about 1/8 inch of insulation from the ends of these wires, and solder the two wires together. Do not solder the wires together between both ESCs. Then place some electrical tape, or heat-shrink tubing, around the bare wires. These switches are no longer needed, since we have a main power switch.

The Robot Brains

At this point, this robot can be made into an R/C robot or an autonomous robot. For an R/C robot, all you need is an R/C receiver. Just mount the receiver in front of the ESCs, and add a 12-inch long piece of hard wire (preferably music/piano wire) to the back of the robot to hold up the antenna. Then use a small piece of double-sided foam tape to hold the receiver on the robot. Chapter 10 provides the details on how to wire the radio-control receiver to the ESCs. Adding a mixing circuit can simplify the steering methods for the robot.

To make an autonomous robot, you need to add some sensors, a microcontroller, and a circuit board. Probably the most open-ended part of the robot-building process is determining what type of sensors and microcontroller to use. Chapters 11 and 12 described many different types of microcontrollers and sensors. All of them will work just fine for this robot. In this example, we will use a Basic Stamp 2 for the microcontroller, since it is small and relatively easy to use.

The Sensors

Deciding which type of sensors to use can be challenging, because the sensors must fit inside the robot body. For this example, we will use the following sensors, which are all discussed in Chapter 12:

- A QRD1114 infrared reflective edge sensor circuit, obtained from Digikey (www.digikey.com). We chose this type because we want the edge sensors as close as possible to the front of the robot, and the wedge shape at the front of the robot doesn't allow a lot of room for other types of sensors.
- A pair of Sharp GP2Y0D02YK infrared object detectors, to look for opponents to the sides of this robot, obtained from Acroname (www.acroname.com). These are easy to install, because there is a lot of empty space on the sides of the robot in front of the wheels.
- A Lynxmotion's IRPD sensor to detect for opponents in front of the robot. We selected this sensor because of its simple control and small size. We cut a small piece of plastic with a matching taper of the wedge, so that the sensor board will sit parallel to the ground. We will place this sensor near the middle of the robot, so that it will be able to detect low-profile robots.

Illustrated here are the edge-detection and side sensors for detecting the edge of the dohyo and the opponents, as viewed from the bottom of the robot.

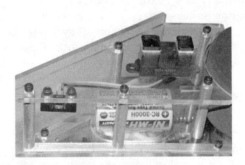

The Circuit Board

After you've selected the sensors, you need to design the circuit board. First, create a schematic drawing of the electrical circuit and connections. You should have a copy of the data sheets for all of the major components, which can be obtained from the component manufacturer. These data sheets provide information about special electrical requirements, such as needing pull-up resistors, 3-volt maximum input signals, and timing sequences. Using these data sheets helps avoid wiring problems that would be difficult to diagnose after the circuit board has been built.

Figure 13-3 shows the schematic drawing for our Viper kit robot using the sensors, ESCs, and a Basic Stamp 2. The following sections explain the circuit board components and assembly.

The Power Circuit   This circuit uses a LM2940CT low drop-out voltage regulator to convert the power from the batteries into a 5-volt power supply for all of the electronics. The visible LED next to the voltage regulator indicates when power is being supplied to the microcontroller. The on/off switch in this circuit is different from the on/off switch used in the main power circuit. That means there are two power switches: one for the main power and one for the microcontroller.

We add the second power switch to the main electronics because, when power is first applied to the Hitec SP-560 ESC, there is a small amount of voltage that goes straight to the motor. This will cause the robot to move forward a couple inches. This characteristic is inherent to this particular ESC. Since the rules of robot sumo state that the robots must not move for 5 seconds at the start of a match, we need to make sure that the robot doesn't move when it is first powered on. This way, we can turn on the main power first, then set the robot down on the sumo ring, and then use the second switch to turn on the microcontroller.

TIP   *It is always a good idea to know how all of the individual components work before designing the circuit board. Then you can design in features that handle the unique quirks of the various components.*

Diagnosis LEDs   Since this circuit only used nine of the sixteen I/O pins of the Basic Stamp, we will add three LEDs to the circuit. These LEDs will be used to indicate the status of the microcontroller. During the programming of each behavior, the different LEDs will turn on and off to indicate what the program is doing. This is especially useful for infrared sensors, since you cannot visually see when the infrared light is on. For this example, we will set two LEDs to light depending on which infrared side sensor is detecting a reflection.

Programming Port   We also add a four-wire serial communication port to this circuit board. Instead of using a traditional DB9 connector, we use a four-pin header to save space. Having an on-board programming port simplifies the program development efforts, since you will not need to remove the Basic Stamp from the circuit to program it.

You will need to make a special adapter cable that connects to the serial cable and to the robot. To make the cable, you need a female DB9 connector, about a foot of 4 conductor cable (or four different colored wires), and a plug that matches the four-pin header on the circuit board.

In this robot, we use a five-pin header strip and plug. Since you will use the programming cable many times during the programming phases, chances of plugging it in incorrectly are pretty high, but you can take a step to prevent this. Cut the number 2 pin off the header strip (see Figure 13-3) and insert it inside the plug. This way, the plug cannot be installed backwards.

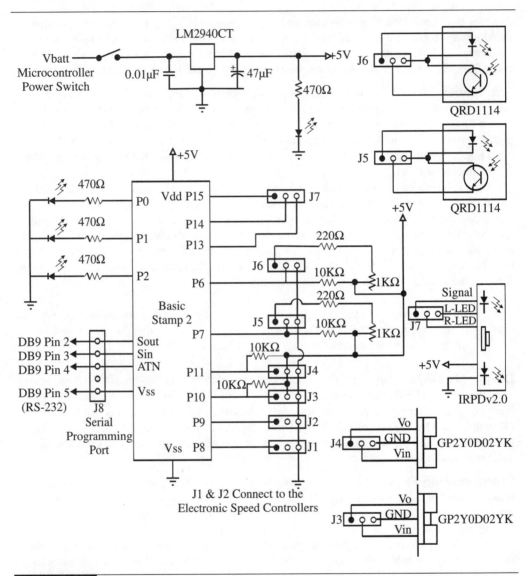

FIGURE 13-3 Schematic drawing for a sumo robot using a Basic Stamp 2

CHAPTER 13: Building the Robot

Circuit Board Assembly   After you've made the schematic drawing, you are ready to assemble the circuit board. You need a prototyping printed circuit board for this. These boards are predrilled with hundreds of holes and have little solder pads on one or both sides of the board. You can obtain these boards at most electronics parts store. Make sure that you get a board that is about 2-3/4 inch square. A smaller board will be too small to add all of the components onto it, and a larger board will need to be cut down to size.

The next step is to lay out all of the parts on the circuit board. This is a trial-and-error process, as you try to get all of the parts to fit together, while minimizing the amount of wires that need to be soldered. There is a lot of art that goes into laying out a circuit board, and no two people will make a circuit board in the same way.

Shown here are the top and bottom sides of this circuit board. All of the components are directly wired to the circuit board except the Basic Stamp. A 24-pin DIP socket is soldered to the circuit board. This way, the Basic Stamp can be easily removed. All the parts that went into this board can be obtained from most electronics parts suppliers, such as Digikey.

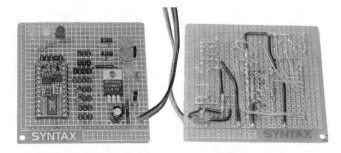

Using Custom Printed Circuit Boards

For those of you who are not comfortable with point-to-point wiring of components on prototyping circuit boards, there is an alternative. Several companies, such as AP Circuits (www.apcircuits.com) and PCB Express (www.pcbexpress.com), will make a custom printed circuit board (PCB) for you. All you need to do is provide a schematic drawing of the circuit to the board manufacturer. (Check with the board manufacturers to determine how the schematic drawing needs to be presented.)

Express PCB (www.expresspcb.com) (yes, PCB Express and Express PCB are different companies) offers a nice software package that you can download so you can draw your own circuit boards. The software package will tell you how much it will cost to make the circuit board. When you are ready to order the circuit board, just e-mail the drawing file that you created with the Express PCB software.

Robot Programming

After you've assembled everything, the next step is to program the Basic Stamp. The following is the working program listing for this robot. The robot will move forward slowly until it reaches the edge of the dohyo. Then it will back up, rapidly spin around, and go forward again. If the side sensors detect an object, the robot will spin 90 degrees to face the robot, and then go forward again. If one of the forward opponent sensors sees a robot in front of it, it will turn toward it. If the sensors detect that the opponent is directly in front of it, the robot will charge forward at maximum speed.

```
'This is a Basic Stamp 2 program running the Viper sumo robot.
'Pins 0, 1, and 2 are for diagnosis purposes to indicate
'program status. The turning variables


'Pin 15 Front opponent object return signal line
'Pin 14 Right front infrared object detection LED
'Pin 13 Left front infrared object detection LED
'Pin 11 Right side sensor signal line
'Pin 10 Left side sensor signal line


i       var     byte        'counter loop variable
Turn    var     byte        'Turn Variable
temp    var     byte        'Tempory counter value
R_Speed var     word        'Right motor speed variable
L_Speed var     word        'Left motor speed variable


R_Motor con     8           'Right Motor ESC signal line is connected
                            ' to pin 8
L_Motor con     9           'Left Motor ESC signal line is connected
                            ' to pin 9
S_Fwd   con     780         'Slow Forward Speed, neutral speed is 750
S_Rev   con     720         'Slow Reverse Speed
F_Fwd   con     1000        'Maximum Forward Speed, 2.0 ms pulse
F_Rev   con     500         'Maximum Reverse Speed, 1.0 ms pulse


Pause 5000                  'Initial 5 second delay per the sumo rules


L_Speed = S_Fwd             'Initialize left motor speed
```

```
R_Speed = S_Fwd           'Initialize right motor speed


Main:              'Main program loop routine
    Gosub Check_Edges     'Check for the edge of the Dohyo
    Gosub Check_Sides     'Check for opponents to sides of the robot
    Gosub Check_Front     'Check for opponents in front of the robot
    Gosub Forward         'Move forward a small amount
Goto Main


Check_Edges:   'Check the right and left edge
    if in6=0 then Right_Edge    'sensors. They are normally high
    if in7=0 then Left_Edge     'but when they detect the white
    Return                      'line, the signal goes low.


Left_Edge:     'Detected the left edge sensor routine
    Gosub Backup          'First back up a small distance
    Turn = 20             'This is the turn increment loop, a larger
    R_Speed = F_Fwd       'value means the robot will turn more.
    L_Speed = F_Rev       'Set the motor speeds to turn to the right
    Gosub Turn_Robot      'Call the turning routine
    L_Speed = S_Fwd       'Reset the forward motor speeds
    R_Speed = S_Fwd
    Return                'Return back to the main routine


Right_Edge:    'Detected the right edge sensor routine
    Gosub Backup
    Turn = 20
    R_Speed = F_Rev
    L_Speed = F_Fwd
    Gosub Turn_Robot
    R_Speed = S_Fwd
    L_Speed = S_Fwd
    Return


Check_Sides:   'Check both sides of the robot for an
    if in10 then LS       'opponent.
    if in11 then RS       'If there is an opponent, the sensor will
```

```
        Return                  'go high.
LS:                             'Left side loop
    Gosub Left_Side             'Call left side routine
    Return
RS:                             'Right side loop
    Gosub Right_Side            'Call right side routine
    Return


Right_Side:     'Detected a robot to the Right side routine
    Turn = 5                'Set the amount of the turning distance
    R_Speed = F_Rev         'Set motor speeds to turn to the right
    L_Speed = F_Fwd
    Gosub Turn_Robot        'Call turning routine
    R_Speed = S_Fwd         'Reset motor speeds for forward motion
    L_Speed = S_Fwd
    Return                  'Return back to the main routine


Left_Side:      'Detected a robot to the Left side routine
    Turn = 5
    R_Speed = F_Fwd
    L_Speed = F_Rev
    Gosub Turn_Robot
    R_Speed = S_Fwd
    L_Speed = S_Fwd
    Return


Check_Front:    'Check for an opponent in towards the front
    Temp = 0                'tempory counter
    High 14                 'Turn on the Right infrared LED for 1 ms
    Pause 1                 'check for a return signal. Store result
    Temp = in15             'in the temp variable. Then repeat for
    Low 14                  'the Left infrared LED. The return signal
    High 13                 'will be high if there is no object
    Pause 1                 'present. Temp = 0 if both sensors
    Temp = Temp + in15 + in15    'returned a signal, Temp = 1 if
    Low 13                  'only the LEFT sensor
    Branch Temp,[Both,Left1,Right1,Nothing]    'detected a signal,
Nothing:                    'Temp = 2 if only
    Low 0                   'RIGHT sensor detected a signal, and
```

```
        Low 2                   'Temp = 3 if no sensors returned a signal
        R_Speed = S_Fwd
        L_Speed = S_Fwd
        Return
    Right1:                     'Have the robot turn slightly to the
        High 2                  'right
        Low 0
        Turn = 2
        R_Speed = F_Rev
        L_Speed = F_Fwd
        Gosub Turn_Robot
        Return
    Left1:                      'Have the robot turn slightly to the
        High 0                  'left
        Low 2
        Turn = 2
        R_Speed = F_Fwd
        L_Speed = F_Rev
        Gosub Turn_Robot
        Return
    Both:                       'Move forward at maximum speed
        high 0
        high 2
        R_Speed = F_Fwd
        L_Speed = F_Fwd
        Return


    Forward:                    'Move the robot forward
        Pulsout R_Motor,R_Speed
        Pulsout L_Motor,L_Speed
        Return


    Backup:                     'Move backwards a fixed amount of
        For i = 1 to 25         'distance set by the counter value
        Pulsout R_Motor, F_Rev  'a larger value will mean the robot
        Pulsout L_Motor, F_Rev  'will move further backwards
        Pause 15
        Next
        Return
```

```
Turn_Robot:      'Turning subroutine for controlling
   For i = 1 to Turn         'the amount of turn and the ESCs
   Pulsout R_Motor, R_Speed  'signals.
   Pulsout L_Motor, L_Speed
   Pause 15
   Next
   Return
```

The Completed Viper Sumo Robot

Shown here is the completed sumo robot. This robot weighs in at a little under 1,400 grams, so it will need some additional weight to compete against other 3-kilogram sumo robots. Lead is an ideal materal for increasin the weight of a robot. The last step that is not shown here is the addition of a thin metal edge that needs to be attached to the front edge of the scoop. This is to protect the plastic edge from becoming damaged when hitting other robots or falling off the edge of the dohyo, and helps to get under the other robot.

This robot turns out to be surprisingly fast. Its sensors are very effective in spotting opponents on the sides and front of the robot. As you've seen, the Viper kit from Lynxmotion is a great 3-kilogram robot sumo platform, but because of its low-profile wedge shape, placing different types of sensors inside the robot can be a challenge.

Terminator Sumo Robot

The Terminator sumo kit from Lynxmotion is a four-wheel-drive sumo kit that has a lot of open space available to add many different types of sensors and microcontrollers. Figure 13-4 shows the general construction steps to build this robot.

This robot uses four of the same motors that are used in the Viper kit. The wheels are a little smaller, at 2-1/2 inches in diameter, thus an 8.4-volt battery will yield a maximum speed of 34 inches per second. The maximum theoretical pushing force is 21 pounds. If the wheels do not slip, this robot should be able to push a weight nearly four times its maximum competition

CHAPTER 13: Building the Robot   **285**

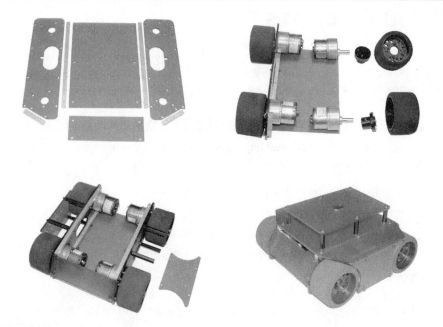

FIGURE 13-4   General assembly steps for the Terminator sumo kit

weight. However, to achieve this kind of pushing capability, you will need to improve the wheel traction (as described in Chapter 15).

Shown here is a clear, plastic version of the Terminator outfitted with a pair of SRF04 ultrasonic range finders, two infrared edge detectors, a pair of ESCs, an 8.4-volt battery pack, and a Next Step carrier board with the Basic Stamp 2p. All of these components can be purchased from Lynxmotion.

Mini Sumo Robots

Mini sumo robots are probably the easiest sumo robots to build, which is one of the reasons why they are so popular. Beginning with a mini sumo robot is a good idea when you are just getting started in robotics. You get mechanics, electronics, and software in a small, low-cost package, that has a simple specific goal of pushing another robot out of a sumo ring.

A Mini Sumo Robot Built from Scratch

We'll go through the procedure for building an autonomous mini sumo robot from scratch. You'll need the following materials to build the core frame of the robot:

- A 3-inch by 3-inch, 1/16- to 1/8-inch thick base plate
- A 4-inch by 1-1/2-inch, 0.025-inch thick aluminum front scoop
- Four AA cell battery holder with 6-inch wire leads
- Four 1-inch long circuit board spacers, with #4-40 threads
- Eight #4-40 by 3/8-inch long screws
- A roll of double-sided foam tape (3M brand heavy-duty mounting tape number 110 works well for this application and can be found at most office supply stores)
- Two standard R/C servos (such as Tower Hobbies TS-53, Futaba FP-S148, Futaba S3003, Hitec HS-422, or Hitec HS-300)
- Two wheels with rubber bands (such as Lynxmotion # SRVT-01 or Acroname # R169)

NOTE *Make sure the spline on the wheels matches the type of servos you use. Different servos have splines that look the same but are not interchangeable. Acroname and Lynxmotion sell a wide variety of different colors and spline options for these types of wheels.*

Base Plate and Front Scoop Creation

The base plate needs to be 3 inches by 3 inches square and approximately 1/8-inch thick. It can be made from any material—plastic, wood, or aluminum. The front scoop can be made from a material other than aluminum, even thicker material, but 0.025-inch thick aluminum is one of the easiest materials to work with using regular hand tools. You can obtain this aluminum at most hardware stores. Figure 13-5 shows a drawing with the dimensions for the base and front scoop.

For the front scoop, you can easily cut the 0.025-inch thick aluminum with a pair of tin snips. Drill the holes into the scoop before you shape the aluminum. To bend the scoop to shape, place the larger flat edge in a vise, so that only the 3-inch wide section is protruding above the vise. Using a piece of flat wood placed flush against the protruding piece of aluminum, bend the

CHAPTER 13: Building the Robot

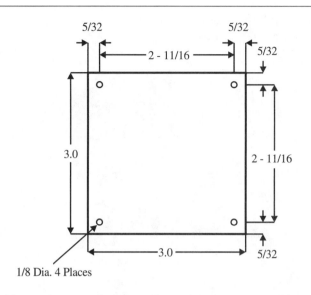

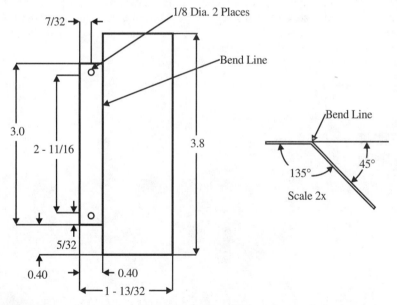

FIGURE 13-5 Geometry dimensions for the mini sumo base plate and front scoop

aluminum to the 45-degree angle by pushing and rotating the wooden board against the aluminum. Shown here is the front scoop before and after bending.

Body Assembly

The next step is to attach the four spacers 1-inch circuit board spaces. These spacers are placed on top of the base board. Use the #4-40 screws to hold the spacers to the base board. Attach the front scoop to the bottom of the base board. The screw and spacer will hold all of the parts together. Figure 13-6 shows the body assembly process.

FIGURE 13-6 Assembling the base frame of the robot

Servo and Wheel Assembly

Now you need to modify the R/C servo into a continuous rotating servo, as described in Chapter 8. Place the rubber bands on the outside diameter of the wheels. (The rubber bands really stretch.) Then press the wheels onto the servo shaft. Remember to make sure you are using a wheel and servo that have matching splines.

Before attaching the servos, use a pencil to mark where the servos should go. The main body of the servo will be toward the front of the robot. This robot was designed this way to move the center of gravity toward the front scoop. The back edge will be approximately 3/4 inch from the back edge of the servo, and there will be about a 0.2-inch gap between the servos. Place the servos with these dimensions on the base plate and measure the width and length of the overall robot. The overall dimensions should be no greater than 3.8 inches wide and long; they can be a little smaller, but not larger. Then mark the base plate where the servos should go.

To attach the servos to the base plate, use two 1/4-inch wide by 1-inch long pieces of double-sided tape for each servo. Make sure the servo case and base plate are clean and dry before adding the tape. Place the tape strips about 1 inch apart, and make sure the wheels are parallel before you press the servos down onto the base plate. Figure 13-7 shows the servos and wheels attached to the base plate.

Battery Box

After you've assembled the parts, the next step is to tape the four AA cell battery box to the bottom of the base board. Use four 1/2-inch square pieces of tape to hold the battery box to the base plate. If you use too much tape, you will damage the parts if you need to take them apart.

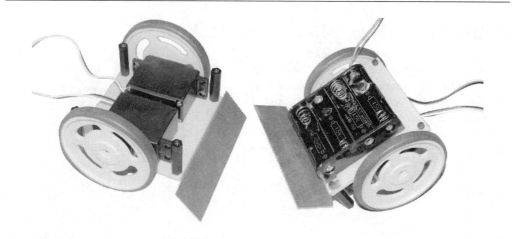

FIGURE 13-7 Attaching the servos, wheels, and battery box to the base plate

Center the battery box toward the front of the robot. Place the tape on the base board (not on the scoop), and then press the battery box onto the base plate. Make sure the wire leads are not pointing to the front of the robot. Figure 13-7 also shows this step.

Electrical Wiring

The last step in the construction process is adding the brains or a remote-control unit to the robot. If you are creating a remote-control mini sumo robot, all you need to do is to make a second 3-inch square base plate and attach it on top of the spacers. Then tape the receiver on top of the base plate. Place the receiver near the front of the base plate to keep the center of gravity as close to the front as possible.

The circuit shown in Figure 13-8 has been used in a mini sumo robot that has been competing for several years and wins most of its contests. This circuit uses the basic infrared edge detectors and modulated opponent detectors, as described in Chapter 12. The brain for this robot is a Basic Stamp 1 from Parallax Inc.

The Circuit Board

The first step in making the circuit board is to cut the prototyping circuit board down to the same size as the base plate, 3 inches square, and drill the four 1/8-inch diameter holes at the corners. Lay out the parts so that the Basic Stamp is at the rear of the circuit board, the 74HC04 is at the center, and the infrared LED and the Panasonic PNA4602M sensor are toward the front.

Do not solder any components in the front 3/4 inches of the circuit board. The 9-volt battery will sit under this area and would interfere with the solder joints. Bend the LED and the infrared sensor leads so that they will come to the front of the circuit board and are on top of the circuit board. Place all three pin headers for the sensors, servos, and the programming port on the sides of the circuit board. Use a single-row, 14-pin socket for mounting the Basic Stamp 1.

Fully Assembled Mini Sumo Robot

Shown here is the completed mini sumo robot. The edge sensors are taped to the corners of the front scoop. To keep the face of the sensor parallel to the ground, a small, 45-degree triangle was cut out of a 1/4-inch thick piece of plastic. Any type of material will work here. Make sure the distance between the sensor and the ground is between 1/32 to 1/8 inch.

CHAPTER 13: Building the Robot

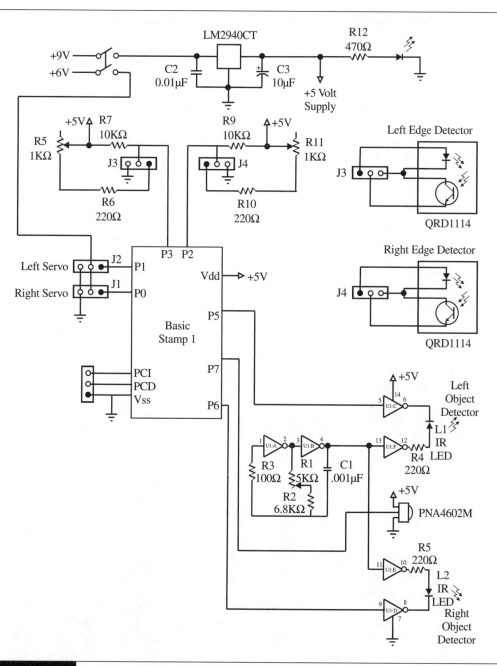

FIGURE 13-8 Mini sumo circuit

Notice how a 9-volt battery fits nicely between the circuit board and the base plate. The three-pin programming port is optional. If you do not add one, you will need to remove the Basic Stamp from the circuit to program it.

Mini Sumo Programming

The following program has been used to win many mini sumo contests.

> **NOTE** *This circuit board and program can also be used in lightweight and 3-kilogram sumo robots. All that you need to do is use larger servos.*

```
'Mini Sumo Program, minisumo.bas Written for a Basic Stamp 1.
'This program is a sample program that uses the IR edge detectors
'to detect the white sumo ring edge and the IR object detector to
'follow its opponent. The mini sumo will move in a straight line
'until it hits the white edge. After the mini sumo hits the white
'edge, it will back up, turn around, then move forward again.
'If the mini sumo sees an object in front of it, it will turn
'towards the object.


'pin0 = Right Servo    These pin(s) are I/O pins, not
'pin1 = Left Servo     physical pins on the Stamp 1
'pin2 = Right Edge Detector
'pin3 = Left Edge Detector
'pin5 = Left Opponent Detector LED
'pin6 = Right Opponent Detector LED
'pin7 = IR Receiver Sensor

dirs=%01100011           'Initialize the I/O pin directions
                         'pin0, pin1, pin5, pin6 are outputs
pause 5000               'Pause 5000 ms (or 5 seconds)


main:           'Main Program Loop
   if pin2 = 0 then Lturn      'Check right edge detector, if the
                               'detector sees the white line, then
                               'goto the left turn routine.
   if pin3 = 0 then rturn      'Check right edge detector, if the
                               'detector sees the white line, then
                               'goto the right turn routine.
   pulsout 0,100               'Send a 1 ms pulse to the right servo
```

```
    pulsout 1,200          'Send a 2 ms pulse to the left servo
    b0 = 0                 'Sample the left object detector for
    for b2 = 1 to 5        '5 times by toggling the IR LED on/off
       pin5 = 1            'The output pin will be high if there
       pin5 = 0            'is no reflected signal. If b0 (or b1)
       b0 = b0 + pin7      'is less than 3 then over 50% of the
    next                   'signals returned back to the receiver.
                           'This give a good indication that
    pulsout 0, 100         'an object was detected, and a
    pulsout 1, 200         'less chance that the signals were
    b1 = 0                 'random noise or false signals
    for b2 = 1 to 5
       pin6 = 1
       pin6 = 0
       b1 = b1 + pin7
    next


    if b0 < 3 then turn    'If a positive object detection
    if b1 < 3 then turn    'was obtained, then goto the turn routine
goto main


turn:                'This routine determines which direction
    b2 = b0 + b1           'to turn. If both detectors return
    if b2 < 5 then main    'equal values, then go straight,
    if b0 < b1 then left   'otherwise turn in the direction that
    if b1 < b0 then right  'had the stronger return probability.
goto main                  'i.e. a lower hit number.


left:            'Make a small left turn move
    pulsout 0, 200         'Send a 2 ms pulse to the right servo
    pulsout 1, 200         'Send a 2 ms pulse to the left servo
    pause 15               'Pause for 15 ms. This delay sets up
goto main                  'the ~50 hz servo update frequency


right:           'Make a small right turn move
    pulsout 0, 100
    pulsout 1, 100
    pause 15
```

```
        goto main


Rturn:              'This is the Right Turn Routine.
    gosub back              'Call the move backwards routine.
    for b2 = 1 to 30        'This loop determines how much the
        pulsout 0, 100      'mini sumo turns. Increasing the
        pulsout 1, 100      'loop value (30) causes the mini
        pause 15            'sumo to turn more to the right,
    next                    'decreasing this value decreases
goto main                   'the amount the mini sumo turns.


Lturn:              'This is the Left Turn Routine.
    gosub back
    for b2 = 1 to 30
        pulsout 0, 200
        pulsout 1, 200
        pause 15
    next
goto main


back:               'This is the backup routine
    for b2=1 to 25          'This loop determines how far the mini
        pulsout 0, 200      'sumo will back up. Increasing the
        pulsout 1, 100      'loop value (25) will increase the
        pause 15            'overall distance. Decreasing the
    next                    'value will cause the mini sumo to
return                      'back up less.
```

Mini Sumo Kits

Because of the popularity of the mini sumo robots, several companies have released (or will soon release) mini sumo kits.

Parallax has a Basic Stamp 2 version of a mini sumo kit called the Sumobot, shown here. This kit comes with all of the parts necessary to build a fully functional mini sumo robot. The main circuit board has a prototyping solderless breadboard for easily adding components and features.

CHAPTER 13: Building the Robot

The Mark III mini sumo robot was a group design effort by members of the Portland Area Robotics Society (www.portlandrobotics.org). This robot was designed to help people get started in the field of robotics, and uses the mini sumo contest as the motivator for building a robot. It uses the Microchip PIC 16F877 microcontroller and a Basic Compiler from Celestial Horizons (www.celestialhorizons.com) for the programming environment.

What makes the Mark III, shown here, a great starter robot is that it was designed from the ground up to be a robot that can be used for many other applications besides the mini sumo contest. With its three edge sensors, it can be easily adapted into a line-following robot or a line-maze-solving robot (a popular contest at the Seattle Robotic Society's Robothon, www.seattlerobotics.org). The main controller board has an expansion port for stacking additional boards to the robot for extra features, such as pyroelectric sensors for detecting flames for the Trinity Fire Fighting contest (www.trincoll.edu/events/robot/index.html), and even an accelerometer option that could be used for inertial navigation systems.

Closing Remarks on Building Robots

At this point, you should have all the information you need to build exciting sumo robots for different weight classes and capabilities. This chapter presented an example of building the Viper 3-kilogram sumo kit from Lynxmotion, adding a set of different components to make it a fully autonomous sumo robot. You also learned how to build a mini sumo robot from scratch.

The rest of this book is devoted to teaching you new tricks to make your robot just a bit better. The next chapters describe various types of competition strategies and ways to improve your robot's pushing capabilities. The final chapter presents a gallery of sumo robot designs, which will give you ideas for what you can do with your robot.

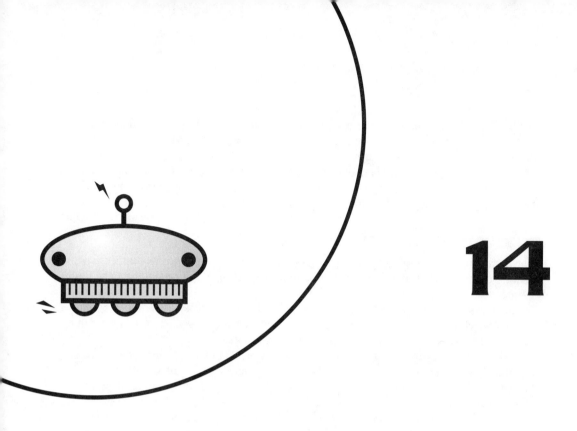

Offensive and Defensive Strategies

Now that you've learned how to build a robot, you're probably wondering what you can do to make your robot a winner. This chapter begins with some general tips and programming tips to help you achieve that goal. Then the rest of this chapter describes various offensive and defensive strategies that you can use in your robot. Some strategies are in software, and some require mechanical designs to incorporate the different strategies.

General Competition Tips

You might be surprised by how many people are still building and programming their robots at 4:00 A.M. the morning before the big event. The most important strategy you can use is to make sure that your robot works before the day of the event. You need to have the robot built long before the event and thoroughly tested days before the event. You need to understand how well your robot performs and know its strengths and weaknesses. And you need to use this information to help you position your robot in the sumo ring.

Along with having your robot prepared for the competition, there are several other things you can do to improve your chances of winning your sumo matches. These range from studying your competition to getting lucky, as described in the following sections.

Knowing Your Competition

All top champions spend a lot of time studying their competition. Winners remember their opponents' strengths and weaknesses, and they exploit them. They don't need to have the best robot—they just need to use better tactics during a match.

Taking a lesson from the champions, you can improve your chances of winning a tournament by carefully watching the other robots during all of the contests. Take some notes on what the robots are doing. Try to identify their strengths and weaknesses.

The appearance of a robot can be very deceiving. Robots that look like they were slapped together the night before the contest, with parts held on by duct tape, might actually be very good robots. Some robots that look like the builders spent $10,000 on them can be all style and no function.

Watching how the robots perform will give you the important information you need to compete. How fast do the robots move? Do they change their speed? Do they move only in a straight line? Where are the sensors? What types of sensors are they using? What do the robots do when they get to the edge of the sumo ring? How far do they back up? Do they have an obvious attack strategy? Do they have any defensive strategies?

Showing Confidence

Something you might not consider is the impact of your demeanor during the match. Try to look confident when you place and control your robot. Everyone will be watching you, as well as watching your robot. If your opponents can see in your eyes that you do not have confidence in your own robot, they will believe that they can beat you, and they will try to intimidate you into making a mistake.

For example, consider an incident that occurred at the final round of the tenth annual Northwest Robot Sumo Tournament. Both robots burned out a motor during a match. The owner of one of these robots set up for the next match, pretending that nothing was wrong. The other builder, unaware that the other guy had the same problem, wanted to concede, thinking his robot didn't have a chance. The crowd talked him into continuing, and he placed his robot back into the ring. He won the match, but he almost lost, because his opponent gave the impression that the other robot was working just fine. Again, the point is that confidence can help you win a match.

The Luck Factor

The best offensive and defensive strategies for your robot really depend on the design of your robot, the design of your opponent's robot, what you want your robot to do, and luck. Yes, as with any contest, luck always plays a factor.

You can be unlucky in a number ways. Did your battery fail during a match? Did you remember to put the batteries in your robot before the match? Is the battery you are running the same battery you have been using for the last two years? Did one of your line sensors stop working, so your robot drives off the ring all by itself? Did a wire break in the power circuit? Did the robot get disqualified because it moved before the 5-second time period? Did an unimportant part that weighs more than 5 grams fall off the robot?

Murphy invented robots so that Murphy's law, "What can go wrong, will go wrong," will be permanently branded onto our foreheads. In every tournament, you will see robots that fail all by themselves due to software, electrical, or mechanical failure. The other robot wins by default. This is the luck factor.

Strategies for Remote-Control Robots

Probably the number one enemy of a remote-control robot is an operator's lack of driving experience. The sumo ring is very small, and it is very easy to drive off the edge of the ring. You can't be thinking about what you are going to do, because the ring is small and the robots are fast. You should plan exactly what you want to do before the match starts, because once it starts, it is pure driving mayhem.

If you are competing with remote-control robots, you need to practice driving, and you need to practice against another sumo robot. You need to learn how to attack, pivot, and avoid. You should learn how to preserve your batteries and motors when the pushing match begins, and how to spin out of a bad situation.

Studying your opponent is also very important in remote-control robot contests, because people have habits that can be used against them. Knowing the weaknesses of the other robot is also very important.

Because of the advanced traction methods used in remote-control robots, getting under the other robot is not very likely. It usually comes down to a pushing war, although you may be able to entice the other robot to make a mistake.

Programming Tips

The first place where a robot wins or loses a tournament begins at home before the match begins. Quite often, people will build the robot, add a bunch of sensors, write a program, and turn on the robot—just to find that it does not work correctly. In most cases, the problem is in the software. Trying to find the problem is often difficult, time-consuming, and frustrating. However, you can design your programs to make them easier to troubleshoot.

Using Subroutines

One of the best ways to program a robot is to use a main program that does nothing but call subroutines, and then repeats itself over and over again. Each subroutine is a very specific function or behavior. Each subroutine calls other subroutines that do specific things.

For example, the main program might check the edge sensors all the time, so it will call a subroutine called Check Edge. The Check Edge subroutine will check the sensors and make a decision. It will either call the Backup and Turn subroutine to get off the white edge of the sumo ring or exit this subroutine and return to the main program. The Backup and Turn subroutine will first call the Backup subroutine, and then call the Turn subroutine.

When programming, you want to break up all of the individual actions and behaviors into stand-alone programs, and start at the lowest level. For example, you might write and test your program in this order:

1. Write a backup subroutine that the main program will call. Program the robot with this backup routine, and then physically test the program with the robot. You want to make sure that this routine works without any problems.

2. Write individual programs for the turning subroutines. The only thing in the robot's memory will be the turning subroutines. Test the routines on the robot. Then combine the two routines and make sure that they work together.

3. After you know the robot is moving the way you want it to, write the algorithm to check the edge sensors, which will call the backup and turning subroutines. Then test the program on the robot. If the robot does not turn the way you want it to at this point, you will know the problem is with the edge-checking routines, not the motion routines.

It becomes easier to debug a program when you program from the lowest level function and build up to the main routine. It is often helpful to first sketch a flowchart with different boxes describing the basic functions. Figure 14-1 shows a simple flowchart of the subroutines described here.

When you test the programs, test them on the actual robot, while the robot is moving. Many programs that work just fine on the workbench don't work correctly on the robot. This is because motors and sensors rarely work exactly as you expect them to work. You will need to make adjustments to your programs to account for how the components actually work.

Subsumptive Programming

Rodney Brooks from MIT developed and coined the term *subsumptive architecture*. Subsumptive architecture uses a series of sensor-based behaviors to produce real-time control over a robot.

In subsumptive programming, some behaviors have higher or lower priority than other behaviors. The robot has a very basic behavior—in the case of robot sumo, moving forward to push anything off the ring. When triggered, the various sensors will initiate a different behavior for the robot. For example, edge sensors trigger a defensive behavior, and opponent sensors trigger an attack behavior. The sensors are assigned different levels of priority. For example, an

CHAPTER 14: Offensive and Defensive Strategies

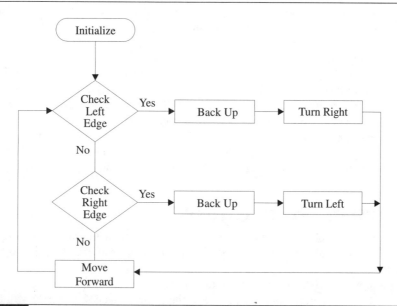

FIGURE 14-1 A flowchart of subroutines for a program

edge sensor has more priority than a side object sensor, because it is more important to avoid driving off the sumo ring than it is to attack the opponent.

These behaviors are like the various subroutines that were described in the previous section. Certain high-priority behaviors can be programmed into low-priority subroutines, so that the appropriate sensors are always monitored. Alternatively, you can use a microcontroller equipped with built-in interrupts. When the interrupt is triggered, the appropriate behavior takes over. This style of programming is very advantageous for a sumo robot, since most of its motion is reaction-based. More information about how to use subsumptive programming can be found in the book *Mobile Robots, Inspiration to Implementation* by Joseph L. Jones and Anita M. Flynn.

Using Multiple Programs

The previous sections provided tips for structuring your programs, which will help you create a good sumo robot. But how can you use software to give your robot a competitive edge? The simple solution is to use multiple programs in your robot.

Remember the suggestion that you should study how your opponents work so you can take advantage of their strengths and weaknesses? There is only so much you can do by positioning your robot on the dohyo. To really take advantage of your knowledge of your opponents, you can have several different programs that use different types of attack strategies.

To let you choose among your programs, your robot will have a set of two or more switches. The rules prohibit reprogramming your robot between the best two out of three rounds, but if you already have the programs on the robot, a flip of a switch changes the attack strategy. When the match starts, the software checks which switch is pressed and executes the program associated with that switch. For example, you might have one program for going slow, one for going fast, another to arc clockwise, another to move in a counter-clockwise arc, and yet a fifth program to have your robot spin wildly across the dohyo.

This type of a strategy will give your robot a big competitive edge, especially against robots that use only one program. For example, if you know that your opponent always attacks straight on, an arcing program will be a good defense. If the other robot needs to move backwards to drop a front edge down onto the ring, you can use a fast attack to get under the other robot before it can get ready. If the other robot has a low-profile wedge that you know can get under your robot, the wildly spinning algorithm may help prevent this attack. Also, your opponent will have no idea what your robot is going to do, and this confusion gives you the psychological edge.

Competition Strategies

The basic strategies for robot sumo are very straightforward: the main offensive strategy is to push the other robot out of the sumo ring, and the main defensive strategy is to not be pushed out of the ring.

Chapter 6 showed that the maximum pushing force a robot has is function of its weight and the coefficient of friction between the wheels and the sumo surface. Chapter 15 will discuss methods to improve the coefficient of friction, but the other part of the equation is weight. You want your robot to be near the maximum weight for the weight class. This helps when you are in a pure pushing battle with another robot. You want to keep the center of gravity on your robot as low as possible. You do not want to be knocked over too easily. Finally, you want to get under the other robot. If you can lift any of the opponent's wheels off the ground, the other robot suddenly loses a lot of pushing power, which makes it easier to push out of the sumo ring.

Offensive Strategies

This section suggests some offensive tactics that you can employ. Some are software-based; others are hardware-based. For the most part, the type of offensive attack strategy you want to use is designed into the robot when you build it.

Front Scoop

Many sumo robots have a wedge-shaped scoop in front. The purpose of the scoop is to try to wedge it under the opposing robot, as shown here.

Some scoops are permanently fixed on the robot. More commonly, the front scoop deploys forward when the match starts. This type of design is very effective. However, if both robots have front scoops, the one whose scoop has the better edge may prevail.

CAUTION *Keep in mind that if the edge of your front scoop is razor sharp, the robot will be disqualified because it poses a safety risk to humans.*

Thin Front Flap

Some robots have a thin (less than 0.030 inch), metal flap that almost lies flat on the surface of the ring, as shown here (on a robot named *Boxter* by Daryl Sandberg). The flap could deploy forward or be permanently fixed.

If the flap is long enough, when the other robot hits the front of your robot, it will drive up the flap. Then it will stop moving. The other robot has now lost its pushing power with respect to the ring. All your robot needs to do is push the other robot off the side of the ring. This technique can burn out the motors of your opponent's robot because they will be stalled on your robot's flap.

If the flap is short, it can be used to get under the other robot as a low-profile front scoop, or it could be combined with a regular front scoop for additional lifting capability.

A variant to this approach is to paint the flap white, so that the other robot will think it is on the white edge of the ring and back away. Now your robot can continue to move forward, and the other robot will back away from it every time it sees the white flap. With this approach, your robot can simply back the other robot off the ring.

Pushing Arm

Some robots have an arm that deploys forward to keep the wedge from the other robot away. It is kind of like you putting your hand on your little brother's head, keeping him at bay. If the pushing arm is used correctly, the other robot will not be able to touch your robot. Another advantage of a pushing arm is that it can easily knock over robots that have a high center of gravity.

When deployed, the arm should be parallel and a couple inches above the ground. The front edge of the arm should have some form of rubber-like material that has a high coefficient of friction, so that it sticks to the other robot.

Corralling Arms and Decoys

There are many robots that use a very tall arm that deploys sideways when the match starts. The robot either spins or drives around in a circle. The arm has a sensor that detects if an opponent has touched the arm, similar to the knitting needle-style edge detector described in Chapter 12. When the robot detects that something has touched its arm, it will either drive forward or rotate the main body toward the opponent. The arm is used to corral the victim off the edge of the ring.

One variant to this approach is to place a flag at the end of the arm to confuse the other robot's sensors into thinking the flag is the robot. Another variant is to use two arms that deploy sideways. If the arms are 30 inches long, they can span the entire width of the sumo ring, thus capturing the other robot. All this robot needs to do is drive forward, and it doesn't need any other sensors. There are several robots in Japan that effectively use two arms.

Brute Force

Some robots don't use any sensors to detect their opponents. They rely on pure power to push the other robot off the sumo ring. These robots use high-traction devices (described in Chapter 15) and high gear-reduction motors. All they do is move back and forth in a slow and deliberate manner. If another robot gets caught in front or behind this type of a robot, it will be pushed out of the sumo ring.

Kinetic Energy

There are two types of coefficients of friction: static and dynamic. Static friction is what prevents objects from starting to slide across the floor. Dynamic friction is the resistance friction when the object is sliding across the floor. Dynamic friction is less than static friction. Up to this point, all of the discussions have been related to static coefficients of friction, because it is the worst case situation. It requires more force to get an object to start to slide across the ground, and less force to keep pushing the object across the ground.

Some robots will use another robot's loss of friction to their advantage. This type of a robot will crash into the other robot at a high rate of speed. The impact causes the other robot to lose static friction because it is momentarily sliding on the surface. Then this robot continues to push. If this is done correctly, a robot with the same weight and traction devices will be pushed out of the sumo ring by the crashing robot.

A variant to this type of an approach is the all-or-nothing attack. The attacking robot will fly across the dohyo with extreme speed and hope that the crashing impact sends the other robot flying out of the ring. If it doesn't, it won't be able to stop fast enough to keep from going out of the ring on its own. The crowd loves this type of attack, since the victim robot is usually launched into the air for a resounding crash on the ground. The robots that rely on speed usually win or lose in the first second of the match.

Driving Along the Perimeter

Another strategy is to have the robot drive along the edge of the sumo ring until its side sensors lock in on the opponent, and then attack the other robot from the side or rear. By driving along the edge of the ring, this robot knows where it is. By using range sensors, it can take multiple readings to determine the position, range, and speed of the opponent. Then you can use this information to calculate an attack path that strikes the opponent on the side or the rear. It is a computationally intensive process, using brains over brawn.

The Sweet Spot

The "sweet spot" on most robots is the rear of the robot. Since most robot designs are focused on the frontal attack, the rear of the robot is often exposed to being easily lifted out of the sumo ring.

In most cases, if your robot can get behind the other robot, the game is over. This is because when your opponent reaches the edge of the ring, it will back up. It will then back up onto your robot, which reduces its pushing traction as it tries to move backwards. At some point, your opponent will start to go forward, and all your robot needs to do is move forward along with it to help it off the edge of the sumo ring.

Path Planning

Another offensive approach is to use multiple preprogrammed motion paths, as described in the "Using Multiple Programs" section earlier in this chapter. By having the robot drive slow, fast, spin around, or drive in circular arcs, you can make it very difficult for the other robot to figure out what your robot is doing. You can also use this approach to ferret out the other robot's capabilities, so you can use a precision strike against that robot in the next two matches.

Disabling Your Opponent

The current rules in Japan do not prohibit intentional damaging of your opponent. As long as your robot does not damage the dohyo or cause injury to humans, it is free to disable your opponent. In essence, autonomous combat robots are legal to use in sumo.

Many local contests may prohibit robots that have features that are there only to damage the opponent. Check the rules of the competition you intend to enter before using this strategy.

Technically, all your robot really needs to do is turn off the other robot. So, if it can flip the other robot's power switch, it automatically wins the match.

Defensive Strategies

To defend against being pushed out of the sumo ring, your robot must be able to detect when a bad situation is occurring and avoid it. This is mainly done with various types of sensors, as described in Chapter 12, but there are some other techniques you can use to protect your robot.

Edge Sensors

At the very least, your robot needs a set of edge sensors so that it won't drive off the sumo ring on its own. Some robots even use redundant sensors to make sure that they don't go off the ring. Other robots will use different types of sensors, so that one sensor type cannot be fooled into thinking that it is on the ring's edge.

Having edge sensors on the rear of the robot can help prevent your robot from backing off the sumo ring. This happens sometimes, especially if there is a battle near the edge of the ring.

Multiple Sensors

Having multiple object sensors around the robot will give your robot a wide field of view, so that the other robot will not be able to easily sneak up behind your robot. Using different types of sensors, such as infrared and ultrasonic sensors, can help you defend against robots that use stealth technology, since you will have different methods for detecting them. Also, at an event, some sensors work differently than they do at home, so having different types of sensors helps reduce the chances of your robot going into a contest blind.

Rear Sensors

Since the back of most robots is their most vulnerable side, you should consider having a sensor that has a high-priority setting at the rear of your robot. When this sensor detects an opponent, it takes control of the robot to make a sudden 180-degree turn to face the opponent trying to sneak in behind your robot.

Stealth Technology

Just like the F117 fighter bomber, stealth technology helps "hide" your robot from other robots. It is hard for the other robot to push your robot out of a ring if it cannot see your robot.

Several robots use stealth technology as their main defensive weapon to win matches. Using flat-angled panels, with sharp corners, on your robot will help deflect any reflective type sensor signal away from your robot and its opponent's sensors. Acoustic absorbing and infrared transparent materials can help. You will need to experiment with different types of geometries for different sensors. The more pyramid-shaped your robot is, the more stealthy it becomes.

Hiding the Dohyo Edge

An interesting defensive technique is to hide the edge of the dohyo from your opponent. Thomas Dickens developed a robot named Blackbird that has a pair of black wings that deploy sideways, as shown here. By covering the white dohyo edges with its wings, this robot can fool the other robot into driving off the edge of the sumo ring. For protection, this robot has a thin, white front flap. This flap confuses an opponent into thinking that it hit the edge of the sumo ring instead of Blackbird. This type of an approach can be effectively used in passive style robots.

Always Driving the Wheels Forward

Most robot builders will reverse one side of the robot's wheels to make a turn. Although this is an effective method for turning, it places your robot at a disadvantage. If another robot is right up against your robot, and one sensor is working but other sensors are not working, your robot will try to turn into the other robot. If the other robot is pushing on your robot, and one set of wheels is going in reverse trying to turn, the other robot will have an easy time walking your robot right off the edge of the sumo ring.

To avoid this type of problem, always drive your wheels forward. Remember from Chapters 5 and 6 that turning with differential-steering robots is related to the relative speeds of the two wheels. So, by having one set of wheels move slowly and another set move fast, your robot will turn—well, actually it will drive in a circle, but it is turning. Doing this can help you prevent the other robot from using your robot's motors against it.

Closing Remarks on Strategy

This chapter presented some ideas on how to incorporate different strategies into your robot. Obviously, there is no one strategy that will win all the time. Each strategy has its advantages and disadvantages. It is up to you to learn how your robot responds to the different types of strategies, and then plan how to exploit the strengths and weaknesses of your robot and your opponents' robots. Probably the best strategy you can have is knowing how well your robot works and how to use it against it opponents. It also helps to be lucky.

Most of the strategy methods must be designed into the robot from the very beginning. If you are new to sumo robots, building a basic wedge-shaped robot is a good place to start. You will learn how to build a robot, and when you compete with it, you will learn how all the types of strategies work against each other. Remember to observe other robots and take notes.

The next chapter will show you how to get a little more traction, which can give your robot more of a competitive edge.

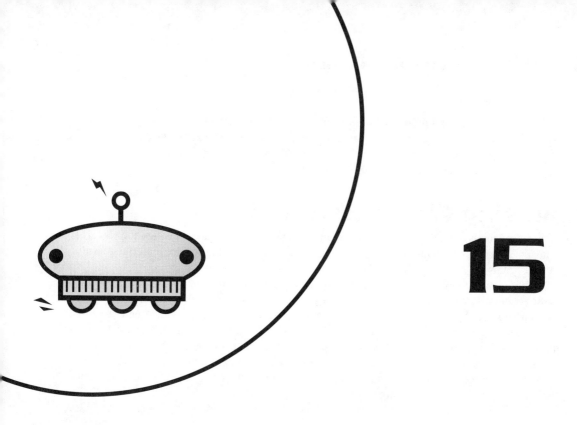

Improving Your Robot's Pushing Capability

Since the goal of robot sumo is to have your robot push the other robot out of the sumo ring, improving your robot's pushing capability can improve your chances of winning matches. In Chapter 6, you learned that the maximum pushing capability of a robot is a function of the weight of the robot and the coefficient of friction between the wheels and the sumo surface (assuming that the motors are strong enough to generate the required torque). Therefore, the methods for improving pushing capability involve traction and robot weight.

This chapter will present several tricks that you can use to improve the maximum pushing capability of your robot. The use of high-traction wheels is a low-cost solution to improving traction, and using vacuums and magnets may give your robot more pushing capability by increasing the apparent weight of your robot.

Before you start building a robot with advanced traction methods, check with your local contest coordinator. Many contests have very specific rules about vacuum systems, magnets, and sticky wheels. There are some contests that specifically ban them or allow only some of these features in certain weight classes.

Improving Wheel Traction

Wheel traction is a function of the wheel materials and type of surface on which the wheels are moving. Since all of the robots are running on the same surface, the surface material doesn't matter in this case.

There are two key aspects of high-traction wheel material: softness and stickiness. Generally, softer materials provide high coefficients of friction, since the weight on the material helps push the material into the imperfections in the surface. So, softer materials provide greater traction. The other factor is how clean the wheels are. If the wheels start collecting dust and dirt, they lose most of their stickiness. The drawback to using really soft materials is that they do not tolerate high forces very well without tearing. Also, the stickier the material, the more dirt is collects. You are not going to get anything for free here. Higher-traction wheels require more maintenance, but they can be the key to a competitive edge.

The following are a couple techniques for improving traction: wheel coating and wheel casting.

Coating the Wheels

The easiest way to improve the traction of your wheels is to coat the surface with a silicone material. You can find silicone gels at any hardware or automotive store. They are commonly used to make gasket seals. Common products are the RTV (room temperature vulcanizing) adhesives used in the automotive world and the silicone adhesives and seals used in kitchen and bathroom applications. The RTV silicones come in several different colors based on their applications, but they all work pretty much the same way for improving wheel traction.

To coat the wheels, spread some of the silicone on a wooden applicator, such as a Popsicle stick, and then apply a thin layer of silicone evenly across the wheel surface that will be in contact with the sumo ring. There is no reason to cover the sides of the wheels. Then let the coated wheels set for 24 hours. Do this in a well-ventilated area, since the fumes can be toxic. Shown here are a couple wheels that have been coated with silicone.

If you have the patience, you can add silicone directly to a wooden, plastic, or metal wheel as it is slowly turning, to build up a thick layer of silicone. If you follow this approach, make sure that each layer is dry before applying the next layer.

Silicone will start to harden over time and lose some of its traction capability. One competition trick is to apply the silicone on the wheels the night before the contest. You can also peel off the old silicone and add some new silicone to wheels. One of the drawbacks to silicone coating is that it may come apart on the sumo ring if the wheels start to spin on the surface.

Casting Wheels

One method used by several robot builders to improve traction is to cast a high-traction tire on their own custom-made wheels. This way, they have complete control of the chemistry that is used to make their materials, including the wheels' color and hardness. When wheel casting is done correctly, it generally yields higher traction than can be obtained with commercially purchased wheels.

Casting Material

The most popular casting material is the two-part, mold-making compounds used in the casting and model-making industries. There are two types of compounds that work well for the sumo robot wheels: silicone and polyurethane. Silicone materials are little stronger than polyurethane materials, but polyurethane materials provide a little more traction capabilities than silicone. Three popular sources for the mold-making compounds are Synair (www.synair.com), TAP Plastics (www.tapplastics.com), and Polytek (www.polytek.com). A search on their websites will show that there are many different options for making custom-casted wheels for your robot.

Both the silicone and polyurethane compounds can be purchased for different Shore A hardnesses, ranging from a soft 10 to a very hard rubber of 80. For high-traction wheels, Shore hardness ranges should be in the 10 to 40 range. One thing to remember is that the softer the material, the easier it is to tear. A Shore A of 10 may be too soft for a powerful 3-kilogram robot, but it could be strong enough for a mini sumo robot. (See Chapter 5 for more information about Shore ratings.)

The casting compounds come in two parts, Part A and Part B, which are mixed together by weight. The Part A compound has a slightly higher density than the Part B compound, so mixing by volume will not result in the precise mixing ratios. You can also adjust the final hardness of the rubber by adding a Part C compound to the Part A and Part B mixture. These compounds are very sensitive to the mixing ratios, so all of the manufacturers recommend that you use scales to weigh the materials before mixing.

Polyurethane compounds are very sensitive to moisture content in the air, and they will suck some of the moisture out of the air and become contaminated. If you live in a moderate to high humidity environment, it is probably better to use a silicone material instead of a polyurethane material.

Custom Wheel Molds

To make a custom wheel, you need a custom mold for the wheel. The mold can be made out of any material. Many people will make the mold out of wood, since it is easy to shape. If you use a porous mold material, like wood, then you need to seal all of the pores, which you can do with wax, shellac, or epoxy.

The mold should be at least a three-part mold, with two sides and a center piece, as shown in Figure 15-1. The center piece should be the same thickness as the wheel. If you have a thick wheel, the center mold piece might need to be two separate C-shaped pieces. In the middle of the center piece is a circular hole, which will be the outside diameter of the tire. The center plate and the back plate should be pinned together (but removable), so that they can be precisely registered with respect to each other.

The back plate of the mold has a hole at the center that will be used to center the wheel inside the mold. The top plate is used to seal the mold. A pair of small holes should be drilled either in the top plate or through the side of the center plate. These holes should line up in the casting area and be close to one another. One hole is where the casting compound is poured, and the other hole is to allow air to escape from the mold.

Prior to pouring the mold material, spray all of the mold surfaces with a mold release. (You can find mold release at the same place where you obtained the mold-making compounds.) Make sure that you do not get the mold release on the wheel surfaces, or the compound will not stick to the surface of the wheel. You may need to roughen the surface of the wheels to get better adhesion.

When you are finished pouring the mold, set it in a dry place, with the pour and vent holes at the highest location so that the compound does not leak out. Allow at least 24 hours for it to cure; full strength usually takes about a week. To remove the wheel, carefully disassemble the mold.

Casting wheels is a messy and time-consuming process that will require a lot of trial-and-error testing of different compounds, tire widths, and thicknesses. You need to test the wheels under full-load conditions to determine if they will work. Custom tuning the mixing ratios can make different wheel hardnesses. Casting wheel designs and compound mixing recipes may be unique to each robot. Most people will not disclose their recipes for their casting methods—it took them a long time to figure out the best mix, and it is their secret for their competitive edge.

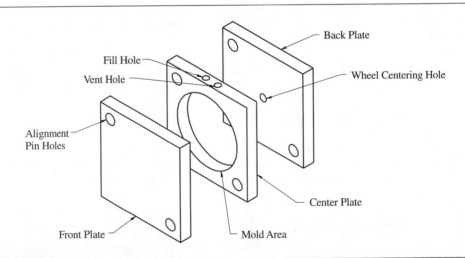

FIGURE 15-1 A custom wheel mold design

> # Increasing the Apparent Weight of Your Robot

It is usually desirable to be at the maximum weight to maximize the theoretical pushing capability of your robot. So what do you do after your robot reaches the maximum weight for the particular robot class? The solution is to increase the *apparent* weight of the robot.

The pushing force of a wheel is a function of the coefficient of friction and the normal force acting on the wheel. Up until now, this normal force has been the weight of the robot. But if you were to push down on your robot, the robot would have more pushing capabilities, because the normal force is now the sum of the robot weight and the force of your hand. Though the robot's weight hasn't increased, the wheels will have a high load on them, which they will see as a higher apparent weight. Pulling down on the robot has the same effect as pushing down on the robot. It follows that all you need to do is figure out a way to either push or pull your robot down onto the sumo ring.

The easiest way for a robot to push down on itself is to place a vacuum between the bottom of the robot and the sumo ring. The vacuum allows the air pressure above the robot to push the robot down onto the sumo ring. The easiest way for a robot to pull itself down onto the sumo ring is to use magnetic-attraction forces (assuming a magnet can be attracted to the sumo surface). Putting a magnet on the bottom of the robot will help increase the apparent weight of the robot more than the weight of the magnet itself. The following sections describe how to implement vacuum and magnetic systems to increase the apparent weight of your robot.

Although there are distinct advantages to increasing the apparent weight of your robot, be aware that your robot structure may not be strong enough to handle the additional weight. The motor bearings may not be able to handle the high loads, axles may bend, the robot frame may bend, or the motors may not be strong enough to turn. Remember that with differential steering, wheels need to slip on the surface in order to turn. As the frictional forces on the wheels increase to improve the pushing capability, the motor must provide enough torque to be able to cause the wheels to slip so that it can turn. Using vacuum or magnetic systems may require you to use stronger motors and gearboxes, which usually increases the cost of making the sumo robot.

Vacuum Systems

There are generally two types of vacuum systems that are used in sumo robots: fan-style vacuums and vacuum pumps. The following sections will describe several different types of vacuum systems, but first, let's see how these systems work to increase the apparent weight of a robot.

How Vacuum Systems Work

Normal atmospheric air pressure is approximately 14.7 pounds per square inch (psi). This means that the surrounding air will exert 14.7 pounds of force on every square inch of surface area. Air pressure acts evenly on all surfaces on an object—the top, sides, bottom, and inside.

A 3-foot by 5-foot window has 2,160 square inches of surface area. The air pressure on one side of the window exerts 31,752 pounds of force. We know a regular window glass cannot support this much load without breaking, but the window doesn't break. This is because the air pressure on both sides of the window is the same, and they cancel each other out. But if the air pressure inside the house is a little different from the pressure outside of the house, the window will shatter. This is why homes in areas that experience tropical storms are not airtight; if they were, their windows would shatter when a storm was nearby because of the air pressure change.

With sumo robots, this air pressure difference can be used to increase the apparent weight of the robot, or the forces on the wheels. You calculate the total downward force (pressure × area) acting on the robot from the air pressure, and then calculate the total upward force due to the air pressure that is acting on the bottom of the robot. The total force acting on the wheels is the weight of the robot, plus the downward force due to the air, minus the upward force due to the air.

For example, suppose you have a 6.6-pound (3-kilogram) robot. The surface area on top of the 8-inch square robot is 64 square inches, which equates to 940.8 pounds (14.7 psi × 64 square inches) of downward force from the 14.7 psi air pressure. The upward force due to the air pressure is also 940.8 pounds, since the area is the same. Hence, the total wheel force is 6.6 pounds (6.6 pounds + 940.8 pounds − 940.8 pounds).

What happens if we can reduce the air pressure under the robot by 2 psi? This will result in an upward force of 812.8 pounds due to the 12.7 psi air pressure under the robot. Now the total reaction forces on the wheels are a whopping 134.6 pounds (6.6 pounds + 940.8 pounds − 812.8 pounds). This is 20 times the original reaction forces on the wheels, so the robot can theoretically push an object that is 20 times its weight. And all of this was done with only a 2 psi pressure difference!

If the coefficient of friction on the wheels is, say, 1.2 and the weight of the robot is 6.6 pounds, the maximum pushing force for this robot is 7.92 pounds. If the coefficient of friction could be doubled to 2.4 (which is an extremely high value that would be difficult to obtain), the maximum theoretical pushing force would be 15.8 pounds. But if we took the original wheels and reduced the air pressure under the robot by only 0.5 psi (down to 14.2 psi), and the cross-sectional area were 64 square inches, the maximum downward force would be 38.6 pounds (6.6 pounds + 940.8 pounds − 14.2 psi × 64 square inches). Thus, the maximum pushing force for this robot is 46.2 pounds (1.2 × 38.6 pounds). It should be obvious that a very small change in pressure will yield greater pushing capabilities than increasing the coefficient of friction.

In reality, you cannot reduce the air pressure under the entire robot. You need a vacuum cup to separate the atmospheric air pressure from the vacuum pressure, as explained a bit later in the "Vacuum Seals" section. So, the upward force will be the sum of the upward forces due to both pressures multiplied over the respective areas on which the pressures are acting. Since the atmospheric air pressure cancels itself out on both sides of the robot, you only need to calculate vacuum pressure multiplied by the area of the vacuum pressure.

A standard pressure gauge will show zero air pressure under atmospheric conditions, even though the absolute air pressure is 14.7 psi. There is no such thing as negative absolute pressure, but a negative gauge pressure is an absolute pressure that is less than 14.7 psi. Vacuum pressure counts backward from atmospheric pressure. For example, a 10 psi absolute air pressure is a negative 4.7 psi gauge pressure. But vacuum pressure will be called 4.7 psi, not −4.7 psi.

For a vacuum sumo robot, the downward force due to the vacuum is vacuum pressure multiplied by the cross-sectional area this pressure is acting on the bottom of the robot. Hence, a 6.6-pound

CHAPTER 15: Improving Your Robot's Pushing Capability

robot with a 4-inch diameter vacuum cup holding 2 psi of vacuum will have a downward force on the wheels of 31.7 pounds (6.6 pounds + $(\pi/4) \times (4 \text{ in.})^2 \times 2$ psi). Thus, a robot with a coefficient of friction of 1.2 will have a maximum pushing force of 38 pounds, although the robot weighs only 6.6 pounds.

Fan-Style Vacuums

A fan-style vacuum consists of an impeller fan, motor, and a housing. The impeller is used to grab the air near the center of the impeller and throw it out to the sides or behind the fan. This creates a high-velocity airflow past the impeller, which draws air into the impeller. By enclosing the area in front of the impeller, a vacuum is created in this area because the air is being sucked out of it and drawn into the impeller blades.

Fan-style vacuums generally do not generate high-level vacuums, but you've seen that small vacuum levels can generate high wheel forces for high-strength pushing. The main advantage to using fan-style vacuums is that they can move a lot of air. Also, they can tolerate a lot of leaking from the vacuum cup seals and still perform well.

Vacuum Fans Impeller design is a fairly complicated process, and fabricating an impeller is extremely difficult and expensive. Fortunately, there is a source for low-cost vacuum impellers: vacuum cleaners. A vacuum cleaner uses this principle to create high-velocity airflow to help suck up dirt. Vacuum cleaner impellers are small and cheap.

One of the best sources for impellers is a small, portable, hand-held vacuum cleaner. Shown here are fan blades from a 2.4-volt cordless Dirt Devil vacuum and a 120 VAC hand-held Dirt Devil vacuum (www.dirtdevil.com). For a robot, the 120 VAC motor will have to be replaced with a high-speed 6-volt DC motor.

When using vacuum cleaner fans, you need to build a special housing so that the air is drawn into the center of the fan blade and exhausted from the sides or behind the fan. The exhausted air must not enter the chamber that is being evacuated; otherwise, you will not create a vacuum.

Also, the gap between the front of the blades and the vacuum chamber should be small, so that the volume of air that enters in the sides of the blades is small when compared to the volume of air that is being drawn into the center of the blades. Figure 15-2 shows this type of fan system.

Shown here is a working vacuum pump based on the layout shown in Figure 15-2. The outside housing is made of clear plastic to show the internal features. The impeller in this fan came from the 2.4-volt cordless Dirt Devil vacuum cleaner. When running at 7.2 volts, it sucks a lot of air and creates a good vacuum.

Ducted Fans   The electric ducted fans used in R/C airplanes to simulate jet engines are great high-velocity fans. These fans are like multibladed propellers. They are mounted inside a duct to precisely draw in air and exhaust it past the electric motor at a high velocity. Shown here is a 2.5-inch diameter electric ducted fan from VASA Model (http://vasamodel.no9.cz). Hobby-Lobby (www.hobby-lobby.com) is a United States distributor for these and many other electric ducted fans.

CHAPTER 15: Improving Your Robot's Pushing Capability

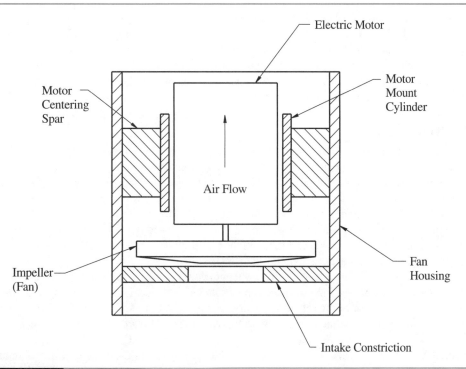

FIGURE 15-2 Layout of a fan system

Electric ducted fans draw the air through the outer blades and exhaust the air directly behind the blades. What makes these fans attractive for sumo robots is that a single kit contains the fan, duct, motor mount, and, sometimes, the electric motor. All of the parts assemble together precisely. All you need is an external mount for the duct.

Diaphragm Pump Vacuums

Another method for creating a vacuum is to use a miniature electric diaphragm pump. This type of pump has many industrial applications. In robotics, diaphragm pumps have become one of the most popular methods for increasing the apparent weight of sumo robots. These pumps can generate very high levels of vacuum—much higher than fan-style vacuums can generate. They are also quieter and draw much less current than fan vacuums do.

Electric diaphragm pumps use a small electric motor to drive a reciprocating arm that pushes and pulls a rubber membrane (a diaphragm). When the membrane is pulled up, its volume increases. To fill the volume, it draws air in from its intake port. There is a small check valve that closes to prevent the air from going back into the intake line, so that air goes out the outlet line when the diaphragm closes. You can can use this pump along with an electrically operated airflow valve, so you can turn on and off the pressure inside the vacuum cup as necessary.

Figure 15-3 shows the 12-volt DP140 diaphragm pump from Medo USA (www.medousa.com), and the 5D1060-101-1035 and 2D1034-101-1035 miniature vacuum pumps from Gast Manufacturing (www.gastmfg.com). The larger pump from Medo can fit inside a 3-1/2 inch square.

The drawback to using diaphragm pumps is that the airflow is very low when compared to a fan-style vacuum. This means that leaks in the vacuum cup area will prevent the vacuum from reaching its full potential. The following section describes how to minimize this problem.

Vacuum Seals

The most important part of a vacuum system on a sumo robot is the vacuum seal between the robot and the sumo ring. The boundary between the bottom of the sumo robot and the dohyo surface that is used to create the vacuum seal is called the *vacuum cup*. A vacuum cup must be able to provide an airtight seal but also allow the robot to move across the surface of the dohyo. The goal is to use as large a cross-sectional area as possible to maximize the vacuum forces.

The problem with traditional cone-shaped vacuum cups used in the material-handling industry is that they are about 1 inch high, which is a difficult size to use in a sumo robot. Most people use custom-made vacuum cups for their robots.

There are many different ways to create a vacuum cup. Figure 15-4 illustrates the components of a good vacuum cup design. First, you need an airtight seal between the vacuum pump (diaphragm or fan-style) and the base of the robot. You can use silicone rubber-sealing compounds to create the seal (the same material that you can use to increase the coefficient of friction on the wheels to improve traction).

The main part of the vacuum cup is the body that goes between the bottom of the robot and the sumo ring surface. You should calculate the height of the body so that, when all of the parts

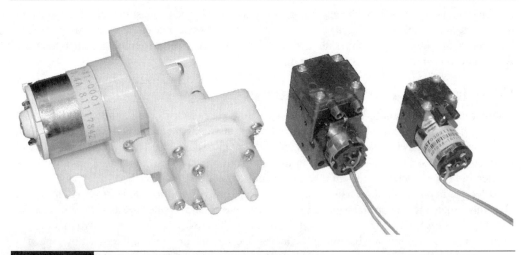

FIGURE 15-3 Miniature diaphragm vacuum pumps

CHAPTER 15: Improving Your Robot's Pushing Capability

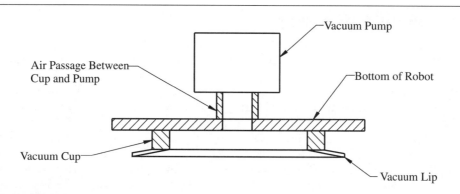

FIGURE 15-4 A vacuum seal design

are put together, the vacuum cup is not supporting any of the weight of the robot. If it is, there will be less weight on the wheels, which reduces the effectiveness of having the vacuum in the first place. The vacuum cup material can be a hard solid material or a soft foam material, as long as it does not allow air to pass through it. Open cell foam (sponge-rubber type) material will leak air through the walls. Closed-cell foam is a better choice. Another approach is to cast a soft rubber frame on the base of the robot, using a silicone or polyurethane compound.

The last part of the seal is a low-friction lip on the edge of the vacuum cup. One approach is to use a thin piece of Teflon sheet (about 0.02-inch thick), cut into the shape of the vacuum cup. Packing tape stuck to the edge also works well. (Sticky side up so it doesn't stick to the sumo ring.) If the tape gets damaged, it can be replaced easily between tournament rounds. This lip should protrude outside the vacuum cup edge, so that the air pressure inside the cup will be lower than the pressure outside the cup. If there is a leak under the lip, there will be a low-pressure spot right under a flexible lip that has the higher air pressure air acting on it. The higher pressure air will push the lip down onto the surface, thus helping to maintain the vacuum.

CAUTION *Remember that the lip is the material that slides along the surface of the dohyo, so it must be a low-friction material. Also, it should not scratch the dohyo, or your robot could be disqualified.*

If you are using a diaphragm pump, you will need a high-quality, very low-leaking vacuum cup. Make sure that the gap between the bottom of the robot and the sumo ring surface inside the vacuum cup is as small as possible, so that the volume of air in this region is minimized. This way, if there are leaks in the seal, there is less volume for the pump to try to pump out. If you are using a fan-style pump, the vacuum cup can tolerate more air leakage and still work.

Vacuum cups will make or break your vacuum system. They will probably be the hardest part to build and get working correctly on your robot. Conceptually, they all work similar to the design shown in Figure 15-3. The exact method of implementation depends on the robot design.

Magnetic Systems

In 2001, the international robot sumo rules changed the requirement for sumo ring construction from aluminum with a vinyl surface to an aluminum ring with a steel surface. The steel surface allows the use of magnets, which can increase the apparent weight of the sumo robot. All you need to do is place some magnets on the bottom of your robot, and then let the magnetic attraction force pull your robot down toward the sumo ring surface. You can get 100 pounds of downward force, without needing any special hardware for your robot.

The main advantages of using permanent magnets is that they are relatively small, have a lot of holding force, and do not need to maintain a vacuum seal to be effective. You may think that magnets are the way to go. But using the magnetic force of attraction is not an easy way to increase the apparent weight of the robot. Magnets must be placed carefully to get any benefit from them and to avoid having your robot get stuck on the dohyo. Also, high-strength magnets are very expensive.

Magnet Types

There are a wide variety of magnets that can be used for this application, including ceramic (ferrite), aluminum-nickel-cobalt (Alnico), rare-earth, and electromagnets. McMaster Carr (www.mcmastercarr.com) sells all of the different types of magnets. A search on the Internet will find many other sources for low-cost magnets.

NOTE *When you are shopping for magnets, realize that the lifting force of the magnet is based on the mild-steel surface that is touching the magnet. When the magnet is advertised to have a lifting force of some amount, it is true only under ideal conditions. In reality, the actual lifting capability will be lower. However, the advertised lifting force does give you an idea on the relative strengths of various magnets.*

Ferrite Magnets The most common magnet types are flexible, disc, and rectangular magnets, as shown in Figure 15-5. These magnets are generally made out of ferrite materials and have a black color. They have the lowest holding strength, as well as the lowest cost, of all of the magnet types. Ferrite magnets can be obtained at just about any hardware, department, or electronics store.

The flexible magnets are the weakest of all of the ferrite magnets, and they are also the least expensive. The flexible magnets that have an adhesive back can be used to mount the magnet to the bottom of a robot. The flexible magnet shown in Figure 15-5 is 1/2 inch wide and has a maximum pull of 6 pounds per linear foot. If the bottom of the robot is 4 by 6 inches, you will need a total of eight 6-inch long strips to cover the bottom of the robot. At 6 pounds per linear foot, each strip will have a pull of 3 pounds, and the total pull for all eight strips will be 24 pounds, almost four times the weight of the robot.

Alnico Magnets The next level of magnets are the Alnico magnets. These magnets have a metallic appearance and come in a wide variety of shapes, such as bars, discs, rings, rectangles, and horseshoes. You can find Alnico magnets at some of the larger hardware stores. They will be called "high-performance magnets" at a hardware store, because they are stronger than ferrite magnets.

CHAPTER 15: Improving Your Robot's Pushing Capability

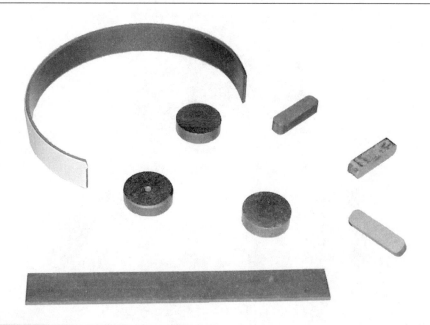

FIGURE 15-5 Various ferrite magnets

Rare-Earth Magnets   The real high-performance magnets are rare-earth magnets, which are also the most expensive type. These include the samarium-cobalt family of magnets and neodymium-iron-boron family of magnets. These magnets are called rare-earth magnets because they contain rare-earth elements. The neodymium magnets are generally stronger than the samarium magnets, but there are some grades of samarium magnets that are stronger than some grades of neodymium magnets.

Rare-earth magnets are usually purchased via mail order or the Internet. The best places to obtain rare earth magnets are from surplus stores such as Electronic Goldmine (www.goldmine-elec.com) and from Wonder Magnet (www.wondermagnet.com).

Figure 15-6 shows a surplus magnet assembly that contains two 1.5-inch long by 0.5-inch wide by 0.1-inch thick neodymium magnets. These magnets have an advertised pull of 15 pounds. You can remove the magnets from their mounts by gently smacking the backside of the mounts with a hammer. Since rare-earth magnets are very brittle, you may break the magnets by doing this, but the pieces work just as well as the unbroken magnets. You can glue about 20 of these magnets to the bottom of a sumo robot, which would result in a downward force of 300 pounds! Of course, with this much force, you would have a really hard time pulling the robot off the sumo ring.

Electromagnets   Electromagnets can be turned on and off, which can be very useful in a sumo match. You can turn the magnets on during pushing and off during high-speed maneuvers. McMaster Carr sells several different types of 12-volt electromagnets that have pulling forces up to a couple hundred pounds.

FIGURE 15-6 Surplus neodymium magnets

The main drawback to electromagnets is that they are fairly large and heavy. You need to decide whether you will use them at the beginning of the robot design process, because all of the other components must be designed around them.

Magnet Placement

In robot sumo, the robots cannot damage the dohyo. Sliding your magnets on the surface of the dohyo could cause unacceptable surface scratches, which would result in disqualification. To avoid this possibility, the magnets must be held above the surface of the dohyo. You will need to place the magnets very carefully, or you will either not get any benefit from them or you will get stuck on the dohyo.

Classical physics states that the force of attraction is inversely proportional to the distance cubed between the magnet and its object of attraction. In other words, if you doubled the distance between the magnet and the metal object, the resulting magnetic force will be approximately one-eighth the original value. This is not exactly true for real magnets. Their actual shape and type, the air between the magnets, and other factors have an effect on how the magnetic force changes as the distance between the magnet and the metal surface increases. The bottom line is that the magnetic force of attraction rapidly drops off to nothing in a very short distance.

CHAPTER 15: Improving Your Robot's Pushing Capability

When placing magnets on the bottom of the robot, you must allow for a small air gap so the magnets do not touch the sumo ring. You can glue the magnets down with a high-strength adhesive such as Loctite or bolt them to the bottom of the sumo robot. You will need to experiment with the air gap between the magnets and the surface of the dohyo. The force of attraction from the magnets will compress the wheels, which causes the magnets to get closer to the metal surface, which increases the magnetic force, which compresses the wheels some more, and so on. If you use high-strength magnets, you may want to use a harder rubber for the wheels, so they are not compressed as easily.

CAUTION *Pinched skin, cuts, and even broken fingers can result from improperly handling high-strength magnets. These magnets should not be able to freely slide along a surface by themselves, because they will find metal or another magnet to slam against. Also, high-strength magnets can damage magnetic storage media, such as floppy disks, computer hard drives, credit cards, computer monitors, and TV screens.*

In Chapter 2, you learned about the Yusei (advantage) point. This type of scoring point was added because of magnetic systems. When a robot gets stuck because one of its magnets touches the sumo ring, the other robot is awarded a yusei point. For example, suppose that one wheel of a four-wheel-drive robot comes off the edge of the sumo ring, but it does not touch the ground around the sumo ring. Robots without magnetic systems can get back on the dohyo in such situations. But if the robot uses magnets, most likely the magnets will rock the robot over, and the magnets will touch the sumo ring edge. The motors on the sumo may not be strong enough to pull the robot back onto the sumo ring. In this case, the other robot gets a yusei point.

Closing Remarks on Traction and Pushing Capabilities

This chapter covered some advanced techniques for sumo robot design. They are generally the most difficult aspect of a sumo robot to implement. If they work correctly, they will give your robot a competitive edge over robots that are not using these techniques, but they don't guarantee a victory. As stated throughout this book, a champion robot will have a good balance of motors, motor controllers, sensors, microcontrollers or radio controllers, traction, strategy, and some luck. But having a lot of traction does help.

If you do successfully improve your robot's pushing capabilities, when people see your robot winning contests, they will notice that you have a unique traction system and bombard you with questions on how you did it. A search on the Internet will show that there are many different websites devoted to every aspect of sumo robots, except for improved traction techniques. There are almost no websites that provide details on how to make high-traction systems. It takes a lot of time and experimentation to determine the right combination of components that will give a robot the competitive edge. Most people do not want to disclose their techniques, and they think their method is the best method. One of the things that makes robot sumo so interesting is that there are still a lot of little secrets about the robots that people are constantly trying to figure out. In this book, you've read about all of the elements that go into making a sumo robot. For the most part, sumo robots are fairly simple to make. You can purchase parts from various sources or build your own robot from the ground up. In robot sumo, there isn't a specific design that always wins, which allows you to use all of your creativity to come up with some wonderful creations. The next chapter shows various robots to give you some ideas on what you can include in your robot. They all look different and perform differently. And most of them are champion robots.

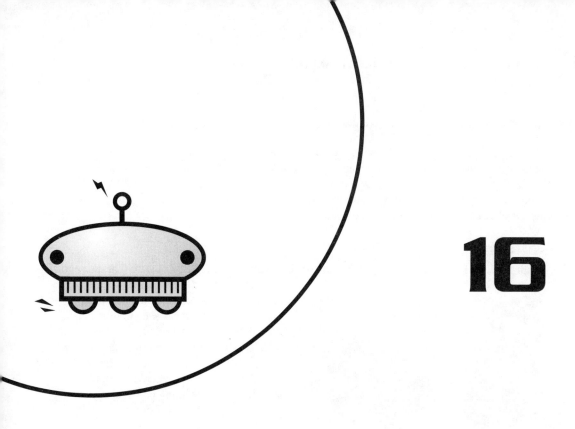

Real-Life Sumo Robots: Lessons from Robot Builders

It is estimated that more than 10,000 robots have competed in robot sumo worldwide since 1990. So far, there hasn't been one sumo robot design that has consistently proven to be superior to other designs. Without a "winning" design to mimic, and with few restrictions on robot design, many different types and styles of sumo robots show up at competitions. This variety makes robot sumo a truly exciting contest, where anyone has a chance to win.

This chapter is a gallery of different sumo robot styles designed by robot builders throughout the world. Most of these robots have placed in the top three positions in various robot sumo tournaments. You'll find examples of robots in different weight classes, including mini sumo, middleweight, and 3-kilogram robots.

Mini Sumo Robots

Mini sumo is one of the fastest growing weight divisions for sumo robots. These robots are small, so you can have your own mini sumo ring in your home. They are easy and inexpensive to build. Figure 16-1 shows a collection of mini sumo robots that competed in the 2002 Seattle Robotics Society Robothon (www.seattlerobotics.org/robothon). You can see that there are a wide variety of designs that can make a contest exciting.

Goliath

Goliath, built by Daryl Sandberg, is more that 10 centimeters long. The rules state that the robot must fit inside a 10-centimeter square box at the start of the contest, so Goliath starts on its back. After the 5-second time delay, it quickly shoots backwards about a quarter inch, which causes the robot to fall forward. Then it drives forward, looking for its hapless victim. Shown here is Goliath in its starting and pushing configurations (courtesy of Daryl Sandberg).

FIGURE 16-1 Collection of mini sumo robots

CHAPTER 16: Real-Life Sumo Robots: Lessons from Robot Builders

Goliath is a very well-built robot. It uses two custom-built wheels that have a polyurethane coating for improved traction. The diameter of the wheels is large so that a pair of Hitec 945MG R/C servo bodies can fit inside the rims. The body is made from a carbon fiber-cloth, fabricated similar to how fiberglass is formed. The images here (courtesy of Daryl Sandberg) show the internal structure of this robot. Goliath uses four Sharp GP2D15 sensors for opponent detectors and a pair of QRB1114 reflective sensors for edge detectors.

Goliath has won many tournaments. One of the things that Goliath has proven is that a robot does not need to be at maximum weight to win a contest. Goliath, a mini sumo, has competed effectively against 3-kilogram sumo robots, giving robots that weigh six times more than it a run for their money.

Nemesis

Pete Burrows has built a very effective mini sumo robot called Nemesis, shown here (courtesy of Pete Burrows), who is truly a nemesis to other robots. At less than 2 inches tall, Nemesis has a very low profile. All of its parts are custom-made components. It uses special 12-volt, 350 rpm micro gearboxes that were obtained from TechMax (www.techmax.com). The battery power for this robot is twelve 180 mAhr NiMH button cells, which were pulled out of a larger battery. This robot has consistently placed in the top three in all of the tournaments it has entered.

Marauder

Marauder, built by Dave Hylands, is a modified Mark III mini sumo kit (described in Chapter 13). Many people like to start with an existing robot platform, such as the Mark III (www.junun.org/MarkIII/Store.jsp) or the SumoBot (www.parallaxinc.com), and then add features to the robot. As you can see in the picture of Marauder (courtesy of Dave Hylands), the builder made some changes to the circuits, added a couple batteries for the motors, and replaced the rubber band with a polyurethane casting. Marauder won the mini sumo contest at the 2002 Western Canadian Robot Games.

Bam

The robots described so far use infrared reflective sensors. Bam, a mini sumo robot developed by Jim Frye, uses ultrasonic sensors. Its design is shown here (courtesy of Jim Frye).

Six-Pack

Six-Pack is a six-wheel drive mini sumo robot also built by Jim Frye. This robot uses LEGO wheels and gears. As shown here (courtesy of Jim Frye), Six-Pack has a front scoop that deploys forward at the start of a match.

Black Widow

Not all robots use sensors to see their opponents. The stealth mini sumo robot named Black Widow, built by Kristina Miles, uses flat surfaces to deflect infrared light away from its opponents, as shown here. The only sensors it uses are infrared edge detectors. This robot runs its servo motors at 9 volts for high-speed maneuvers. It uses silicone-coated wheels for improved traction and pushing capabilities. Black Widow also uses multiple internal programs for determining the type of attack to use on its opponents. It randomly moves across the sumo ring and eventually finds and overpowers its opponents. This robot has beaten Nemesis and Goliath in some of their meetings.

Walking Robots: Black Marauder the Biped and Pete's Folly the Hexapod

Most people prefer to build robots that use wheels for locomotion, but some builders like the mechanical challenge of designing walking sumo robots. Shown here are Black Marauder, a two-legged robot, and Pete's Folly, a six-legged robot, built to compete in mini sumo contests.

Although these robots are easily knocked over by their opponents, they are great crowd-pleasers. When Black Marauder's silicone-coated feet are on the ground, hitting them will cause many robots to come to a stop. This indicates that a better design for a walking robot could be used to win sumo tournaments.

Remote-Control Mini Sumo: Remote Possibility

Not all mini sumo robots are autonomous. Shown here is a mini sumo robot named Remote Possibility. This robot uses the Marvin Slyder kit from Sine Robotics (www.sinerobotics.com) and a Microchip PIC microcontroller to interpret the radio-control receiver commands, to control the modified servos. Although this robot is not pretty, its silicone-coated wheels and good driving controls have enabled it to go undefeated in the remote-control class of the Northwest Robot Sumo Tournament since 2000.

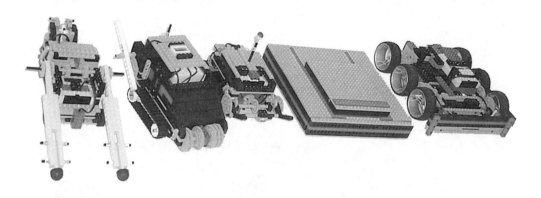

FIGURE 16-2 Various LEGO sumo robots (courtesy of Steve Hassenplug)

LEGO Sumo Robots

LEGO sumo robots are rapidly gaining popularity throughout the world. There are many sumo contests that allow only robots made of LEGOs. All of the other sumo contests allow any robot constructed of LEGOs. Many of these robots use the RCX brick from the Lego Mindstorms Robotics Invention System (discussed in Chapter 11). Figure 16-2 shows several different designs for LEGO sumo robots.

Middleweight Sumo Robots

Middleweight robot sumo class follows all of the same rules as the 3 kg sumo class, except the maximum weight of the sumo is 1.5 kg instead of 3.0 kg. The lightweight sumo class is the same as the middle and 3 kg sumo classes except that it has a 1.0 kg max weight. The middleweight sumo class is one of the popular weight classes that are used at many robot contests, such as the sumo events at the Central Illinois Robot Club (http://circ.mtco.com).

Too Much

Sometimes, a bidirectional robot that can function in forward and reverse works better than a robot that needs to back up and turn around to continue a battle. Peter Campbell designed a unique robot named Too Much, who has the same attacking capability going forwards or backwards. This robot, shown here (courtesy of Peter Campbell), was handmade using the expanded PVC called Sintra.

Too Much uses a Basic Stamp 2 on the Next Step developer board from Lynxmotion to monitor the four infrared proximity detectors, and a modified life-tracker board from Lynxmotion to control two Hitec HS-805BB servos. The high-traction wheels for this robot come from the lint pickup rollers mentioned in Chapter 5. Because of the symmetry of this robot, a carrying handle was built into the top center of the robot.

Since this robot doesn't need to turn around when it comes to the edge of the sumo ring, other robots cannot sneak behind it. This robot's unique design has resulted in it finishing in the top two positions in the middleweight sumo class at the Central Illinois Robot Club's tournaments since 2000 and winning the middleweight sumo class at ITT Tech in Fort Wayne, Indiana.

Full-Sized Sumo Robots

From a spectator's viewpoint, 3-kilogram sumo robots are the most exciting to watch—they move fast and hit hard. Because of their larger size, these robots can handle more options and are fun to build. Figure 16-3 shows two robots competing at the 2002 International Robot Sumo Tournament, sponsored by Fuji Soft ABC Inc.

Sometimes, robots are not the winners in robot sumo tournaments. During the first two years of the Northwest Robot Sumo Tournament, a rock (shown here) was the main target used to generate interest in the event. The first year, no robot was able to move this small rock out of the sumo ring. In the second year, the rock won most of its contests, but some robots gave it a good push; other robots broke down after hitting the

CHAPTER 16: Real-Life Sumo Robots: Lessons from Robot Builders

FIGURE 16-3 Two 3-kilogram robots competing at the 2002 International Robot Sumo Tournament

rock. The rock no longer competes, but the rock will always be known as the champion of the beginning years of the Northwest Robot Sumo Tournament.

Big TX

Mr. Mato Hattori built Big TX, shown here, who competed in robot sumo contests in Japan. Because of his excitement for robot sumo, Mr. Hattori made a video about how Big TX was built and the Japanese robot sumo events, and then distributed it to several robot clubs throughout the United States. Since then, Mr. Hattori has been known as the person who introduced robot sumo to the United States and helped robot sumo spread internationally.

Leo

Leo, built by Tom Dickens, is a long-time competitor and champion sumo robot in the Seattle area. This robot uses a pair of cordless drill motors to drive both of the two wheels. What is interesting about this robot is that the motors come from two different types of cordless drills. This results in the robot not driving in a straight line; it actually drives in an arc, which turns out to be a very effective drive method. The motor speed controller for this robot is a simple relay-based control with a single MOSFET transistor (described in Chapter 9). The sides of this robot are hinged. Under the sides are switches that are used to detect if the robot has hit anything. At the top of the robot is a plastic character named Leonardo, one of the Teenage Mutant Ninja Turtles, as shown here, which makes a great handle for picking up the robot.

Student Engineering Projects

Shown here are two autonomous sumo robots from Singapore that competed in the 2002 International Robot Sumo Tournament in San Francisco. These robots were part of an engineering project by various students from the Mechanical Engineering Division and InnoHub (Alpha Centre), School of Engineering, Ngee Ann Polytechnic, Singapore.

Skeleton

Skeleton is a remote-control, 3-kilogram sumo robot, built by Mr. Benedict Brandon Phay Yee How, Ms. Jenny Abdul Jabbar, and Mr. Kwa Zhaowei, from Singapore. This robot, shown here, competed at the 2002 International Robot Sumo Tournament. Skeleton uses 24 AAA batteries to drive a pair of Maxon RE-25 gearhead motors, which use a pair of bevel gears to create a right-angle drive. The five devices that look like large transistors are L6203 4-amp motor controllers from ST Microelectronics (www.st.com). The main microcontroller on this robot is an NEC μPD78310A, which is used to interpret edge-detection signals and radio-control commands to control the motors.

Boxter

Some robots are not very impressive looking on the outside, but looks aren't everything. As you can see here (courtesy of Daryl Sandberg), Boxter is a unimposing 3-kilogram sumo robot with a big, thin flap in front. This robot, built by Daryl Sandberg, sits on its back at the start of the match, and then flips forward after the match begins. It slowly moves around the sumo ring, looking for its opponent. Boxter's main feature is its strong pushing motors, controlled by standard R/C ESCs. The flap in front is a thin piece of very flexible spring steel, which sits nearly flat on the sumo ring. Boxter uses the Basic Stamp 1 as its brains. Many people in the amateur robotics community believe that a Basic Stamp 1 isn't powerful enough to win in robot sumo. Boxter proves them wrong by winning more than 90 percent of its matches!

The photo shown here was taken at the tenth annual Northwest Robot Sumo Tournament. It looks like the other robot (built by Pete Burrows) is pushing Boxter out of the sumo ring. In fact, the other robot is actually stuck on top of Boxter's front flap. Boxter simply and slowly pushed this robot out of the sumo ring, winning the match. Although Boxter isn't one of the best looking robots, it is one of the best performing robots.

Overkill

Overkill is a six-wheel-drive, 3-kilogram sumo robot, built by Jim Frye. As shown here (courtesy of Jim Frye), Overkill has a front scoop that deploys forward at the start of a match. This robot uses two Basic Stamp 2 modules to control all of its functions, including its six ultrasonic range sensors, bump switches, and edge detectors. Overkill has consistently placed in the top three in all of the tournaments it has entered.

OctoBot

OctoBot is a unique robot, built by Jeff Loitz, that competes in the Central Illinois Robot Club's sumo tournaments. This robot is made of black expanded PVC. It uses two BotBoard I microcontrollers to monitor a pair of infrared line detectors, object detectors, mini SSC (serial servo controller) II, and bump switches. This robot uses a total of 13 different R/C servos. Each servo controller can control eight different R/C-style servos using serial communications.

The photo shown here illustrates how OctoBot starts with its eight lint-roller wheels folded up, so that the original starting body will fit inside a 20-centimeter square box, as required by the robot sumo rules. At the start of the tournament, four servos are used to lower the eight wheels and the front plow. Each wheel is powered by its own 49 ounce-inch standard R/C servo. When the wheels and front plow are deployed, this 5.5-pound robot is 14 inches wide by 13.5 inches long. The high-traction wheels on this sumo robot allow it to climb an incredible 65-degree slope!

During one of OctoBot's competitions, it went against a robot that has a white front scoop to confuse other robots' edge detectors. The white scoop did exactly what it was designed to do. OctoBot started moving backwards, as it was programmed to do in this situation. Although it has an edge detector on its rear to help prevent it from being pushed backwards off the sumo ring, the software was programmed to ignore the rear sensor if the forward sensor was triggered. The end result was that OctoBot backed right off the dohyo. After OctoBot hit the floor, its front object sensors detected the edge of the sumo ring, and OctoBot went after the sumo ring. To everyone's amazement, OctoBot started to push the sumo ring across the floor, along with the other robot on top of it (that was randomly moving along the surface). The crowd loved it, and OctoBot won the Coolest Robot award for its unexpected demonstration of pure strength.

Which Robot Is Better?

Shown here is a match between a pair of robots built by Bill Harrison of Sine Robotics (www.sinerobotics.com). The core design of these four-wheel-drive robots is the same, but the type of sensors and the programming software are different. One robot is using an arm that deploys forward with a flag to confuse the opponents. The other robot uses ultrasonic sensors to detect its opponent. These robots use sealed-lead acid (SLA) batteries for their main power supply and custom-made wheels silicone-casted wheels for high traction. Except for the sensors, these robots are identical. This time ultrasonics won over the arm sensor, but the competition results are 50-50 with these robots.

Now Go Build a Robot!

This chapter showed many different styles of sumo robots built by people throughout the world. There are many other designs that have won contests or will win future contests. This gallery of robots should give you ideas for different types things you may want to include in your robot designs. As long as you are not violating one of the few rules for robot design, you are limited only by your own imagination.

This book provided a lot of information about the various components that go into sumo robots—for just about any type of robot you want to build. By now, you should have learned everything you need to know in order to build a champion, exciting, and fun robot. The appendix at the end of this book lists all of the related sources and references mentioned throughout the book. For the beginning robot builder, robot sumo gives you a specific goal to achieve with your robot as you learn. For the experienced robot builder, the many tips and tricks presented in this book will help you make better sumo robots. At the time this book was published, Japanese competitors were undefeated in the International Robot Sumo Tournaments. Perhaps the secrets presented in this book can be used to end that reign.

Now it is time to have fun—go build a sumo robot!

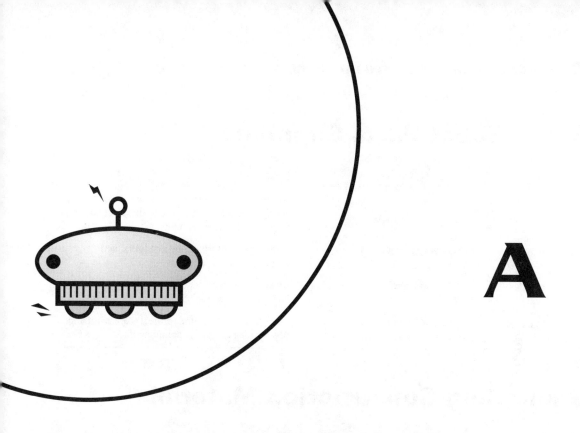

Robot Resources and References

This appendix lists sources for suppliers of items associated with robot sumo, robot sumo references, robot sumo resources, major robot sumo tournaments, and robotics clubs. Because most companies and organizations now advertise and do business via the Internet, only web addresses are shown here. You can find other contact information for the companies and organizations at their websites.

Sumo Robot Parts Suppliers

Company	Web Address	Comments
Lynxmotion	www.lynxmotion.com	Number one robot sumo parts and kits supplier in North America. Also sells a wide variety of other types of robot parts and kits.
Acroname	www.acroname.com	Wide variety of robot parts, sensors, and microcontrollers
Parallax, Inc.	www.parallaxinc.com	Makers of SumoBot and Basic Stamp products, and great source for other robot components
Mark III Project	www.junun.org/MarkIII/store.jsp	The Mark III mini sumo project store, with a lot of very nice electronic components

Sumo Ring Construction Materials

Company	Web Address	Comments
Lonseal Company	www.lonseal.com	Vinyl surface material

Wheels, Hubs, and Mounting Hardware

Company	Web Address	Comments
Acroname	www.acroname.com	Omni-directional wheels, mini sumo wheels, and splined servo mounting hubs
Dave Brown Products Inc.	www.dbproducts.com	Excellent foam wheels
Lynxmotion	www.lynxmotion.com	Wide selection of wheels and mounting hubs
McMaster-Carr	www.mcmastercarr.com	General-purpose parts supplier
Parallax, Inc.	www.parallaxinc.com	Mini sumo wheels
Team Associated	www.teamassociated.com	Foam wheels, hubs, and mounting hardware

Appendix A: Robot Resources and References

Company	Web Address	Comments
Traxxas Corporation	www.traxxas.com	Foam wheels, hubs, and mounting hardware
DuraTrax	www.duratrax.com	Foam wheels, hubs, and mounting hardware
Tower Hobbies	www.towerhobbies.com	Well-stocked hobby store
Team R/C Racing Components	www.tm-rc-racingcomponents.com	Wide variety of soft foam R/C car racing wheels and tires

Motor and Gearmotor Sources

Company	Web Address	Comments
Lynxmotion	www.lynxmotion.com	Good selection of sumo class gearmotors
Astro Flight, Inc.	www.astroflight.com	Extremely powerful DC motors
Tamiya	www.tamiya.com	Excellent gearmotors; available only in Japan
Jameco Electronics	www.jameco.com	Wide variety of motors and gearmotors, including Hsiang Neng gearmotors
Graupner	www.graupner.com	Wide selection of high-speed electric motors
Trinity Products	www.teamtrinity.com	High-powered R/C car electric motors
Mabuchi Motor Co.	www.mabuchi-motor.co.jp	Motor manufacturer
Johnson Electric	www.johnsonmotor.com	Large Motor Manufacturer
Maxon	www.maxonmotorusa.com	Motor manufacturer
Hsiang Neng DC Micro Motor Manufacturing Corp.	www.hsiangnengmotors.com.tw	Motor manufacturer
Pitman Motors	www.pittmannet.com	Motor manufacturer
MicroMo Electronics	www.micromo.com	Motor manufacturer
Colman Motor Products	www.merkle-korff.com	Motor manufacturer

Surplus Motors and Gearmotors

Company	Web Address	Comments
Servo Systems	www.servosystems.com	Wide selection of motors; the catalog has more motors than those shown on the website
Marlin P. Jones & Assoc., Inc.	www.mpja.com	Good source for low-cost surplus motors and gearmotors
The Electronic Goldmine	www.goldmine-elec.com	Good source for low-cost surplus motors and gearmotors
TechMax	www.techmax.com	Excellent source for low-cost miniature electric motors and gearmotors

R/C Servo Motors

Company	Web Address
Futaba	www.futaba.com
Hitec	www.hitecrcd.com
Airtronics	www.airtronics.net
FMA Direct	www.fmadirect.com
Cirrus	http://cirrus.globalhobby.com
JR	www.jrradios.com
Tower Hobbies	www.towerhobbies.com

Gears and Drive Components

Company	Web Address	Comments
Stock Drive Products	www.sdp-si.com	Excellent source for any type of drive component
PIC Design	www.pic-design.com	Source for small gears, belts, pulleys, and clutches
W.M. Berg, Inc.	www.wmberg.com	Source for all types of gears

Batteries

Company	Web Address
Panasonic	www.panasonic.com/industrial/battery/industrial
Sanyo	www.sanyo.com/industrial/batteries/index.html
Eveready	www.eveready.com
Duracell	www.duracell.com
Power-Sonic	www.power-sonic.com
Duratrax	www.duratrax.com
Team Trinity	www.teamtrinity.com
Tower Hobbies	www.towerhobbies.com

Motor Controllers

Company	Web Address	Comments
Solutions Cubed	www.solutions-cubed.com	Excellent 10-volt and higher motor controllers that accept serial, analog, and 1 to 2 ms R/C servo-style pulse inputs
Vantec	www.vantec.com	Number one source for high-current motor controllers
Innovation First	www.innovationfirst.com	Maker of the 60-amp Victor 883 motor controller
Lynxmotion	www.lynxmotion.com	Wide selection of motor controllers for sumo robots
Acroname, Inc.	www.acroname.com	Several nice motor controllers including the Magnavision motor controller
Sozbots	www.sozbots.com	Source of miniature 5-amp dual-motor controllers
HVW Technologies	www.hvwtech.com	Nice selection of motor controllers and parts
Mondo-tronics	www.robotstore.com	Good selection of motor controller kits
HiTec RCD	www.hitecrcd.com	R/C car electronic speed controllers (ESCs)
Futaba	www.futaba.com	R/C car ESCs

Company	Web Address	Comments
Novak	www.teamnovak.com	R/C car ESCs
Traxxas	www.traxxas.com	R/C car ESCs
LRP Electronic	www.lrp-electronic.de	R/C car ESCs
Duratrax	www.duratrax.com	R/C car ESCs
Scott Edwards Electronics	www.seetron.com	Eight R/C servo motor controller

Remote-Control Equipment

Company	Web Address	Comments
HiTec RCD	www.hitecrcd.com	Major R/C equipment manufacturer
Futaba	www.futaba.com	Major R/C equipment manufacturer
Airtronics	www.airtronics.net	Major R/C equipment manufacturer
Innovation First	www.innovationfirst.com	Maker of the Issac robot controller system
Graupner	www.graupner.com	V-tail mixing circuts

Microcontrollers

Company	Web Address	Comments
Parallax, Inc.	www.parallaxinc.com	Maker of Basic Stamps, SumoBot, and a wide variety of support kits, components, and instructional documentation
Basic Micro, Inc.	www.basicmicro.com	Maker of Atom 24, 28, and 40 microcontrollers and excellent basic compilers
Acroname, Inc.	www.acroname.com	Creator of the BrainStem
Javelin Stamp	www.javelinstamp.com	Dedicated website for Parallax's Javelin Stamp
Gleason Research	www.gleasonresearch.com	Distributors for the Handy Board

Company	Web Address	Comments
BotBoard Plus	www.kevinro.com	BotBoard products and support hardware
Savage Innovations	www.oopic.com	Maker of the OOPic microcontroller
NetMedia, Inc.	www.basicx.com	Maker of the BasicX microcontroller
Solarbotics	www.solarbotics.com	Maker of BEAM neuro-networks
Microchip	www.microchip.com	Maker of the PIC microcontroller
Atmel	www.atmel.com	Maker of the AVR microcontrollers
Ubicom	www.ubicom.com	Maker of the SX microcontrollers
Cypress Microsystems	www.cypressmicro.com	Maker of the PSoC microcontrollers
RoboMinds	www.robominds.com	Makers of the powerful Motorola 68332 based robot controller boards
Celestial Horizons	www.celestialhorizons.com	A basic compiler

Robot Sensors

Company	Web Address	Comments
Lynxmotion	www.lynxmotion.com	Wide variety of robot sensors
Acroname, Inc.	www.acroname.com	Excellent selection of robot sensors
HVW Technologies	www.hvwtech.com	Wide variety of robot sensors
Parallax, Inc.	www.parallaxinc.com	Wide variety of robot sensors
Milford Instruments, Ltd.	www.milinst.com	Ultrasonic sensors
Mondo-tronics	www.robotstore.com	Wide variety of robot sensors
Devantec (Robot Electronics)	www.robot-electronics.co.uk	Maker of the SRF04 ultrasonic range finders
Sharp	www.sharp.co.jp	Optical sensor manufacturer
Panasonic	www.panasonic.co.jp/semicon/e-index.html	Optical sensor manufacturer
Lite-On Optoelectronics	www.liteon.com/opto.index.html	Optical sensor manufacturer
QT Optoelectronics	www.qtopto.com www.fairchildsemi.com	Infrared reflective sensor manufacturer

Electronic Parts Suppliers

Company	Web Address
Digikey	www.digikey.com
Mouser Electronics	www.mouser.com
Jameco Electronics	www.jameco.com
Allied Electronics	www.alliedelec.com
Radio Shack	www.radioshack.com

Mechanical Parts Suppliers

Company	Web Address	Comments
McMaster Carr	www.mcmastercarr.com	Has just about everything you could imagine
Stock Drive Products	www.sdp-si.com	Excellent source for any type of drive component
Small Parts, Inc.	www.smallparts.com	Small parts, fasteners, and metal stock materials
Grainger	www.grainger.com	Another great place that sells just about anything

Vacuum Pump Suppliers

Company	Web Address	Comments
MEDO USA, Inc.	www.medousa.com	Maker of miniature electric diaphragm vacuum pumps
Gast Manufacturing, Inc.	www.gastmfg.com	Maker of miniature electric diaphragm vacuum pumps

High-Strength Magnets

Company	Web Address	Comments
ForceField	www.wondermagnet.com	Excellent source for low-cost, high-strength magnets
Magnet Sales and Manufacturing Inc.	www.magnetsales.com	Thousands of different types and sizes of magnets
The Electric Goldmine	www.goldmine-elec.com	Good source for low-cost magnets
Australian Magnet Technology	www.magnet.au.com	Good source for a wide variety of magnets
Dura Magnetics, Inc.	www.duramag.com	Great source for a wide variety of magnets

Silicon and Polyurethane High Traction Wheel Casting Compounds

Company	Web Address
Synair, Inc.	wwww.synair.com
Tap Plastics	www.tapplastics.com
Polytek	www.polytek.com

Printed Circuit Board Manufacturers

Company	Web Address
AP Circuits	www.apcircuits.com
PCB Express	www.pcbexpress.com
Express PCB	www.expresspcb.com

Hobby Stores

Company	Web Address	Comments
Tower Hobbies	www.towerhobbies.com	Well-stocked hobby store that offers most of the R/C parts and equipment sold by the major manufacturers
Hobby-Lobby	www.hobby-lobby.com	Wide selection of parts; major distributor for European equipment from companies such as Graupner
Hobby People	www.hobbypeople.com	Wide selection of parts, and the distributor for the Cirrus brand servos.

Tools

Company	Web Address	Comments
Lee Valley Tools	www.leevalley.com	Exellent woodworking tools
Dremel	www.dremel.com	Various hand-held power tools and accessories—a must for any hobbyist
Optimum Designs Inc.	www.optascope.com	A really nice, low-cost PC-based oscilloscope

References

Robot Building

Applied Robotics, by Edwin Wise (Prompt Publications, 1999)   A good overview of basic experimental robotics.

Build Your Own Combat Robot, by Pete Miles and Tom Carroll (McGraw Hill, 2002) A great book on how to build combat robots, including sumo robots.

Build Your Own Robot, by Karl Lunt (A K Peters, 2000)   A great all-round reference for advanced small robot building.

Mobile Robots, Inspiration to Implementation by Joseph Jones and Anita Flynn (A K Peters, 1999)   A good intermediate book for mobile robot building.

Robots, Androids & Animatrons, by John Lovine (McGraw-Hill, 1997)   A good introduction to experimental robotics.

The Robot Builder's Bonanza, by Gordon McComb (McGraw-Hill, 2000)   A great first book on experimental robotics.

Electronics

The Art of Electronics, by Paul Horowitz and Winfield Hill (Cambridge University Press, 1989)   This is the electronics bible—a must for building electronic circuits.

Motor Control Electronics Handbook, by Richard Valentine (McGraw-Hill, 1998)   An excellent book on electric motor circuits.

Electric Motor Handbook, by Robert Boucher (Astroflight, 2001)   An excellent book on electric motors.

Machine Design

Mechanical Engineering Design, by Joeseph Shigley (McGraw-Hill, 1988)   Mechanical engineers' bible for machine design.

Machinery's Handbook, 26th Ed., by Erik Oberg (Industrial Press, 2000)   A must-have for all machinists.

Fundamentals of Machine Component Design, by Robert Juvinall and Kurt Marshek (Wiley & Sons, 1999)   An excellent book on machine design.

Programming

Programming and Customizing the Pic Microcontroller, by Myke Predko (McGraw-Hill, 1998)   An excellent book on using and programming the PIC Microcontroller.

Programming and Customizing the Basic Stamp, by Scott Edwards (McGraw-Hill, 1998)   An excellent book on using and programming the Basic Stamp.

Other Robot References and Resources

Resource	Web Address	Comments
Robot Books	www.robotbooks.com	This is a great store for robotics books, kits, and BattleBot class motors.
Open Source Motor Controller (OSMC) project	www.groups.yahoo.com/OSMC	This project involves open-source development efforts for motor controllers. There is a lot of good information about how motor controllers work.
Nuts & Volts Magazine	www.nutsvolts.com	A must-have magazine; if you don't have it already, subscribe to it. It covers excellent real-world use of electronics and many robotic applications.
Build Your Own Remote-Controlled Sumo-Bot	www.mcgraw-hill.com	A new robot sumo kit/book, by Mike Predko and TAB Electronics (McGraw-Hill), which is a nice kit for getting started in the world of robot sumo.
FindChips.com	www.findchips.com	This is a search engine that will tell you what companies are selling a specific integrated circuit.
Sine Robotics	www.sinerobotics.com	
Larry Barello's AVR and Robotics web page	www.barello.net	Great source for learing about closed loop motor control, and balancing robots. You can also get some nice motor controllers here.

Major Robot Sumo Tournaments

The following are the major robot sumo tournaments:

- Robothon (www.seattlerobotics.org/robothon)
- Northwest Robot Sumo Tournament (www.sinerobotics.com/sumo)
- Portland Area Robotics Society (www.portlandrobotics.org)
- Central Illinois Robot Club (http://circ.mtco.com)
- All Japan Robot Sumo Tournament (www.fsi.co.jp/sumo)
- International Robot Sumo Tournament (www.fsi.co.jp/sumo and www.robots.org)
- Western Canadian Robot Games (www.robotgames.com)

You can find the official robot sumo rules at www.fsi.co.jp/sumo, the website of the International Robot Sumo Tournament and All Japan Robot Sumo Tournament organizers and rules committee.

Robotics Clubs

The following are some of the many robotics clubs in the United States and Canada, and many of which have annual sumo contests:

- Atlanta Hobby Robot Club (www.botlanta.org)
- Central Illinois Robot Club (http://circ.mtco.com)
- Chicago Area Robotics Group (www.chibots.org)
- Connecticut Robotics Society (www.ctrobots.org)
- Dallas Personal Robotics Group (www.dprg.org)
- Ottawa Robotics Enthusiasts (www.ottawarobotics.org)
- Phoenix Area Robotics Experimenters (www.parex.org)
- Portland Area Robotics Society (www.portlandrobotics.org)
- Robot Society of America (www.robots.org)
- Rockies Robotics Group (www.rockies-robotics.com)
- Seattle Robotics Society (www.seattlerobotics.org)
- Southern Oregon Robotics Club (www.1sorc.com)
- Triangle Amateur Robotics (www.triangleamateurrobotics.org)
- Twin Cities Robotics Group (www.tcrobots.org)
- Vancouver Robotics Club (www.vancouverroboticsclub.org)
- Winnipeg Area Robotics Society (www.winnipegrobotics.com)

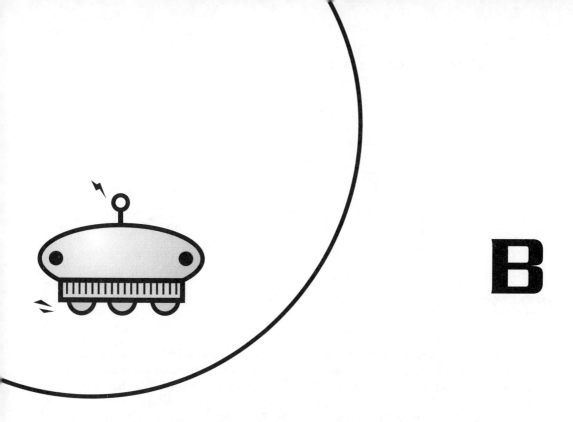

Robot Power Supplies

Most electronic circuits require 5 volts to operate properly, but batteries do not output 5 volts by themselves. In order to use batteries with electronic circuits, you need a voltage regulator to reduce the voltage from its normal rating to 5 volts.

One of the most popular 5-volt regulators is the LM7805 voltage regulator from National Semiconductor (www.national.com). The LM7805 is a 3-pin component that looks like a large transistor. This regulator can supply up to one amp of current before shutting down. If you try to draw too much current, you may damage the regulator. Figure B-1 shows a schematic drawing on how to wire up this voltage regulator. To use this circuit, all you need is the voltage regulator and two capacitors, C1 and C2. The capacitor values shown are the minimum recommended values from the manufacturer, and they can be simple ceramic capacitors. The voltage rating of these capacitors should be at least twice as high as the battery voltage. The resistor and the LED are optional. They are included in this circuit so you would have an indicator that will let you know that the circuit is being powered.

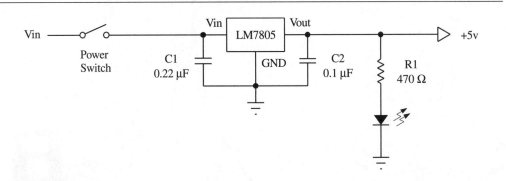

FIGURE B-1 Voltage regulator circuit

When working with these voltage regulators, you need to understand what voltage drop out means. A simple way to understand how the voltage drop out affects the circuit is to subtract the drop out voltage from the input voltage. If the result is greater than 5 volts, then the output voltage will be 5 volts. If the result is less than 5 volts, then the output voltage will be the result. The drop out voltage for the LM7805 is 2 volts. So for this circuit to work properly, the input voltage must be greater than 7 volts. If the input voltage is less than 7 volts, then the output voltage from the regulator will be less than 5 volts. So this means that a 4 AA cell battery pack cannot be used with this voltage regulator to provide a regulated 5 volts to the rest of the robot's electronics. But a 9-volt battery will work.

One additional thing to keep in mind when working with these voltage regulators is the current draw. Let say we have a 9-volt battery, and the motors are connected directly to the battery, and a voltage regulator is connected to the same battery to provide power to the microcontroller. Now this 9-volt alkaline battery has an internal resistance of 1.8 Ω, and there was a sudden surge of current being drawn by the motors that is equal to 1 amp. This results in a 1.8-volt drop in the battery, so the battery is now outputting 7.2 volts. Under this condition the voltage regulator will work just fine. But when the battery is used up a little bit, say, down to 8.5 volts, this 1 amp surge will drop the voltage down to 6.7 volts, which will result in the voltage regulator outputting 4.7 volts. Many microcontrollers will start to reset when the voltage gets much lower than this. This is one of the reasons why you don't power the motors and electronics with the same battery.

You will run into the same type of problems if you have a lot of sensors and components that add up to a high current draw from the batteries. To get around this problem is to either use a higher voltage battery or a low drop out voltage regulator such as the LM2940CT-5. This voltage regulator has a drop out voltage down to 0.5 volts. Thus the input voltage can be as low as 5.5 volts for the circuit to continue to work properly. Though more expensive than the LM7805, it could save you some power headaches when you are pushing the limits. The last thing you want to see happen to your robot is the microcontrollers resetting during a match because the batteries are getting low.

Appendix B: Robot Power Supplies

Table 1 shows a list of three popular voltage regulators, along with their maximum output current rating and the manufacturers minimum recommended values for the two capacitors. The LM78L05AC looks like a small transistor, and is used for low current demand applications. All of the capacitors are ceramic capacitors except for the LM2940CT-5 regulator. The C2 capacitor must be a polarized capacitor with the positive terminal connected to the Vout on the regulator. The manufacturer recommends a Tantalum capacitor, but an Aluminum electrolytic capacitor will work.

Regulator	Current	Max. Input Voltage	Drop Out Voltage	C1	C2
LM78L05AC	0.1 amp	20	2.0	0.33 µF	0.1 µF
LM7805C	1.0 amp	25	2.0	0.22 µF	0.1 µF
LM2940CT-5	1.0 amp	26	0.5	.47 µF	22 µF

A search on the National Semiconductor website using the key words "voltage regulator" will yield hundreds of different types of voltage level regulators to choose from. They are not just limited to 5 volts. They have many different output voltage levels and maximum current limits for a wide variety of applications.

Index

Numbers

3kg sumo weight class, 24
27MHz radio frequency band, 190
50MHz radio frequency band, 191
72MHz radio frequency band, 190
75MHz radio frequency band, 191

A

Abdul Jabbar, Jenny, 335
Accidents, 18
Ackerman steering, 69–70, 91
Acroname Inc., 72, 82, 225, 241, 276, 340
Acrylics, 49
Adhesives, 53
Advantage points, 17, 19, 323
Alignment, 87
Alkaline batteries, 120–121
All Japan Robot Sumo Tournament, 350, 351
All-or-nothing attack, 305
Alnico magnets, 320
Aluminum sumo robots, 50–51
AM radio systems, 192
American Wire Gauge (AWG), 133
Amp-hour (Ahr) battery rating, 123–126
Amplitude modulation (AM), 192–193
Angular contact bearings, 111
Antennas, 196–198
    length of, 197
    orientation of, 197–198
AP Circuits, 279
Apparent weight, 313–323
    magnetic systems for increasing, 320–323

vacuum systems for
    increasing, 313–319
  See also Weight
Artificial intelligence (AI), 212
Assembly language, 215
Atmel, 230
Autonomous sumo robots, 5

B

Backdriving, 109
Ball bearings, 111
Bam sumo robot, 328
Barello, Larry, 350
Base plate, 286, 287
Base-loaded antenna, 197
Basic Atom modules, 223–224
Basic Compiler, 224–225
Basic Micro, Inc., 223
Basic programming language, 216
Basic Stamp modules, 217–221
    features, 218–219
    mixing circuits, 204–206
    programming, 219–221
    servo calibration, 162
Batteries, 119–136
    alkaline, 120–121
    capacity of, 123–126
    comparison chart of, 129
    connecting in packs, 128–130, 131
    effects on motors, 148–149
    internal resistance of, 126–128
    lithium-ion, 122–123
    memory-effect problems, 121
    nickel-cadmium, 121
    nickel-metal-hydride, 121–122
    placement of, 132–133
    recharging, 130, 132
    sealed-lead-acid, 122
    suppliers of, 343
    types of, 120–123
    voltage regulators for, 353–355
    wiring requirements, 133–134
Battery eliminator circuit (BEC), 182, 199
Battery packs, 128–130
    attaching, 289–290
    illustrated, 130, 131
    multiple, 135
Baum, Dave, 227
BEAM robots, 212
Bearings, 109–111
    rolling-element, 111
    sliding, 109–111
Belt-drive systems, 105–106
Bertocchini, Carlo, 64
Big TX sumo robot, 7, 333
Bipolar transistors, 171–172
Black & Decker gearboxes, 115
Black Marauder sumo robot, 330
Black Widow sumo robot, 268–269, 329
Body assembly, 288
BoeBot robotics kit, 7
Book resources, 348–349, 350
Border-marking tool, 43
BotBoard Plus microcontroller, 228–229
Boxter sumo robot, 335–336
BrainStem microcontroller, 225–226
Brass sumo robots, 51
Brooks, Rodney, 300

Brushless motors, 139
Brute force, 304
Build Your Own Remote-Controlled Sumo-Bot (Predko), 350
Building sumo rings, 30–46
    metal rings, 36–38
    mini rings, 38–40
    painting process, 42–45
    practice rings, 30
    top surface preparation, 40–42
    wooden rings, 30–36
Building sumo robots, 271–296
    autonomous robots, 275–276
    base plate for, 286, 287
    battery box for, 289–290
    body assembly process, 288
    circuit board for, 276–279, 290
    electrical wiring for, 274–275
    fasteners for, 53–55
    front scoop for, 286–288
    full-sized robots, 271–285
    kits for, 271–285, 294–295
    materials for, 48–52
    mini robots, 286–295
    motor controller for, 273–274
    performance specifications for, 272–273
    programming process, 280–284, 292–294
    radio-control robots, 275
    recommended books on, 348–349
    robot weight and, 48
    sensors for, 276
    servo assembly process, 289
    tools used for, 55–59
    wheel assembly process, 289
Bullet-proof glass, 49
Burrows, Peter, 62, 327, 336

C

Cable chains, 100–101
Calibrating sensors, 268
Campbell, Peter, 331
Capacitors, 194–195
Capacity of batteries, 123–126
Carbon-zinc batteries, 123
Casting wheels, 311–312
Celestial Horizons, 295
Center distance calculations
    belt-drive systems, 106
    chain-drive systems, 101–103
Center frequency, 244
Chain-drive systems, 100–104
    center distance calculations, 101–103
    roller and ladder assembly, 103
    sprocket-to-shaft attachment, 103–104
    types of chains, 100–101
Channel number, 190
Channels, radio-control, 189, 190
Checksum test, 193
Circuit boards, 347
    assembly of, 279
    custom-printed, 279
    full-sized robot, 276–279

    mini robot, 290, 291
    schematic drawings of, 278, 291
Clock speed, 213–214
Clubs, robotics, 351
Coating wheels, 310–311
Competition strategies, 297–308
    defensive, 306–308
    general tips, 297–299
    offensive, 302–306
    programming tips, 299–302
    remote-control robots, 299
    sensorless, 268–269
Computer radios, 193
Confidence, 298
Constants of motors, 143–146
    calculating, 145–146
    internal resistance, 145
    no-load current, 145
    speed, 144
    torque, 144–145
Constructing sumo rings. *See* Building sumo rings
Constructing sumo robots. *See* Building sumo robots
Construction set wheels, 76–77
Cooling motors, 150–151
Cordless screwdriver/drill motors, 114–115
Corralling arms, 304
CPU (central processing unit), 213
Cross-communication interference, 265
Custom gearboxes, 116–117
Custom wheels, 77–79
    design of, 77–78
    material for, 79
    molds for, 311–312
    O-ring, 80–82

Cylindrical adhesives, 104
Cypress Microsystems, 230

D

Dave Brown Products Inc., 73
Dead band, 203
Decoys, 304
Defensive strategies, 306–308
    See also Competition strategies
Devantech, 259
Diagnosis LEDs, 277
Diaphragm pump vacuums, 317–318
Dickens, Tom, 228, 334
Differential steering, 70–71
    kinematics of, 91–93
Digikey, 276
Diodes, 195
Direct-drive method, 62, 99
Direction control, 166–174
    H-bridge circuits and, 166
    integrated circuits and, 173–174
    R/C ESCs and, 181
    relay control and, 167–168
    solid-state H-bridges and, 171–172
    switch combinations and, 166–167
    transistor control and, 168–171
Dirt Devil vacuum, 315
Disabling opponents, 306
Disqualification, 18
Distance-measuring sensors

calculating distance, 255–258
proximity detectors, 249–253
range detectors, 253–258
sharp infrared, 249–258
ultrasonic, 258–261
Dohyo. *See* Sumo rings
Donuts, 76
Double-elimination tournament, 25–26
Dremel MultiPro, 56
Drive shafts
mounting wheels to, 83–88
parts suppliers for, 342
Driving along the perimeter, 305
Driving with wheels forward, 308
Ducted fans, 316–317
DuraTrax, 87, 130
Duty cycle, 176
Dynamic friction, 305

E

Edge-detection sensors, 234–242, 306
infrared, 241, 253
lever-action switch, 234, 235
long-range, 234–237
mechanical, 234–238
optical, 238–242
photoresistor-based, 238–239
phototransistor-based, 240–241
placement of, 261
See also Opponent-detection sensors
EEPROM, 213
Effective points, 16–17, 19

Electric motors. *See* Motors
Electrical wiring. *See* Wiring process
Electromagnets, 321–322
Electronic Goldmine, 113, 321
Electronic speed controllers (ESCs), 178–185
advanced, 184–185
commercial H-bridges, 179
performance specifications, 181–182
power supply, 182–183
radio-control, 179–184
speed and direction control, 181
Electronic tools, 56–59
Electronics
parts suppliers for, 346
recommended books on, 349
Elevon mixers, 201
Epoxies, 53
ESCs. *See* Electronic speed controllers
Expanded foam PVC, 49–50
Exponential control, 203
Express PCB, 279

F

Failsafe control, 208–209
Fan-style vacuums, 315–317
Fasteners, 53–55
adhesives, 53
screws and bolts, 53–55
Federal Communications Commission (FCC), 189
Ferrite magnet motors, 138

Ferrite magnets, 320, 321
FindChips search engine, 350
Flat belts, 66
Flyback diode, 170
Flynn, Anita M., 301
FM radio systems, 193
Foam wheels, 74–76
Four-wheeled robots, 63
Frequency bands, 189–192
Frequency crystals, 191–192
Frequency modulation (FM), 193
Friction, 93–94
Front flap, 303–304
Front scoop, 286–288, 302–303
Frye, Jim, 63, 328, 329, 336
Full-sized sumo robots, 271–285
    autonomous, 275–276
    circuit board for, 276–279
    electrical wiring for, 274–275
    examples of, 332–338
    kits for building, 271–285
    motor controller for, 273–274
    performance specifications, 272–273
    programming process, 280–284
    radio-controller for, 275
    sensors for, 276
    See also Mini sumo robots; Sumo robots
Futaba Corporation, 130

G

Game points, 16–17, 19
Game principles, 14–15
Game procedures, 15–16
Gast Manufacturing, 318
Gearboxes, 112–117
    custom, 116–117
    electric motors and, 113
    modified, 114–116
Gearmotors, 113
    R/C servos for, 154–163
    selection process for, 152–154
    suppliers of, 341–342
    See also Motors
Gears, 107–109
    gear-to-shaft attachment, 109
    parallel-shaft applications, 107–108
    parts suppliers, 342
    perpendicular-shaft applications, 108–109
Goliath sumo robot, 326–327
Graupner, 202
Green, Marvin, 228

H

Hand tools, 55–56
Handy Board microcontroller, 228
Hardware communications, 215
Harrison, Bill, 7, 24, 28, 53, 63, 65, 207, 235, 337
Hattori, Mato, 7, 333

H-bridge circuits, 166
    commercial, 179
    relay-controlled, 168
    solid-state, 171–173
Heat generation, 146–148, 151
Helical gears, 108
Hex nut-style mounting, 86–87
Hiding the dohyo edge, 307
High-strength magnets, 347
Hitec RCD, 199
Hobby stores, 348
Holonomic motion control, 71–72
Honda robots, 68
HVW Technologies, 173, 179, 246
Hylands, David, 328

Integrated development environment (IDE), 221
Integrated-circuit-based motor controllers, 173–174
Intelligence, robotic, 211
Interference problems, 194–196
    cross-communication interference, 265
    electrical noise from motors, 194–196
    transmissions from other radios, 196
Internal resistance of batteries, 126–128
International Robot Sumo Tournament, 332, 333, 350, 351

I

I/O pins, 213
Indirect-drive method, 99
Information resources. *See* Resources
Infrared sensors
    circuit schematic, 245
    edge-detection, 241, 253
    opponent-detection, 242–261
    proximity, 249–253
    range, 253–258
    reflective, 243–249
    sharp, 249–258
    versions of, 248–249
Injury and accidents, 18
Innovation First, 183–184
Input pin voltage, 237–238

J

Jameco Electronics, 113
Japan
    origins of robot sumo in, 7
    R/C frequency bands used in, 191
Japanese official rules, 11–19
    See also Rules of robot sumo
Javelin Stamp module, 221–223
    features, 221–222
    programming, 222–223
Jones, Joseph L., 301
Jorgenson, Robert, 7
Joysticks, 189

K

Keyways, 104
Kinematics, 91–99
    differential steering and, 91–93
    pushing force and, 93–95
    speed reduction and, 95–98
    wheel torque and, 98–99
Kinetic energy, 305
Knitting needle sensor, 234–237

    steering methods and, 68–72
    styles of, 61–68
Loitz, Jeff, 336
Long-range edge-detection sensor, 234–237
Lonseal Company, 29
Lonstage vinyl material, 29, 41
Luck factor, 299
Lunt, Karl, 228, 229
Lynxmotion, 54, 76, 83, 199, 241, 248, 271–272, 273, 340

L

Ladder chains, 100–101, 103
Lee Valley Tools, 34
Lego Mindstorms RCX brick, 226–227
LEGO sets
    tank treads from, 65
    wheels from, 76–77
LEGO sumo robots, 331
LegOS programming language, 227
Leo sumo robot, 334
Lever-action switch, 234, 235
Light-emitting diodes (LEDs), 238, 277
Lightweight sumo weight class, 24
Lint pickup tubes, 79, 80
Lithium batteries, 123
Lithium-ion (Li-ion) batteries, 122–123, 130
Location detection, 263, 265–266
Locomotion, 61–88
    mounting wheels for, 83–88
    selecting wheels for, 73–88

M

Machine design, 349
Machine screws, 54–55
Machining aluminum, 51
Magnetic systems, 320–323
    placement of magnets, 322–323
    suppliers of, 347
    types of magnets, 320–322
Marauder sumo robot, 328
Mark III Project, 340
Marlin P. Jones & Associates, 113
Martin, Fred, 226, 228
Masking tape, 43–44
Match, definition of, 12
Match drilling, 85, 115
Materials for robots, 48–52
    aluminum, 50–51
    brass, 51
    comparison of, 51–52
    plastic, 49–50
    wood, 48–49

Index

Maxon, 113
MBasic compiler, 224–225
McMaster-Carr, 79, 88, 320
Mechanical parts suppliers, 346

Mechanical sensors
    for edge detection, 234–238
    for opponent detection, 242–243
Medo USA, Inc., 318
Memory, microcontroller, 215
Memory-effect battery problems, 121
Mercury batteries, 123
Meshed gears, 107
Metal sumo rings, 36–38
    design of, 36–38
    metal surface for, 40–41
    schematic diagram of, 37
    steps in constructing, 38
    vinyl surface for, 41–42
Metal surface
    adding to sumo rings, 40–41
    material used for, 30
    painting, 46
Microchip Corporation, 213, 214, 230
Microcomputers, 212
Microcontrollers, 211–231
    Basic Atom modules, 223–224
    Basic language compiler, 224–225
    Basic Stamp modules, 217–221
    BotBoard Plus, 228–229
    BrainStem, 225–226
    clock speed of, 213–214
    explanation of, 212–213
    features of, 213–215
    Handy Board, 228
    hardware communications for, 215
    Javelin Stamp module, 221–223
    Lego RCX brick, 226–227
    memory of, 215
    programming environments for, 215–216
    single-chip solutions, 229–230
    suppliers of, 344–345
    types of, 216–229
Middleweight sumo robots, 331–332
Miles, Kristina, 329
Milford Instruments, Ltd., 258
Mini sumo rings, 38–39, 40
Mini sumo robots, 286–295
    base plate for, 286, 287
    battery box for, 289–290
    body assembly process, 288
    building from scratch, 286–294
    circuit board for, 290, 291
    electrical wiring for, 290
    examples of, 326–330
    front scoop for, 286–288
    kits for building, 294–295
    programming process, 292–294
    remote-control, 200–201, 330
    servo assembly process, 289
    walking, 330
    weight class for, 24
    wheel assembly process, 289

See also Full-sized sumo robots; Sumo robots
Mini sumo wheels, 79–83
    custom-machined, 83
    O-ring, 79, 80–82
    sources of, 82
MIT Media Laboratory, 226, 228
Miter gears, 108
Mixing circuits, 204–206
Mobile Robots, Inspiration to Implementation (Jones & Flynn), 301
Model airplane wheels, 73
Modified gearboxes, 114–116
    cordless screwdriver/drill motors, 114–115
    motor replacement, 115–116
    remote-control servo motors, 116
MOFSET transistors, 171–172
Mondo-tronics, 179, 259
Motion studies. *See* Kinematics
Motor Control Electronics Handbook (Valentine), 172
Motor controllers, 165–186
    advanced, 184–185
    building your own, 172
    commercial products as, 178–185
    direction control and, 166–174
    electronic speed controllers and, 178–185
    full-sized robots and, 273–274
    H-bridge circuits and, 166
    integrated-circuits and, 173–174
    pulse-width modulation and, 175, 176–178
    relays and, 167–168
    solid-state H-bridges and, 171–173
    speed control and, 175–185
    suppliers of, 343–344
    switch combinations and, 166–167
    transistors and, 168–171
    variable resistors and, 175
Motorola, 230
Motors, 137–163
    battery effects on, 148–149
    brushless, 139
    characteristics of, 139
    constants of, 143–146
    cooling, 150–151
    efficiency of, 142
    ferrite magnet, 138
    finding, 152–154
    heat generated by, 146–148
    internal resistance of, 145
    no-load current of, 140, 145
    performance specifications for, 143–144, 151–152, 153
    PMDC, 137, 138–139
    power of, 142, 146
    pushing the limits of, 149–151
    R/C servos used as, 154–163
    rare-earth, 138–139
    reducing electrical noise from, 194–196
    selection process for, 151–154
    sources of, 163
    speed of, 140, 144

stall current of, 141–142
suppliers of, 341–342
torque of, 140–141, 144–145
types of, 138–139
voltage increases and, 149–150
Mounting wheels, 83–88
alignment issues, 87
hex nut-style mounting, 86–87
parts suppliers for, 340–341
pinned shaft method, 85
quick-disconnect coupling, 87–88
quick-release pinning, 86
semipermanent pinning, 85–86
set screw mounting, 84
Movement. *See* Locomotion
Multiple programs, 301–302
Multiple sensors, 306
Multi-wheeled robots, 63

N

Narrow-field-of-view sensors, 265–266
National Semiconductor, 239, 249, 353–355
Nemesis sumo robot, 327
Nickel-cadmium (NiCd) batteries, 121, 130, 132
Nickel-metal-hydride (NiMH) batteries, 121–122, 130, 132
Noga, Markus, 227
No-load current, 140, 145
Northwest Robot Sumo Tournament, 332, 350
Nozawa, Hiroshi, 7
NQC programming language, 227
Nuts & Volts magazine, 350

O

Object-detection sensors. *See* Opponent-detection sensors
OctoBot sumo robot, 336–337
Offensive strategies, 302–306
See also Competition strategies
Olsen, John, 82
Omni-directional roller wheels, 71–72
Open Source Motor Controller (OSMC) project, 172, 350
Opponent-detection sensors, 242–261
mechanical, 242–243
multiple, 262–263, 265–266
panning, 265
placement of, 261, 262, 263
reflected infrared, 243–249
sharp infrared, 249–258
ultrasonic, 258–261
See also Edge-detection sensors
OPTAscope, 58–59
Optical edge-detection sensors, 238–242
infrared, 241
photoresistor-based, 238–239
phototransistor-based, 240–241
O-ring wheels, 79, 80–82
Oscilloscope, 57–59
Overkill sumo robot, 336

P

Painting sumo rings, 42–46
    border, 44–45
    metal surface, 46
    starting lines, 45
    vinyl surface, 46
    wooden surface, 42–45
Panning sensors, 265
Parallax Inc., 8, 82, 204, 217, 221, 249, 294, 340
Parallel-shaft applications, 107–108
Parts suppliers. *See* Resources
Path planning, 306
PBasic programming language, 216, 217
PCB Express, 279
Penalties, 17
Performance specifications
    for building sumo robots, 272–273
    for electronic speed controllers, 181–182
    for motors, 143–144, 151–152, 153
Perpendicular-shaft applications, 108–109
Pete's Folly sumo robot, 330
Phay Yee How, Benedict Brandon, 335
Photocells, 238
Photoresistor-based edge detectors, 238–239
Phototransistor-based edge detectors, 240–241
Pinned shaft method, 85
Pistol-grip configuration, 189
Pitch, 100
Plain bearings, 109–111
Plastic sumo robots, 49–50
PMDC motors, 137, 138–139
Polycarbonates, 49
Polytek, 311
Polyurethane, 311, 347
Portland Area Robotics Society, 295
Position feedback method, 160–162
Power circuit, 277
Power supplies
    electronic speed controllers and, 182–183
    voltage regulators and, 353–355
Power tools, 56
Power transmission, 89–118
    bearings, 109–111
    belt-drive systems, 105–106
    chain-drive systems, 100–104
    gearboxes, 112–117
    gears, 107–109
    kinematics and, 91–99
    making choices for, 90
    methods of, 99–100
    system components, 99–117
Power-Sonic, 130
Practice sumo ring, 30, 40
Predko, Mike, 350
Printed circuit boards, 347
Program System on a Chip (PSoC), 230
Programming microcontrollers, 215–216
    Basic Atom, 224
    Basic Stamp, 219–221
    BotBoard Plus, 228–229
    BrainStem, 226
    Handy Board, 228
    Javelin Stamp, 222–223
    Lego RCX, 227

Programming port, 277
Programming strategies, 299–302
    examples of, 280–284, 292–294
    full-sized robots, 280–284
    mini robots, 292–294
    multiple programs, 301–302
    recommended books on, 349
    subroutines, 300, 301
    subsumptive programming, 300–301
Proximity detectors, 249–253
    long-range edge detection with, 253
    wiring and testing, 250–252
Published resources, 348–349, 350
Pulse-code modulation (PCM), 193
Pulse-position modulation (PPM), 157–158
Pulse-width modulation (PWM), 175, 176–178
Pushing arm, 304
Pushing capability, 309–324
    improving wheel traction, 310–312
    increasing apparent weight, 313–323
Pushing force, 93–95
    friction and, 93–94
    torque and, 94–95
PVC plastic, 49–50

Q

QT Optoelectronics, 241
Quick-disconnect coupling, 87–88
Quick-release pinning, 86

R

R/C servo motors, 154–163
    calibrating, 162
    color codes, 157
    comparison chart, 156
    connector styles, 157
    control signals, 156–158
    disassembling, 159–160
    modifying, 158–159
    parts suppliers, 342
    position feedback method, 160–162
    reassembling, 163
    using as robot motors, 155–156
Racing slicks, 74
Radial-load bearings, 111
Radio Shack, 130
Radio-control (R/C) systems, 188–206
    antenna setup, 196–198
    channel capabilities, 189
    configuring, 198–206
    control signals, 194
    electronic speed controllers, 179–184
    frequency bands, 189–192

interference problems, 194–196
major subsystems, 188
transmission methods, 192–193
Radio-controlled car motors, 138
RAM (random access memory), 213
Range sensors, 264
    sharp infrared, 253–258
    ultrasonic, 258–261
Rare-earth magnets, 321
Rare-earth motors, 138–139
Real-life sumo robots, 325–338
    full-sized robots, 332–338
    LEGO robots, 331
    middleweight robots, 331–332
    mini robots, 326–330
Rear sensors, 262, 264, 307
Recharging batteries, 130, 132
References, 348–349, 350
    See also Resources
Reflective sensors, 243–249
    circuit schematic, 245
    limitations of, 266–267
    variations of, 248–249
Relay-controlled motors, 167–168
Rematches, 15
Remote Possibility sumo robot, 330
Remote-control car wheels, 74–76
    donuts for making, 76
    foam wheels, 74–76
    racing slicks, 74
Remote-control servo motors, 116
Remote-control sumo robots, 5, 187–209
    antenna setup, 196–198
    competition strategies, 299

configuring, 198–206
failsafe control, 208–209
interference problems, 194–196
mini sumo robots, 200–201
parts suppliers, 344
radio-control systems, 188–206
steering controls, 201–206
tether-control systems, 206–208
Resources, 339–351
    batteries, 343
    electronic parts, 346
    gears and drive components, 342
    high-strength magnets, 347
    hobby stores, 348
    major tournaments, 350
    mechanical parts, 346
    microcontrollers, 344–345
    motor controllers, 343–344
    motors and gearmotors, 341–341
    printed circuit boards, 347
    publications, 348–349, 350
    remote-control equipment, 344
    robot sensors, 345
    robotics clubs, 351
    sumo ring construction materials, 340
    sumo robot parts suppliers, 340
    tools, 348
    vacuum pumps, 346
    wheel casting compounds, 347

wheel mounting hardware, 340–341
wheels and hubs, 340–341
Restarting a stopped match, 23
Restrictions on robots, 22–23
Richards, Steve, 72
Rings. *See* Sumo rings
Robot Books online store, 350
Robot sumo
    description of, 1–3, 4
    history of, 7
    major tournaments, 350
    resources and references, 339–351
    rules of, 11–26
Robothon, 295, 326, 350
Robotics clubs, 351
Robots. *See* Sumo robots
Roller bearings, 111
Roller chains, 100–101, 103
Rolling-element bearings, 111
Ross, Kevin, 228
Round-robin tournament, 25
RTV adhesives, 310
Rules of robot sumo, 11–26
    definition of a match, 12
    dohyo specifications, 12–13
    game principles, 14–15
    game procedures, 15–16
    injury and accidents, 18
    interpretation of, 19–24
    modifications of, 19, 23–24
    objections to, 18
    official Japanese rules, 11–19
    robot markings, 18
    robot specifications, 13–14
    tournament styles, 25–26
    violations and penalties, 17–18

weight classes, 24
yuko (effective) points, 16–17

S

Safety guidelines, 6
Sandberg, Daryl, 303, 326–327, 335
Sargent, Randy, 226
SBasic programming language, 229
Scenix microcontrollers, 218
Schottky diodes, 195
Screws, for robot assembly, 53–55
Sealed-lead-acid (SLA) batteries, 122, 130, 132
Seattle Robotics Society, 295, 326
Semipermanent pinning, 85–86
Sensors, 233–270
    accuracy issues, 266–269
    calibrating, 268
    competitive strategies without, 268–269
    distance calculation, 255–258
    edge-detection, 234–242
    full-sized robot, 276
    interference problems, 265
    mechanical, 234–238, 242–243
    multiple, 262–263, 265–266
    opponent-detection, 242–261, 263
    optical, 238–242
    panning, 265
    placement of, 261–265
    proximity, 249–253
    range, 249–253, 258–261, 262
    reflective infrared, 243–249

sharp infrared, 249–258
suppliers of, 345
types of, 233–234, 264
ultrasonic, 258–261
Servo horns, 81–82, 159
Servo mixing, 201–202
Servo motors, 154–163
Servo speed, 154–155
Servo Systems Company, 113
Set screw mounting, 84
Shaft attachments
belt-drive systems and, 106
chain-drive systems and, 103–104
gear systems and, 109
Sharp infrared sensors, 249–258
distance calculation, 255–258
edge detection, 253
proximity detection, 249–253
range detection, 253–258
wiring and testing, 250–252
Short circuit switch positions, 167
Shunts, 146
Side sensors, 264
Silicone, 310–311, 347
Silver batteries, 123
Silverman, Brian, 226
Sine Robotics, 24, 82, 200, 207, 330, 350
Single-elimination tournament, 25
Six-legged robots, 67
Six-Pack sumo robot, 329
Size
of sumo rings, 28–29
of sumo robots, 4–5, 19–20
Skeleton sumo robot, 335
Sliding bearings, 109–111

Software mixing, 204–206
Solarbotics, 212
Solid-state H-bridges, 171–173
Solutions Cubed, 83, 184
Sony robots, 68
Sozbots, 183
Specifications
for sumo rings, 12–13, 28–29
for sumo robots, 13–14, 21–22
Speed
kinematics of, 91
motor, 140, 144
Speed control, 175–185
commercial products for, 178–185
pulse width modulation and, 176–178
variable resistors and, 175
Speed reduction, 95–98
SPI serial communication systems, 215
Sprocket-to-shaft attachment, 103–104
Spur gears, 107
Stall current, 141–142
Starting lines
masking, 43–44
painting, 45
Starting the match, 15, 20–21
restarting a stopped match, 23
Static friction, 305
Stealth technology, 268, 307
Steel surface. *See* Metal surface
Steering methods, 68–72
Ackerman steering, 69–70
differential steering, 70–71
holonomic motion control, 71–72

Index

servo mixing, 201–202
synchro drive system, 71
Sticky wheels, 79
Student engineering projects, 334
Subroutines, 300, 301
Subsumptive programming, 300–301
Sumo, 4
Sumo rings, 27–46
    building, 30–46
    colors of, 28–29
    metal rings, 36–38
    mini rings, 38–39
    painting, 42–46
    practice rings, 30
    size of, 28–29
    specifications for, 12–13, 28–29
    top surface of, 29–31, 40–42
    vinyl material for, 29–30, 340
    weight of, 46
    wooden rings, 30–36
Sumo robots
    batteries for, 119–136
    building, 271–296
    competition strategies for, 297–308
    design of, 5–6
    fasteners for assembling, 53–55
    full-sized, 271–285, 332–338
    locomotion styles of, 61–68
    markings on, 18
    materials used for, 48–52
    microcontrollers for, 211–231
    mini, 286–295, 326–330
    motor controllers for, 165–186
    motors for, 137–163
    mounting wheels on, 83–88
    parts suppliers for, 340
    power transmission in, 89–118
    pushing capability of, 309–324
    R/C configurations for, 198–206
    real-life examples of, 325–338
    remotely controlling, 187–209
    resources for, 339–351
    restrictions on, 22–23
    safety guidelines for, 6
    sensors for, 233–270
    sizes of, 4–5, 19–20
    specifications for, 13–14, 21–22
    steering methods of, 68–72
    tools for constructing, 55–58
    voltage regulators for, 353–355
    weight of, 19, 48, 313
    wheels for, 73–83, 310–312
Superglue, 53
Suppliers. *See* Resources
Surface materials
    metal, 30, 40–41
    vinyl, 29–30, 41–42
    wood, 35–36, 40
Surplus motors, 341–341
Sweet spot, 305
Switch combinations, 166–167
Synair, 311
Synchro drive system, 71
Synchronous belts, 66, 105–106

T

Tamiya Inc., 74, 112
Tank-treaded robots, 64–66
    tread supports for, 66
    types of tank treads, 64–66
TAP Plastics, 311
TEA programming language, 226
Team Associated, 83
Team Trinity, 75
Temperature issues, 146–148
Temporary adhesives, 53
Tensile load specifications, 101
Terminal blocks, 134–135
Terminator sumo robot kit, 284–285
Tether-control systems, 206–208
    general specifications for, 207
    schematic diagram of, 208
    simple circuits for, 207–208
Thin front flap, 303–304
3kg sumo weight class, 24
Thrust bearings, 111
Tilden, Mark, 212
Timing belts, 66, 105–106
Too Much sumo robot, 331–332
Tools for robot construction, 55–58
    electronic tools, 56–59
    hand tools, 55–56
    power tools, 56
    suppliers of, 348
Torque
    increasing, 98–99
    motor, 140–141, 144–145
    pushing force and, 94–95
Tournaments
    list of major, 350
    styles of, 25–26
Tower Hobbies, 130
Traction, 310–312
Trammels, 33–34
Transducers, ultrasonic, 258–261
Transistor-controlled relays, 168–171
Transit time, 154–155
Transmission methods, 192–193
Traxxas Corporation, 87
Trickle charging, 130, 132
Two-wheeled robots, 62

U

UART serial communication systems, 215
Ubicom Inc., 218, 230
Ultrasonic range sensors, 258–261
Uninterruptible power supply (UPS), 122
United Kingdom R/C frequency bands, 191
United States R/C frequency bands, 190–191

V

Vacuum systems, 313–319
    diaphragm pump, 317–318
    fan-style, 315–317
    how they work, 313–315
    suppliers, 346
    vacuum cups, 318–319
Valentine, Richard, 172
Vantec, 183

Variable resistors, 175
Variable-speed control, 175–178
    commercial electronic products for, 178–185
    pulse width modulation and, 176–178
    variable resistors and, 175
Vinyl surface
    adding to sumo rings, 41–42
    material used for, 29
    painting of, 46
    supplier for, 29, 340
Violations, 17–18
Viper sumo robot kit, 272–284
Virtual peripherals (VPs), 222
Voltage
    effects of increasing, 149–150
    input pin, 237–238
    microcontroller operating range, 215
Voltage regulators, 353–355
V-tail mixers, 201, 202–203

W

Walking robots, 67–68
    design of, 68
    examples of, 330
    future of, 68
    six-legged, 67
Walter, Grey, 211
Warnings, 17
Weesner, Dana and David, 83

Weight
    of sumo rings, 46
    of sumo robots, 19, 24, 48, 313
    See also Apparent weight
Weight classes, 24
Western Canadian Robot Games, 350
Wheel torque, 98–99
Wheel traction, 310–312
    casting wheels for, 311–312
    coating wheels for, 310–311
Wheeled robots, 62–63
Wheels, 73–88
    casting, 311–312
    coating, 310–311
    construction set, 76–77
    custom, 77–79, 311–312
    mini sumo, 79–83
    model airplane, 73
    molds for, 311–312
    mounting to drive shafts, 83–88
    omni-directional roller, 71–72
    O-ring, 79, 80–82
    parts suppliers for, 340–341
    remote-control car, 74–76
    sources of, 73–77
    traction of, 310–312
Wiring process
    diagram, 274
    full-sized robots, 274–275
    mini robots, 290
    requirements, 133–134
Wonder Magnet, 321
Wood sumo robots, 48–49

Wooden sumo rings, 30–36
    design of, 30–31
    metal surface for, 40–41
    schematic diagram of, 31
    steps in constructing, 31–36
    trimming and finishing, 35–36
    vinyl surface for, 41–42
    wooden surface for, 35–36, 40
Wooden surface
    finishing, 35–36, 40
    painting, 42–45
Worm gears, 108–109

Y

Yuko (effective) points, 16–17, 19
Yusei (advantage) points, 17, 19, 323

Z

Zhaowei, Kwa, 335
Zinc-air batteries, 123

INTERNATIONAL CONTACT INFORMATION

AUSTRALIA
McGraw-Hill Book Company Australia Pty. Ltd.
TEL +61-2-9415-9899
FAX +61-2-9415-5687
http://www.mcgraw-hill.com.au
books-it_sydney@mcgraw-hill.com

CANADA
McGraw-Hill Ryerson Ltd.
TEL +905-430-5000
FAX +905-430-5020
http://www.mcgrawhill.ca

GREECE, MIDDLE EAST, NORTHERN AFRICA
McGraw-Hill Hellas
TEL +30-1-656-0990-3-4
FAX +30-1-654-5525

MEXICO (Also serving Latin America)
McGraw-Hill Interamericana Editores S.A. de C.V.
TEL +525-117-1583
FAX +525-117-1589
http://www.mcgraw-hill.com.mx
fernando_castellanos@mcgraw-hill.com

SINGAPORE (Serving Asia)
McGraw-Hill Book Company
TEL +65-863-1580
FAX +65-862-3354
http://www.mcgraw-hill.com.sg
mghasia@mcgraw-hill.com

SOUTH AFRICA
McGraw-Hill South Africa
TEL +27-11-622-7512
FAX +27-11-622-9045
robyn_swanepoel@mcgraw-hill.com

UNITED KINGDOM & EUROPE (Excluding Southern Europe)
McGraw-Hill Education Europe
TEL +44-1-628-502500
FAX +44-1-628-770224
http://www.mcgraw-hill.co.uk
computing_neurope@mcgraw-hill.com

ALL OTHER INQUIRIES Contact:
Osborne/McGraw-Hill
TEL +1-510-549-6600
FAX +1-510-883-7600
http://www.osborne.com
omg_international@mcgraw-hill.com

Books that kick BOT!

Don't miss these innovative releases–
ideal for today's robot fans and thrill-seekers...

Robot Invasion: 7 Cool & Easy Robot Projects
Dave Johnson
ISBN: 0-07-222640-4

BugBots, JunkBots, and Bots on Wheels: Building Simple Robots with BEAM Technology
Dave Hrynkiw & Mark W. Tilden
ISBN: 0-07-222601-3

Build Your Own Combat Robot
Pete Miles & Tom Carroll
ISBN: 0-07-219464-2

BattleBots® The Official Guide
Mark Clarkson
ISBN: 0-07-222425-8

Robot Sumo: The Official Guide
Pete Miles
ISBN: 0-07-222617-X

McGraw-Hill/Osborne products are available worldwide wherever quality computer books are sold. For more information visit www.osborne.com.

OSBORNE DELIVERS RESULTS! OSBORNE

BUILD A POWERFUL SUMO-BOT DESIGNED TO WITHSTAND METAL AGAINST METAL COMPETITION!

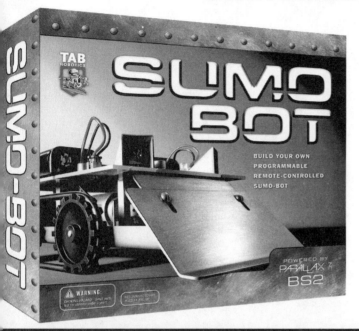

MORE POWER!
MORE SOPHISTICATION!
MORE FUN!

Here's a fun and affordable way for hobbyists to take their robot building skills to the next level and be part of the HOTTEST NEW CRAZE in amateur robotics.

SUMO-BOT is PRE-PROGRAMMED with the following behaviors so you can have fun right away:

Random Movement—moves about randomly, reversing direction when an obstacle is in its way

Photovore—Seeks out light

Photophobe—Avoids light

Maze Solver—Follows a wall

Program it further! Instructions for programming are on the CD-ROM.

GREAT FOR AGES 14+, THE KIT COMES COMPLETE WITH:

✔ **Pre-assembled PCB**

✔ **Multi-function, dual channel remote control**

✔ **Robot hardware including collision-sensing infra-red LED and receivers**

✔ **CD-ROM with programming instructions and file chapters of robot building tips and tricks**

Features:

- The **POWER** of the Parallax BASIC Stamp 2 controller
- A **TRACKED DRIVE** for maneuverability over all types of terrain
- **"FLIPPING ABILITY"** to put the competition on its back
- A **STEEL FRAME** that encourages **CUSTOMIZATION**
- Mechanical features designed to **WITHSTAND** the shock of combat
- **AA BATTERY POWERED** for longer operation (batteries not included)

2003 • ISBN: 0-07-141193-3 • $99.95
Available at bookstores everywhere!